Advances in Mathematical Analysis and its Applications

Advances in Mathematical Analysis and its Applications is designed as a reference text and explores several important aspects of recent developments in the interdisciplinary applications of mathematical analysis (MA) and highlights how MA is now being employed in many areas of scientific research. It discusses theory and problems in real and complex analysis, functional analysis, approximation theory, operator theory, analytic inequalities, the Radon transform, nonlinear analysis, and various applications of interdisciplinary research; some topics are also devoted to specific applications such as the three-body problem, finite element analysis in fluid mechanics, algorithms for difference of monotone operators, a vibrational approach to a financial problem, and more.

- The book encompasses several contemporary topics in the field of mathematical analysis, their applications, and relevancies in other areas of research and study.
- It offers an understanding of research problems by presenting the necessary developments in reasonable details.
- The book also discusses applications and uses of operator theory, fixed-point theory, inequalities, bi-univalent functions, functional equations, and scalar-objective programming, and presents various associated problems and ways to solve such problems.
- Contains applications on wavelets analysis and COVID-19 to show that mathematical analysis has interdisciplinary as well as real-life applications.

The book is aimed primarily at advanced undergraduates and postgraduate students studying mathematical analysis and mathematics in general. Researchers will also find this book useful.

Advances in Mathematical Analysis and its Applications

Edited by
Bipan Hazarika
Santanu Acharjee
H. M. Srivastava

CRC Press is an imprint of the
Taylor & Francis Group, an **informa** business

A CHAPMAN & HALL BOOK

First edition published 2023
by CRC Press
6000 Broken Sound Parkway NW, Suite 300, Boca Raton, FL 33487-2742

and by CRC Press
4 Park Square, Milton Park, Abingdon, Oxon, OX14 4RN

CRC Press is an imprint of Taylor & Francis Group, LLC

© 2023 selection and editorial matter, Bipan Hazarika, Santanu Acharjee and H. M. Srivastava; individual chapters, the contributors

Reasonable efforts have been made to publish reliable data and information, but the author and publisher cannot assume responsibility for the validity of all materials or the consequences of their use. The authors and publishers have attempted to trace the copyright holders of all material reproduced in this publication and apologize to copyright holders if permission to publish in this form has not been obtained. If any copyright material has not been acknowledged please write and let us know so we may rectify in any future reprint.

Except as permitted under U.S. Copyright Law, no part of this book may be reprinted, reproduced, transmitted, or utilized in any form by any electronic, mechanical, or other means, now known or hereafter invented, including photocopying, microfilming, and recording, or in any information storage or retrieval system, without written permission from the publishers.

For permission to photocopy or use material electronically from this work, access www.copyright.com or contact the Copyright Clearance Center, Inc. (CCC), 222 Rosewood Drive, Danvers, MA 01923, 978-750-8400. For works that are not available on CCC please contact mpkbookspermissions@tandf.co.uk

Trademark notice: Product or corporate names may be trademarks or registered trademarks and are used only for identification and explanation without intent to infringe.

ISBN: 978-1-032-35804-8 (hbk)
ISBN: 978-1-032-36227-4 (pbk)
ISBN: 978-1-003-33086-8 (ebk)

DOI: 10.1201/9781003330868

Typeset in SFRM1000
by KnowledgeWorks Global Ltd.

Publisher's note: This book has been prepared from camera-ready copy provided by the authors.

Contents

Preface		xi
Editors		xv
Contributors		xvii

1 Some applications of double sequences   1
Dragan Djurčić and Ljubiša D.R. Kočinac
- 1.1 Introduction 1
  - 1.1.1 Double sequences 1
  - 1.1.2 Selection principles 4
  - 1.1.3 Asymptotic analysis 5
- 1.2 S_1 selection principle and double sequences 6
- 1.3 α_2-selection principle and double sequences 9
- 1.4 Double sequences and the exponent of convergence 14
- Bibliography 17

2 Convergent triple sequences and statistical cluster points   21
Mehmet Gürdal and Mualla Birgül Huban
- 2.1 Introduction 21
- 2.2 Known definitions and properties 22
- 2.3 $\mathcal{I}_3$-statistical cluster points of triple sequences 23
  - 2.3.1 $\Gamma^{\mathcal{I}_3}$-statistical convergence 28
- 2.4 Lacunary $\mathcal{I}_3$-statistical cluster points 31
- Bibliography 35

3 Relative uniform convergence of sequence of positive linear Functions   39
Kshetrimayum Renubebeta Devi and Binod Chandra Tripathy
- 3.1 Introduction 39
- 3.2 Preliminaries and definitions 40
- 3.3 Relative uniform convergence of single sequence of functions 42
- 3.4 Statistical convergence of sequence 44
- 3.5 Double sequences 48
- 3.6 Relative uniform convergence of difference double sequence of positive linear functions 52
- Bibliography 54

Contents

4 Almost convergent sequence spaces defined by Nörlund matrix and generalized difference matrix — 57
Kuldip Raj and Manisha Devi
- 4.1 Introduction and preliminaries ... 57
- 4.2 Main results ... 62
- Bibliography ... 68

5 Factorization of the infinite Hilbert and Cesàro operators — 71
Hadi Roopaei
- 5.1 Introductions and preliminaries ... 71
- 5.2 Hilbert matrix ... 72
- 5.3 Hausdorff matrix ... 73
- 5.4 Cesàro matrix of order n ... 73
- 5.5 Copson matrix ... 75
- 5.6 Gamma matrix of order n ... 75
- 5.7 Factorization of the infinite Hilbert operator ... 76
  - 5.7.1 Factorization of the Hilbert operator based on Cesàro operator ... 77
  - 5.7.2 Factorization of the Hilbert operator based on gamma operator ... 81
  - 5.7.3 Factorization of the Hilbert operator based on the generalized Cesàro operator ... 86
- 5.8 Factorization of the Cesàro operator ... 90
- Bibliography ... 93

6 On theorems of Galambos-Bojanić-Seneta type — 95
Dragan Djurčić and Ljubiša D.R. Kočinac
- 6.1 Introduction ... 95
- 6.2 Known results ... 97
  - 6.2.1 Classes ORV_s and ORV_f and their subclasses ... 97
  - 6.2.2 Rapid and related variations ... 100
- 6.3 New result ... 106
- Bibliography ... 109

7 On the spaces of absolutely p-summable and bounded q-Euler difference sequences — 113
Taja Yaying
- 7.1 Introduction ... 113
  - 7.1.1 Euler matrix of order 1 and sequence spaces ... 114
  - 7.1.2 Quantum calculus ... 115
- 7.2 q-Euler difference sequence spaces $e_p^q(\nabla)$ and $e_\infty^q(\nabla)$ ... 117
- 7.3 Alpha-, beta-, and gamma-duals ... 119
- 7.4 Matrix transformations ... 122
- Bibliography ... 125

Contents   vii

8 Approximation by the double sequence of LPO based on multivariable q-Lagrange polynomials   129
Behar Baxhaku and P. N. Agrawal
- 8.1 Introduction 129
- 8.2 Double sequence of $\mathfrak{K}_{n,q}^{\beta^{(1)},\cdots,\beta^{(r)}}(.)(x)$ 131
- 8.3 Approximation by using power series summability method (p.s.s.m) 132
  - 8.3.1 Illustrative example 136
- 8.4 $\mathcal{A}$-statistical convergence of operators $\mathfrak{K}_{n_1,q_{n_1}}^{n_2,q_{n_2}}(.)(\mathbf{x})$ 139
  - 8.4.1 Application of Theorem 8.4.4 143
- 8.5 $\mathcal{A}$-statistical convergence by GBS operators 146
- Bibliography 151

9 Results on interpolative Boyd-Wong contraction in quasi-partial b-metric space   155
Pragati Gautam and Swapnil Verma
- 9.1 Introduction and preliminaries 155
- 9.2 Main results 159
- Bibliography 165

10 Applications of differential transform method on some functional differential equations   169
Anil Kumar, Giriraj Methi, and Sanket Tikare
- 10.1 Introduction 169
- 10.2 Preliminaries 170
  - 10.2.1 Definition of differential transform 171
  - 10.2.2 Faà di Bruno's formula and Bell polynomials 171
  - 10.2.3 Description of the method 173
  - 10.2.4 Convergence results 174
  - 10.2.5 Error estimate 175
- 10.3 Applications 175
  - 10.3.1 Example 1 176
  - 10.3.2 Example 2 178
  - 10.3.3 Example 3 181
  - 10.3.4 Example 4 184
  - 10.3.5 Example 5 186
- Bibliography 188

11 Solvability of fractional integral equation via measure of noncompactness and shifting distance functions   191
Bhuban Chandra Deuri and Anupam Das
- 11.1 Introduction 191
  - 11.1.1 Some notations 192
  - 11.1.2 Measure of noncompactness 192
- 11.2 Main result 193

11.3 Application	196
Bibliography	201

12 Generalized fractional operators and inequalities integrals   205
Juan E. Nápoles Valdés and Florencia Rabossi

12.1 Introduction	205
12.2 Integral inequalities with some integral operators	208
12.2.1 Generalized integral operators	208
12.2.2 Generalized fractional integral operators	212
12.2.3 Weighted integral operators	220
12.3 A general formulation of the notion of convex function	222
12.4 Integral inequalities and fractional derivatives	223
Bibliography	225

13 Exponentially biconvex functions and bivariational inequalities   229
Muhammad Aslam Noor and Khalida Inayat Noor

13.1 Introduction	229
13.2 Preliminary results	231
13.3 Properties of exponentially biconvex functions	234
13.4 Bivariational inequalities	244
Bibliography	248

14 On a certain subclass of analytic functions defined by Bessel functions   251
B. Venkateswarlu, P. Thirupathi Reddy, and Shashikala A

14.1 Introduction	251
14.2 Coefficient bounds	254
14.3 Neighborhood property	256
14.4 Partial sums	258
Bibliography	262

15 A note on meromorphic functions with positive coefficients defined by differential operator   265
B. Venkateswarlu, P. Thirupathi Reddy, and Sujatha

15.1 Introduction	265
15.2 Coefficient inequality	268
15.3 Distortion theorem	269
15.4 Integral operators	271
15.5 Convex linear combinations and convolution properties	272
15.6 Neighborhood property	275
Bibliography	276

16 Sharp coefficient bounds and solution of the Fekete-Szegö problem for a certain subclass of bi-univalent functions associated with the Chebyshev polynomials — 279

Amol Bhausaheb Patil

16.1 Introduction 279
    16.1.1 Bi-univalent function 280
    16.1.2 Subordination 281
    16.1.3 Chebyshev polynomials 282
    16.1.4 The function class $\mathcal{CH}_\Sigma(\lambda,\mu,x)$ 282
16.2 Coefficient estimates for the class $\mathcal{CH}_\Sigma(\lambda,\mu,x)$ 284
    16.2.1 Some immediate consequences of the theorem 287
Bibliography 288

17 Some differential sandwich theorems involving a multiplier transformation and Ruscheweyh derivative — 291

Alb Lupaş Alina

17.1 Differential subordination and superordination 291
17.2 Strong differential subordination and superordination 300
Bibliography 308

18 A study on self similar, nonlinear and complex behavior of the spread of COVID-19 in India — 311

Dibakar Das, Sankalpa Chowdhury, Gourab Das, Anuska Chanda, Swapnesh Khamaru, and Koushik Ghosh

18.1 Introduction 312
18.2 On the importance of the tests performed 314
18.3 Theory 315
    18.3.1 Calculation of moving averages 315
    18.3.2 Calculation of Hurst exponent by finite variance scaling method 316
    18.3.3 Estimation of fractal dimension by Higuchi's method . 317
    18.3.4 Multifractal analysis by multifractal detrended fluctuation analysis 318
    18.3.5 Analysis for non-linearity using delay vector variance method 320
    18.3.6 0-1 test for chaos detection 322
    18.3.7 Mathematical aspects of self-organized criticality ... 323
18.4 Data 323
18.5 Results 324
Bibliography 330

Index — 335

Preface

Mathematical analysis is one of the main branches of pure mathematics. But, its recent interdisciplinary applications have placed it into a common domain of both pure and applied mathematics. It is need of the hour to establish a mode of communication between researchers of different areas who frequently use mathematical analysis. Moreover, the crisis of a book having cutting-edge research trends of mathematical analysis can be felt. Thus, we have written this book. It has chapters on the following topics: double and triple sequences, convergence of double and triple sequences, sequence spaces, summability theory, matrix transformations, Hilbert matrix operators, quantum Euler matrix, multivariate Lagrange polynomials, multivariate Lagrange-Hermite polynomials, contraction in metric space, application of fixed point theory, application of measure of non-compactness in fractional integral equations, Bessel's functions, bi-univalent functions, Ricci solitons, and complex behavior of COVID-19. The chapters are reflecting advances in mathematical analysis as well as applications in other mathematical areas along with real life situations, which include COVID-19. They are written by several experts of mathematical analysis and allied areas.

Chapter 1 demonstrates theories of double sequences in asymptotic analysis of divergent processes (Karamata theory of regular variation and de Haan theory of rapid variation). Moreover, it shows applications of double sequences in selection principle theory, mainly in S_1-type and α_2-type selection principles.

Chapter 2 investigates some problems concerning lacunary statistical cluster and λ-statistical cluster points of triple sequences via ideals in finite dimensional spaces, and some of its properties in finite dimensional Banach spaces.

A systematic development of the notion of relative uniform convergence in the field of functions of real variables is discussed in Chapter 3.

In Chapter 4, the Nörlund matrix, generalized difference matrices, and Orlicz function are applied to generate some spaces of almost convergent sequences over n-normed spaces. Further, various topological and algebraic properties are investigated in newly defined sequence spaces.

Chapter 5 discusses factorization for infinite Hilbert and Cesàro operators based on Cesàro and Gamma operators. Generalized versions of two well-known Hilbert and Hardy's inequalities are obtained.

In Chapter 6, several results of the Galamnos-Bojanić-Seneta type in the theory of regularly, rapidly varying functions and sequences are obtained. These results give a unified approach to the study of functional and sequential directions in the regular variation (in the sense of Karamata) and rapid variation (in the sense of de Haan). It also presents several new results for the classes Pl_f^* and Pl_s^* of functions and sequences.

Chapter 7 introduces generalized Banach sequence spaces e_p^q and e_∞^q by using quantum analog of Euler matrix of order 1 in ℓ_p and ℓ_∞, respectively. Moreover, basis for e_p^q, Köthe duals of spaces e_p^q and e_∞^q are obtained. Theorems characterizing some classes of compact operators on e_p^q and e_∞^q using Hausdorff measure of non-compactness are obtained.

Chapter 8 discusses construction of linear positive operators which solely depend on the generating functions of the multivariate polynomials viz. multivariate Lagrange polynomials, multivariate q (or basic)-analogue of the Lagrange polynomials, q (or basic)-Lagrange-Hermite polynomials, etc.

Chapter 9 introduces interpolative Boyd-Wong contraction mapping on quasi-partial b-metric space and applies it in sensitivity analysis of experimental signals and synthesis of scientific data.

Chapter 10 aims to investigate differential transformation on some functional differential equations. It is applied on some proportional linear and nonlinear delay differential equations, which in turn converts the problem into a system of ordinary differential equations. To deal with non-linearity, Faà di Bruno's formula is applied.

Chapter 11 proposes some new fixed point theorems involving measure of non-compactness and shifting distance functions. Further, the existence of solution of fractional integral equation is obtained in Banach space.

Chapter 12 presents some fractional and generalized integral operators, and analyzes their relationships with different local differential operators. This chapter also establishes new Hermite-Hadamard inequalities for h-convex functions.

In Chapter 13, new classes of the exponentially biconvex functions are introduced. Some relationships between various concepts of exponentially biconvex functions are established. It is shown that optimality conditions of differentiable exponentially biconvex functions can be characterized by exponentially variational inequalities.

A new class $k\text{-}\tilde{U}S_s(\tau, \gamma, t)$ of analytic functions in the open unit disc U with negative coefficients is introduced in Chapter 14. This chapter also determines coefficient estimates, neighborhood properties, and partial sums for functions f belonging to this class.

Chapter 15 introduces a new subclass of meromorphically uniformly convex functions with positive coefficients defined by a differential operator and obtains coefficient estimates, growth and distortion theorem, radius of convexity, integral transforms, convex linear combinations, convolution properties, and neighborhoods for the class $\sigma_\chi^+(\alpha, \beta)$.

Chapter 16 gives a new subclass $\mathcal{CH}_\Sigma(\lambda, \mu, x)$ of bi-univalent functions that are defined in the open unit disc $\mathcal{D}$ using the Chebishev polynomials. Also, the sharp initial two Taylor-Maclaurin coefficient bounds and the sharp estimate on the Fekete-Szegö functional functions are obtained in $\mathcal{CH}_\Sigma(\lambda, \mu, x)$.

In Chapter 17, the operators $IR_{\lambda,l}^{m,n}$ is defined as the Hadamard product of the multiplier transformation $I(m, \lambda, l)$. Moreover, Ruscheweyh derivative R^n given by $IR_{\lambda,l}^{m,n} : \mathcal{A} \to \mathcal{A}$, $IR_{\lambda,l}^{m,n} f(z) = (I(m, \lambda, l) * R^n) f(z)$ and $\mathcal{A}_n = \{f \in \mathcal{H}(U) : f(z) = z + a_{n+1}z^{n+1} + ..., z \in U\}$ are introduced. They are classes of normalized analytic functions with $\mathcal{A}_1 = \mathcal{A}$. Also, certain subordination and superordination results involving the operator $IR_{\lambda,l}^{m,n}$ are derived.

Chapter 18 investigates monofractal dimension and multifractal dimension by using Higuchi method, multifractal detrended fluctuation analysis, etc. To examine the non-linearity, delay vector variance analysis is considered. It is shown that the spread of COVID-19 in Indian context is yet not capable of self repair after the possible damages caused due to the intervention of vaccine drive.

Bipan Hazarika, Gauhati University, India
Santanu Acharjee, Gauhati University, India
H. M. Srivastava, University of Victoria, Canada

Editors

Bipan Hazarika is a Professor of Department of Mathematics, Gauhati University, Guwahati, Assam. Earlier, he worked at Rajiv Gandhi University, Arunachal Pradesh, India from 2005 to 2017. He was Professor at Rajiv Gandhi University upto 2017. He received an M.Sc. and Ph.D. from Gauhati University, Guwahati, India. His main research areas are sequences spaces, summability theory, applications of fixed point theory, fuzzy analysis, and non-absolute integrable function spaces. He published more than 160 research articles in several renowned international journals. He is a regular reviewer of more than 50 different journals published by Springer, Elsevier, Taylor and Francis, Wiley, IOS Press, World Scientific, American Mathematical Society, IEEE, De Gruyter, etc. He published books on Differential Equations, Differential Calculus and Integral Calculus. Recently he edited following books: "Sequence Spaces Theory and Applications, Chapman and Hall/CRC" and "Fixed Point Theory and Fractional Calculus: Recent Advances and Applications, Springer". He is an editorial board member of more than 6 international journals and guest editor of a special issue named "Sequence spaces, Function spaces and Approximation Theory" of Journal of Function Spaces.

Santanu Acharjee is an Assistant Professor of Department of Mathematics, Gauhati University, Guwahati, Assam, India. He pursued an M.Sc. and Ph.D. in Mathematics from Gauhati University in 2011 and 2016 respectively. His areas of specialization are topology, soft computing, artificial intelligence, mathematical social science, mathematical economics, social networks, human trafficking and anti-terrorism research, etc. He has published more than twenty five research articles and one book chapter. He has been collaborating with several eminent researchers of various well-known institutes viz. University of Oxford, Creighton University, University of Auckland, Kuwait University, University of California-Riverside, Russian Academy of Science, etc. He is a member of American Mathematical Society (USA), Life member of Indian Science Congress Association (India) and a member of International Association of Engineers (Hong Kong). He is a reviewer of more than 38 different journals, some of them are: Psychological Methods, Journal of Theoretical and Philosophical Psychology, Chaos, Solitons and Fractals, Information Sciences, International Journal of Fuzzy System, Soft Computing, Journal of Mathematics and Computer Science, Journal of Logic, Language and Information, Filomat, Archive for Mathematical Logic, The B.E. Journal of

Theoretical Economics, Theoretical Computer Sciences, Journal of Mathematical Psychology, ACM Transactions on Computational Logic, Topology and Its Applications, Fuzzy Sets and Systems, Journal of Artificial Intelligence Research, International Journal of Economic Theory, Philosophical Studies, Theory and Decision, Mathematics in Computer Science, Algebra: Logic and Number Theory, Social Choice and Welfare, Mathematical Social Sciences, Journal of Advanced Studies in Topology, Fuzzy Optimization and Decision making, International Journal of Intelligent Systems, etc.

Moreover, he is a regular reviewer for Mathematical Reviews (AMS), Zb-MATH Open (Germany) and several journals of American Psychological Association. He is an editorial board member of several journals. He has delivered seven invited talks in international conferences. Moreover, he was invited as Visiting Researcher by The Fields Institute for Research in Mathematical Sciences, Canada. NBHM (Govt. of India) granted travel grant for his visit to Canada. He jointly introduced a new area of mathematical research named "Bitopological dynamical system" in the year 2020.

H. M. Srivastava is a Professor Emeritus of Department of Mathematics and Statistics, University of Victoria, Canada. He joined the University of Victoria in 1969 as Associate Professor (1969–1974) and then became Full Professor (1974–2006). Professor Srivastava has held numerous visiting positions including those at West Virginia University in the U.S.A (1967–1969), Universite Laval in Canada (1975), and the University of Glasgow in the U.K. (1975–1976) and indeed also at many other universities and research institutes in different parts of the world. Professor Srivastava has published 21 books (monographs and edited volumes), 30 book chapters, 45 papers in international conference proceedings, and over 1,000 scientific research journal articles on various topics of mathematical analysis and applied mathematics.

Currently, he is actively associated (as editor, honorary editor, senior editor, associate editor, or editorial board member) with over 200 international scientific research journals. His biographical sketches have appeared in various issues of more than 50 international biographies, directories, etc. Professor Srivastava was considered as highly-cited researcher for the years 2015, 2017, 2018, and 2020 by Clarivate Analytics (Web of Science). He has also been listed and ranked in the twelfth place in General Mathematics among the top two percent.

Contributors

P. N. Agrawal
Department of Mathematics, Indian Institute of Technology Roorkee
Roorkee, India
https://orcid.org/0000-0003-3029-6896

Alb Lupaş Alina
Department of Mathematics and Computer Science, University of Oradea, Universitatii street, 410087
Oradea, Romania
https://orcid.org/0000-0002-2855-7535

Behar Baxhaku
Department of Mathematics, University of Prishtina
Prishtina, Kosovo
https://orcid.org/0000-0002-8990-1440

Anuska Chanda
Department of Electrical Engineering, University Institute of Technology, The University of Burdwan
Golapbag (North), Burdwan, West Bengal, India
https://orcid.org/0000-0003-3257-786X

Sankalpa Chowdhury
Department of Computer Science Engineering, University Institute of Technology, The University of Burdwan
Golapbag (North), Burdwan, West Bengal, India
https://orcid.org/0000-0002-3673-1812

Anupam Das
Department of Mathematics, Cotton University
Guwahati, Assam, India
https://orcid.org/0000-0002-1529-9266

Dibakar Das
Department of Electrical Engineering, University Institute of Technology, The University of Burdwan
Golapbag (North), Burdwan, West Bengal, India
https://orcid.org/0000-0003-1624-1231

Gourab Das
Department of Electrical Engineering, University Institute of Technology, The University of Burdwan
Golapbag (North), Burdwan, West Bengal, India
https://orcid.org/0000-0001-6502-0764

Bhuban Chandra Deuri
Department of Mathematics, Rajiv
  Gandhi University, Rono Hills,
  Doimukh
Arunachal Pradesh, India
https://orcid.org/0000-0002-2966-5220

Kshetrimayum Renubebeta Devi
Department of Mathematics, Tripura
  University, Agartala
Tripura, India
https://orcid.org/0000-0003-2819-3193

Manisha Devi
Department of Mathematics, Shri
  Mata Vaishno Devi University
J & K, India
https://orcid.org/0000-0001-5150-475X

Dragan Djurčič
University of Kragujevac, Faculty of
  Technical Sciences in Čačak
Čačak, Serbia
https://orcid.org/0000-0002-7141-4724

Pragati Gautam
Department of Mathematics, Kamala
  Nehru College, University of
  Delhi, August Kranti Marg
New Delhi, India
https://orcid.org/0000-0002-4031-3672

Mehmet Gürdal
Department of Mathematics,
  Suleyman Demirel University
Isparta, Turkey
https://orcid.org/0000-0003-0866-1869

Koushik Ghosh
Department of Mathematics,
  University Institute of Technology,
  The University of Burdwan
Golapbag (North), Burdwan, West
  Bengal, India
https://orcid.org/0000-0003-2138-6320

Mualla Birgül Huban
Isparta University of Applied
  Sciences
Isparta, Turkey
https://orcid.org/0000-0003-2710-8487

Swapnesh Khamaru
Department of Applied Electronics
  and Instrumentation
  Engineering,University Institute
  of Technology, The University of
  Burdwan
Golapbag (North), Burdwan, West
  Bengal, India
https://orcid.org/0000-0003-2244-9522

Ljubiša D.R. Kočinac
University of Niš, Faculty of Sciences
  and Mathematics
Niš, Serbia
https://orcid.org/0000-0002-4870-7908

Anil Kumar
Department of Mathematics &
  Statistics, Manipal University
  Jaipur
Rajasthan, India
https://orcid.org/0000-0002-8063-2109

Giriraj Methi
Department of Mathematics & Statistics, Manipal University Jaipur
Rajasthan, India
https://orcid.org/0000-0001-7702-5183

Muhammad Aslam Noor
Department of Mathematics, COMSATS University
Islamabad, Pakistan
https://orcid.org/0000-0001-6105-2435

Khalida Inayat Noor
Department of Mathematics, COMSATS University
Islamabad, Pakistan
https://orcid.org/0000-0002-5000-3870

Amol B. Patil
Department of Mathematics, AISSMS College of Engineering, Pune
Maharashtra, India
https://www.orcid.org/0000-0002-2907-8376

Florencia Rabossi
UNNE, FaCENA, Ave. Libertad 5450
Corrientes 3400, Argentina
https://orcid.org/0000-0001-9482-3734

Kuldip Raj
Department of Mathematics, Shri Mata Vaishno Devi University
J & K, India
https://orcid.org/0000-0002-2611-3391

P. Thirupathi Reddy
Department of Mathematics, School of Engineering, NNRESGI, Medichal-Dist-50088, Telengana, India
https://orcid.org/0000-0002-0034-444X

H. Roopaei
Department of Mathematics, University of Alberta Edmonton
Alberta, Canada
https://orcid.org/0000-0001-7190-3387

Shashikala A
Department of Mathematics, GSS GITAM University, Doddaballapur
Bengaluru North, Karnataka, India
https://orcid.org/0000-0002-4533-2987

Sujatha
Department of Mathematics, GSS GITAM University, Doddaballapur
Bengaluru North, Karnataka, India
https://orcid.org/0000-0002-2109-3328

Sanket Tikare
Department of Mathematics, Ramniranjan Jhunjhunwala College
Ghatkopar (West), Mumbai, India
https://orcid.org/0000-0002-9000-3031

Binod Chandra Tripathy
Department of Mathematics, Tripura University
Agartala, Tripura, India
https://orcid.org/-0002-0738-652

Juan E. Nápoles Valdes
UNNE, FaCENA, Ave. Libertad 5450
Corrientes 3400, Argentina
https://orcid.org/0000-0003-2470-1090

B. Venkateswarlu
Department of Mathematics, GSS, GITAM University, Doddaballapur
Bengaluru North, Karnataka, India
https://orcid.org/0000-0003-3669-350X

Swapnil Verma
Department of Mathematics, Kamala Nehru College, University of Delhi, August Kranti Marg
New Delhi, India
https://orcid.org/0000-0002-2041-3334

Taja Yaying
Department of Mathematics, Dera Natung Government College
Itanagar, Arunachal Pradesh, India
https://orcid.org/0000-0003-3435-8417

Chapter 1

Some applications of double sequences

Dragan Djurčić

Ljubiša D.R. Kočinac

1.1	Introduction	1
	1.1.1 Double sequences	1
	1.1.2 Selection principles	4
	1.1.3 Asymptotic analysis	5
1.2	S$_1$ selection principle and double sequences	6
1.3	α_2-selection principle and double sequences	9
1.4	Double sequences and the exponent of convergence	14
	Bibliography	17

1.1 Introduction

In this chapter we discuss relationships between the theory of real double sequences and selection principles theory, a field of mathematics having nice and deep relations with various mathematical disciplines: game theory, combinatorics, function spaces, and so on. By $\mathbb{N}$ and $\mathbb{R}$, we denote the set of natural numbers and the set of real numbers, respectively. Single sequences will be denoted by $\mathbf{x} = (x_n)_{n \in \mathbb{N}}$, $\mathbf{y} = (y_n)_{n \in \mathbb{N}}$, and so on, while double sequences will be denoted by $\mathbf{X} = (x_{m,n})_{m,n \in \mathbb{N}}$, $\mathbf{Y} = (y_{m,n})_{m,n \in \mathbb{N}}$, and so on. We use the symbol c_2 to denote the set of real double sequences; $c_{2,+}$ denotes the set of double sequences of positive real numbers.

1.1.1 Double sequences

In 1900, Alfred Israel Pringsheim introduced the concept of convergence of real double sequences:

1. A double sequence $\mathbf{X} = (x_{m,n})_{m,n\in\mathbb{N}}$ converges to $a \in \mathbb{R}$ (notation P-lim $\mathbf{X} = a$ or P-lim $x_{m,n} = a$), if $\lim_{\min\{m,n\}\to\infty} x_{m,n} = a$, i.e., if for every $\varepsilon > 0$ there is $n_0 \in \mathbb{N}$ such that $|x_{m,n} - a| < \varepsilon$ for all $m,n > n_0$ (see [29], and also [14,31]). The limit a is called the *Pringsheim limit* of $\mathbf{X}$.

   In this chapter we denote by $c_2^{a,P}$ the set of all double real sequences converging to a point $a \in \mathbb{R}$ in Pringsheim's sense, and similarly for $c_{2,+}^{a,P}$. We also consider the following two kinds of convergence of double sequences.

2. A double sequence $\mathbf{X} = (x_{m,n})_{m,n\in\mathbb{N}}$ *max-converges* to $a \in \mathbb{R}$ (notation max-lim $\mathbf{X} = a$ or max-lim $x_{m,n} = a$), if $\lim_{\max\{m,n\}\to\infty} x_{m,n} = a$, i.e., if for every $\varepsilon > 0$ there is $n_0 \in \mathbb{N}$ such that $|x_{m,n} - a| < \varepsilon$ for all $m > n_0$ or for all $n > n_0$.

   $c_2^{a,\max}$ denotes the class of real double sequences such that $\lim_{\max\{m,n\}\to\infty} x_{m,n} = a$.

3. A double sequence $\mathbf{X} = (x_{m,n})_{m,n\in\mathbb{N}}$ *sum-converges* to $a \in \mathbb{R}$ (notation sum-lim $\mathbf{X} = a$ or sum-lim $x_{m,n} = a$), if $\lim_{m+n\to\infty} x_{m,n} = a$, i.e., if for every $\varepsilon > 0$ there is $n_0 \in \mathbb{N}$ such that $|x_{m,n} - a| < \varepsilon$ for all $m,n \in \mathbb{N}$ such that $m + n > n_0$.

   $c_2^{a,\mathrm{sum}}$ denotes the set of real double sequences sum-converging to a.

Observe
$$c_2^{a,\mathrm{sum}} \subsetneq c_2^{a,\max} \subsetneq c_2^{a,P}.$$

The most investigated convergence of double sequence is P-convergence. A considerable number of papers which appeared in recent years study mostly the set $c_2^{a,P}$ and its subsets from different points of view (see, for instance, the papers [1,10,11,19–21,23–27,32,33] and the books [22,34]). Some results in this investigation generalize known results concerning single sequences to certain classes of double sequences, while other results reflect the specific nature of the Pringsheim convergence (for example, a double sequence may converge without being bounded).

In [13], Hardy introduced the notion of *regular convergence* for double sequences: a double sequence $\mathbf{X} = (x_{m,n})_{m,n\in\mathbb{N}}$ regularly converges to a point $a \in \mathbb{R}$ if it P-converges to a and for each $m \in \mathbb{N}$ and each $n \in \mathbb{N}$ there exist the following two limits:
$$\lim_{n\to\infty} x_{m,n} = R_m, \quad \lim_{m\to\infty} x_{m,n} = C_n.$$

$\bar{c}_2^{a,P}$ denotes the set of elements $(x_{m,n})_{m,n\in\mathbb{N}}$ in $c_2^{a,P}$ which are bounded, regular and such that $\lim_{m\to\infty} x_{m,n} = \lim_{n\to\infty} x_{m,n} = a$.

A double sequence $\mathbf{X} = (x_{m,n})_{m,n\in\mathbb{N}}$ is *bounded* if there is $M > 0$ such that $|x_{m,n}| < M$ for all $m,n \in \mathbb{N}$.

Notice that a P-convergent double sequence need not be bounded.

If P-lim $|\mathbf{X}| = \infty$, (equivalently, for every $M > 0$ there are $n_1, n_2 \in \mathbb{N}$ such that $|x_{m,n}| > M$ whenever $m \geq n_1$, $n \geq n_2$), then $\mathbf{X}$ is said to be *definitely divergent*.

We give now a few facts which will be used in the sequel without special mention.

Fact 1. To each double sequence $\mathbf{X} = (x_{m,n})_{m,n \in \mathbb{N}}$, the following single sequences are assigned:

1.1. The *Landau-Hurwicz sequence* $\omega(\mathbf{X}) = (\omega_n(\mathbf{X}))_{n \in \mathbb{N}}$, where for each $n \in \mathbb{N}$

$$\omega_n(\mathbf{X}) = \sup\{|x_{k,l} - x_{p,q}| : k \geq n, l \geq n, p \geq n, q \geq n\}.$$

1.2. The *diagonal sequence* $d(\mathbf{X}) = (d_n(\mathbf{X}))_{n \in \mathbb{N}}$, where

$$d_n(\mathbf{X}) = \sum_{k=1}^{n}\left(\sum_{l=1}^{n} x_{k,l}\right).$$

If there is $D(\mathbf{X}) \in \mathbb{R}$ such that $\lim_{n \to \infty} d_n(\mathbf{X}) = D(\mathbf{X})$, one says that $\mathbf{X}$ has the *finite diagonal sum*, denoted by D-$\Sigma\mathbf{X}$.

ℓ_2^D denotes the class of double sequences $\mathbf{X}$ from c_2 with finite diagonal sum D-$\Sigma\mathbf{X}$.

1.3. The sequence $v(\mathbf{X}) = (v_n(\mathbf{X}))_{n \in \mathbb{N}}$, where

$$v_n(\mathbf{X}) = \sum_{i=1}^{n-1}(x_{i,n} + x_{n,i}) + x_{n,n}.$$

Fact 2. To each double sequence $\mathbf{X} = (x_{m,n})_{m,n \in \mathbb{N}}$, one assigns the double sequence

$$\mathbf{S}(\mathbf{X}) = (s_{m,n}(\mathbf{X}))_{m,n \in \mathbb{N}}, \text{ where } s_{m,n}(\mathbf{X}) = \sum_{k=1}^{m}\left(\sum_{l=1}^{n} x_{k,l}\right).$$

If there is a number $T(\mathbf{X}) = P - \lim \mathbf{S}(\mathbf{X})$, then we say that $\mathbf{X}$ has the *finite Pringsheim sum*, denoted by P-$\Sigma\mathbf{X}$.

ℓ_2^P denotes the class of double sequences $\mathbf{X} \in \mathbb{S}_2$ with finite Pringsheim sum P-$\Sigma\mathbf{X}$.

Fact 3. If we have a double sequence $(\mathbf{X}^{k,l})_{k,l \in \mathbb{N}}$ of double sequences $\mathbf{X}^{k,l} = (x_{m,n}^{k,l})_{m,n \in \mathbb{N}}$, then we can arrange it in a sequence $(\mathbf{X}^i = (x_{m,n}^i)_{m,n \in \mathbb{N}} : i \in \mathbb{N})$ of double sequences.

The following proposition shows a connection between P-converges of double sequences and their Landau-Hurwicz sequences.

Proposition 1.1.1. ([2]) *A double sequence* $\mathbf{X} = (x_{m,n})_{m,n \in \mathbb{N}}$ *belongs to the class* $c_2^{a,P}$, $a \in \mathbb{R}$, *if and only if* $\lim_{n \to \infty} \omega_n(\mathbf{X}) = 0$.

Proof. ($\Rightarrow$) Assume that the double sequence $\mathbf{X}$ belongs to $c_2^{a,P}$ for an arbitrary and fixed $a \in \mathbb{R}$. Let $\varepsilon > 0$ be given. There is $n_0 \in \mathbb{N}$ such that $|x_{j,k} - a| \leq \varepsilon/2$ for each $j \geq n_0$ and each $k \geq n_0$. Therefore, we have

$$|x_{j,k} - x_{r,s}| = |x_{j,k} - a + a - x_{r,s}| \leq |x_{j,k} - a| + |x_{r,s} - a| \leq \varepsilon/2 + \varepsilon/2$$

for all $j, k, r, s \geq n_0$. This implies that for each $n \geq n_0$ we have

$$0 \leq \omega_n(\mathbf{X}) \leq \sup\{|x_{j,k} - x_{r,s}| : j \geq n_0, k \geq n_0, r \geq n_0, s \geq n_0\} \leq \varepsilon,$$

i.e., $\lim_{n \to \infty} \omega_n(\mathbf{X}) = 0$.

($\Leftarrow$) Let $\mathbf{X} = (x_{m,n})_{m,n \in \mathbb{N}}$ be a double sequence with $\lim_{n \to \infty} \omega_n(\mathbf{X}) = 0$. For a given $\varepsilon > 0$, there is $n_1 \in \mathbb{N}$ such that $0 \leq |x_{j,k} - x_{r,s}| \leq \varepsilon/2$ for $j \geq n_1$, $k \geq n_1$, $r \geq n_1$, $s \geq n_1$, because

$$0 \leq \omega_n(\mathbf{X}) = \sup\{|x_{j,k} - x_{r,s}| : j \geq n_1, k \geq n_1, r \geq n_1, s \geq n_1\} \leq \varepsilon/2$$

for $n \geq n_1$. Since for all $j, r \geq n_1$ it holds $|x_{j,j} - x_{r,r}| \leq \varepsilon/2$, it follows that the sequence $(x_{t,t})$ is convergent (as a Cauchy sequence), i.e., there is $A \in \mathbb{R}$ such that $\lim_{t \to \infty} x_{t,t} = A$. This implies there is $n_2 \in \mathbb{N}$ such that $|x_{t,t} - A| \leq \varepsilon/2$ for each $t \geq n_2$. Therefore, for $n_0 = \max\{n_1, n_2\}$ and all $j, k \geq n_0$ we have

$$|x_{j,k} - A| \leq |x_{j,k} - x_{j,j}| + |x_{j,j} - A| \leq \varepsilon.$$

$\square$

1.1.2 Selection principles

Selection principles theory is an old theory with roots in 1920s and 1930s. Nowadays, it is one of the most investigated areas of mathematics. For more details concerning this theory, see, for example, [15]. In this chapter we will discuss the selection principles related to collections of single or double sequences.

Let $\mathcal{A}$ and $\mathcal{B}$ be (not necessarily distinct) subfamilies of c_2. Then:

1. $\mathsf{S}_1(\mathcal{A}, \mathcal{B})$ denotes the selection hypothesis: for each sequence $(A_n : n \in \mathbb{N})$ of elements in $\mathcal{A}$, there is a sequence $(a_n : n \in \mathbb{N})$ such that for each n, $a_n \in A_n$ and $(a_n : n \in \mathbb{N}) \in \mathcal{B}$ [15].

2. $\mathsf{S}_1^{(d)}(\mathcal{A}, \mathcal{B})$ denotes the selection hypothesis: for each double sequence $(\mathbf{A}_{m,n} : m, n \in \mathbb{N})$ of elements of $\mathcal{A}$, there are elements $a_{m,n} \in \mathbf{A}_{m,n}$ such that the double sequence $(a_{m,n})_{m,n \in \mathbb{N}}$ belongs to $\mathcal{B}$ [8].

3. Consider now an order on the set $\mathbb{N} \times \mathbb{N}$. Let $\varphi : \mathbb{N} \times \mathbb{N} \to \mathbb{N}$ be a bijection. Set $(m_1, n_1) \leq_\varphi (m_2, n_2) \Leftrightarrow \varphi(m_1, n_1) \leq \varphi(m_2, n_2)$, where $\leq$ is the natural order in $\mathbb{N}$.

    $\mathsf{S}_1^\varphi(\mathcal{A}, \mathcal{B})$ denotes the selection hypothesis: for each sequence $(\mathbf{A}_n : n \in \mathbb{N})$ of elements of $\mathcal{A}$, there is an element $\mathbf{B} = (b_{\varphi^{-1}(n)})_{n \in \mathbb{N}}$ in $\mathcal{B}$ such that $b_{\varphi^{-1}(n)} \in \mathbf{A}_n$ for all $n \in \mathbb{N}$ [8].

4. $\alpha_2(\mathcal{A}, \mathcal{B})$ denotes the selection hypothesis: for each sequence $(\mathbf{A}_n : n \in \mathbb{N})$ of elements of $\mathcal{A}$, there is an element $\mathbf{B}$ in $\mathcal{B}$ such that $\mathbf{B} \cap \mathbf{A}_n$ is infinite for all $n \in \mathbb{N}$ [16, 17].

5. $\alpha_2^{(d)}(\mathcal{A}, \mathcal{B})$ denotes the selection hypothesis: for each double sequence $(\mathbf{A}_{m,n} : m, n \in \mathbb{N})$ of elements of $\mathcal{A}$, there is an element $\mathbf{B}$ in $\mathcal{B}$ such that $\mathbf{B} \cap \mathbf{A}_{m,n}$ is infinite for all $(m, n) \in \mathbb{N} \times \mathbb{N}$ [8].

Notice that to each of these selection principles one associates, in a natural way, an infinitely long two person game.

1.1.3 Asymptotic analysis

In [5] (see also [18] about asymptotic analysis of divergent processes) the class $\mathsf{Tr}(\mathsf{R}_{\mathsf{s},-\infty})$ of translationally rapidly varying (single) sequences was introduced and studied: a sequence $\mathbf{x} = (x_n)_{n \in \mathbb{N}}$ of positive real numbers belongs to the class $\mathsf{Tr}(\mathsf{R}_{\mathsf{s},-\infty})$ of *translationally rapidly varying sequences* if

$$\lim_{n \to \infty} \frac{x_{[n+\alpha]}}{x_n} = 0$$

for each $\alpha \geq 1$. Here $[r]$ denotes the integer part of $r \in \mathbb{R}$.

By $\ell_{2,\mathsf{Tr}(\mathsf{R}_{\mathsf{s},-\infty})}^{D}$, we denote the subclass of ℓ_2^D consisting of double sequences $\mathbf{X}$ such that $\omega(\mathbf{S}(\mathbf{X})) \in \mathsf{Tr}(\mathsf{R}_{\mathsf{s},-\infty})$.

A generalization of this notion to double sequences is given in the following definition.

Definition 1.1.2. ([2]) A double sequence $\mathbf{X} = (x_{m,n})_{m,n \in \mathbb{N}} \in c_{2,+}$ belongs to the class $\mathsf{Tr}(\mathsf{R}_{\mathsf{s}_2,-\infty})$ of *translationally rapidly varying double sequences* if

$$\lim_{\min\{m,n\} \to \infty} \frac{x_{[m+\alpha],[n+\beta]}}{x_{m,n}} = 0$$

for each $\alpha \geq 0$ and each $\beta \geq 0$ such that $\max\{\alpha, \beta\} \geq 1$. Here $[x]$ denotes the integer part of $x \in \mathbb{R}$.

Example 1.1.3. The class $\mathsf{Tr}(\mathsf{R}_{\mathsf{s}_2,-\infty})$ is nonempty. The double sequence $(x_{m,n})_{m,n \in \mathbb{N}}$ is defined by

$$x_{m,n} = \frac{1}{(m+n)!}, \quad m \in \mathbb{N}, n \in \mathbb{N},$$

belongs to this class.

Proposition 1.1.4. ([2]) $\mathsf{Tr}(\mathsf{R}_{\mathsf{s}_2,-\infty}) \subset c_{2,+}^{0,P}$.

Proof. Let $\mathbf{X} = (x_{m,n})_{m,n \in \mathbb{N}} \in \mathsf{Tr}(\mathsf{R}_{\mathsf{s}_2,-\infty})$ and let $\varepsilon = \frac{1}{2}$, $\alpha = \beta = 1$. There is $n_0 = n_0(1/2, 1, 1) \in \mathbb{N}$ such that

$$\frac{x_{m+1,n+1}}{x_{m,n}} \leq \frac{1}{2}$$

for all $m, n \geq n_0$. For $m = n \geq n_0$, we have $x_{n+1,n+1} \leq \frac{1}{2}x_{n,n}$. Therefore, $\lim_{n\to\infty} x_{n,n} = 0$. Similarly, for $\varepsilon = \frac{1}{2}$ and $\alpha = 1$, $\beta = 0$, there is $n_1 = n_1(1/2, 1, 0) \in \mathbb{N}$ such that $x_{m+1,n} \leq \frac{1}{2}x_{m,n}$ for all $m, n \geq n_1$ which implies that for $n \geq n_1$, $\lim_{m\to\infty} x_{m,n} = 0$. Finally, for $\varepsilon = \frac{1}{2}$, $\alpha = 0$, $\beta = 1$ there is $n_2 = n_2(1/2, 0, 1) \in \mathbb{N}$ such that $x_{m,n+1} \leq \frac{1}{2}x_{m,n}$ for all $m, n \geq n_2$. From here we get $\lim_{n\to\infty} x_{m,n} = 0$, for each $m \geq n_2$.

Let now $\varepsilon > 0$ be arbitrary (and fixed). Then there is $n_\varepsilon \in \mathbb{N}$ such that $x_{n,n} \leq \varepsilon$ for each $n \geq n_\varepsilon$. Set $n^* = \max\{n_\varepsilon, n_1, n_2\}$. Then $x_{m,n} \leq \varepsilon$ for each $m, n \geq n^*$, which means that $\mathbf{X} \in c_{2,+}^{0,P}$. $\square$

The following example shows that the inclusion in the above proposition is proper.

Example 1.1.5. The double sequence $\mathbf{X} = (x_{m,n})_{m,n \in \mathbb{N}}$ defined by

$$x_{m,n} = \begin{cases} \frac{1}{m}, & \text{for } m \in \mathbb{N}, n \in \{1, 2, \cdots, m\}; \\ \frac{1}{n}, & \text{for } n \in \mathbb{N}, m \in \{1, 2, \cdots, n\}. \end{cases}$$

evidently belongs to the class $c_{2,+}^{0,P}$. However, it does not belong to $\mathsf{Tr}(\mathsf{R}_{\mathsf{s}_2,-\infty})$ because for $\alpha = \beta = 1$ and $m = n$ we have

$$\lim_{n\to\infty} \frac{x_{m+1,n+1}}{x_{m,n}} = \lim_{n\to\infty} \frac{n}{n+1} = 1.$$

1.2 S$_1$ selection principle and double sequences

In this section we present a few results which show applications of the selection principle S$_1$ to double sequences.

Theorem 1.2.1. ([8]) *For $a \in \mathbb{R}$ the selection principle* $\mathsf{S}_1^{(d)}(c_2^{a,P}, \overline{c}_2^{a,P})$ *is true.*

Proof. Let $(\mathbf{X}^{j,k} : j, k \in \mathbb{N})$ be a double sequence of elements in $c_2^{a,P}$, and assume that for all $j, k \in \mathbb{N}$, $\mathbf{X}^{j,k} = (x_{m,n}^{j,k})_{m,n \in \mathbb{N}}$. Construct the double sequence $\mathbf{Y} = (y_{m,n})_{m,n \in \mathbb{N}}$ in the following way:
1. $y_{1,1} = x_{m_1,m_1}^{1,1} \in \mathbf{X}^{1,1}$, where $m_1 \in \mathbb{N}$ is such that $\left|x_{m,n}^{1,1} - a\right| \leq \frac{1}{2}$ for each $m \geq m_1$ and each $n \geq m_1$.

For $s, t \in \mathbb{N}$ and $q = \max\{s, t\} \geq 2$, we select $y_{s,t}$ to be $x_{m_p, m_p}^{s,t} \in \mathbf{X}^{s,t}$, where
$$p = \begin{cases} (q-1)^2 + t, & \text{if } q = s; \\ (q-1)^2 + 2t - s, & \text{if } q = t. \end{cases}$$
and $|x_{m,n}^{s,t} - a| \leq \frac{1}{2^q}$ for each $m \geq m_p$ and each $n \geq m_p$.

We prove $\mathbf{Y} = (y_{m,n})_{m,n \in \mathbb{N}} \in c_2^{a,P}$. Let $\varepsilon > 0$ be given. Choose $r \in \mathbb{N}$ such that $\frac{1}{2^r} < \varepsilon$. For each $m \geq r$ and each $n \geq r$, by construction of $\mathbf{Y}$, we have $|y_{m,n} - a| \leq \frac{1}{2^r} < \varepsilon$, i.e., $\mathbf{Y} \in c_2^{a,P}$. From the of $\mathbf{Y}$, we also easily conclude that $\mathbf{Y}$ actually belongs to $\overline{c}_2^{a,P}$. The theorem is proved. □

Corollary 1.2.2. *For $a \in \mathbb{R}$ the selection principle $\mathsf{S}_1^{(d)}(c_2^{a,P}, c_2^{a,P})$ is satisfied.*

An improvement of this corollary is the following result.

Theorem 1.2.3. ([9]) *For a given $a \in \mathbb{R}$, the selection principle $\mathsf{S}_1^{(d)}(c_2^{a,P}, c_2^{a,\mathrm{sum}})$ holds.*

Proof. Let $(\mathbf{X}^{j,k} = (x_{m,n}^{j,k})_{m,n \in \mathbb{N}} : j, k \in \mathbb{N})$ be a double sequence of elements in $c_2^{a,P}$. We define a sequence $\mathbf{Y} = (y_{j,k})_{j,k \in \mathbb{N}}$ in the following way.

For fixed $j, k \in \mathbb{N}$ pick an element $x_{m,n}^{j,k} \in \mathbf{X}^{j,k}$ such that $|x_{m,n}^{j,k} - a| \leq \left(\frac{1}{2}\right)^{j+k-1}$. In this way we get the double sequence $\mathbf{Y}$. We prove $\mathbf{Y} \in c_2^{a,\mathrm{sum}}$. For each $\varepsilon > 0$ find $i_0 \in \mathbb{N}$ so that $\frac{1}{2^i} < \varepsilon$ for all $i \geq i_0$. Then for all $j, k \in \mathbb{N}$ such that $j + k - 1 \geq i_0$, it holds $|y_{j,k} - a| \leq \varepsilon$. Therefore, $\mathbf{Y} \in c_2^{a,\mathrm{sum}}$ which completes the proof. □

Theorem 1.2.4. ([12]) *The selection principle $\mathsf{S}_1^{(d)}(c_{2,+}^{0,P}, \ell_{2,\mathrm{Tr}(\mathsf{R}_{s,-\infty})}^{D})$ is satisfied.*

Proof. Let $(\mathbf{X}^{k,l} : k, l \in \mathbb{N})$ be a double sequence of double sequences $\mathbf{X}^{k,l} = (x_{m,n}^{k,l})_{m,n \in \mathbb{N}}$ in $c_{2,+}^{0,P}$. We construct the double sequence $\mathbf{Y} = (y_{k,l})_{k,l \in \mathbb{N}}$ in the following way ($n \in \mathbb{N}$):

[Step 1: $(k,l) = (1,1)$] Pick $y_{1,1}$ in the double sequence $\mathbf{X}^{1,1}$ such that $y_{1,1} \leq 1$;

[Step 2: $(k,l) \in \{(1,2),(2,1),(2,2)\}$] Choose $y_{k,l} \in \mathbf{X}^{k,l}$ so that $y_{k,l} \leq \frac{1}{2^2} \cdot \frac{v_1(\mathbf{Y})}{3}$;

[Step n: $(k,l) \in \{(i,n),(n,i) : i \leq n\}$] Choose $y_{k,l} \in \mathbf{X}^{k,l}$ so that $y_{k,l} \leq \frac{1}{n^2} \cdot \frac{v_{n-1}(\mathbf{Y})}{2n-1}$. And so on.

In this way we obtain that $\mathbf{Y}$ is a double sequence of positive real numbers.

Claim 1. $\mathbf{Y} \in \ell_2^D$.

Observe that for every $n \in \mathbb{N}$, $d_n(\mathbf{Y}) \leq \sum_{i=1}^n \frac{1}{i^2}$, since $d_n(\mathbf{Y}) = \sum_{i=1}^n v_i(\mathbf{Y})$.

Therefore, the sequence $(d_n(\mathbf{Y}))_{n \in \mathbb{N}}$ converges, i.e., $Y \in \ell_2^D$.

Claim 2. $\omega(\mathbf{S}(\mathbf{Y})) = (\omega_n(\mathbf{S}(\mathbf{Y})))_{n \in \mathbb{N}} \in \mathsf{Tr}(\mathsf{R}_{s,-\infty})$.

First, notice that for each $n \in \mathbb{N}$, $\omega_n(\mathbf{S}(\mathbf{Y})) = D - \Sigma \mathbf{Y} - d_n(\mathbf{Y})$. For sufficiently large $n \in \mathbb{N}$ we have

$$\frac{\omega_{n+1}(\mathbf{S}(\mathbf{Y}))}{\omega_n(\mathbf{S}(\mathbf{Y}))} = \frac{D - \Sigma \mathbf{Y} - d_{n+1}(\mathbf{Y})}{D - \Sigma \mathbf{Y} - d_n(\mathbf{Y})} = 1 - \frac{d_{n+1}(\mathbf{Y}) - d_n(\mathbf{Y})}{D - \Sigma \mathbf{Y} - d_n(\mathbf{Y})}$$

$$= 1 - \frac{v_{n+1}(\mathbf{Y})}{v_{n+1}(\mathbf{Y}) + v_{n+2}(\mathbf{Y}) + \ldots}$$

$$= 1 - \frac{1}{1 + \frac{v_{n+2}(\mathbf{Y})}{v_{n+1}(\mathbf{Y})} + \frac{v_{n+3}(\mathbf{Y})}{v_{n+1}(\mathbf{Y})} + \ldots}$$

$$= 1 - \frac{1}{1 + \frac{v_{n+2}(\mathbf{Y})}{v_{n+1}(\mathbf{Y})} + \frac{v_{n+3}(\mathbf{Y})}{v_{n+2}(\mathbf{Y})} \cdot \frac{v_{n+2}(\mathbf{Y})}{v_{n+1}(\mathbf{Y})} + \ldots}$$

$$\leq 1 - \frac{1}{1 + \frac{v_{n+2}(\mathbf{Y})}{v_{n+1}(\mathbf{Y})} + \frac{v_{n+3}(\mathbf{Y})}{v_{n+2}(\mathbf{Y})} + \ldots}.$$

Since the series $\sum_{n=1}^{\infty} \frac{v_{n+1}(\mathbf{Y})}{v_n(\mathbf{Y})}$ is convergent, we have that $\frac{v_{n+2}(\mathbf{Y})}{v_{n+1}(\mathbf{Y})} + \frac{v_{n+3}(\mathbf{Y})}{v_{n+2}(\mathbf{Y})} + \ldots$ tends to 0 for $n \to \infty$. Thus we conclude

$$\lim_{n \to \infty} \frac{\omega_{n+1}(\mathbf{S}(\mathbf{Y}))}{\omega_n(\mathbf{S}(\mathbf{Y}))} = 0$$

which means that $\omega(\mathbf{S}(\mathbf{Y})) \in \mathrm{Tr}(R_s, -\infty)$, i.e., $\mathbf{Y} \in \ell_{2,\mathrm{Tr}(R_s,-\infty)}^D$. This completes the proof of the theorem. □

Remark 1.2.5. The double sequence $\mathbf{Y}$ in the proof of the previous theorem P-converges to 0, i.e., $\mathbf{Y} \in c_{2,+}^{0,P}$. Indeed, since $\mathbf{Y} \in \ell_2^D$, for each $\varepsilon > 0$ there is $n_0 \in \mathbb{N}$ such that $v_n(\mathbf{Y}) \leq \varepsilon$ for all $n \geq n_0$. It follows that for all $p, q \in \mathbb{N}$, $y_{n_0+p,n_0+q} \leq \varepsilon$, hence $\mathbf{Y} \in c_{2,+}^{0,P}$.

Below, we give another result involving translational rapid variability.

Because the selection property S_1 is monotone in the second coordinate, by Proposition 1.1.4, the following theorem is an improvement of Corollary 1.2.2 (and Theorem 1.2.1).

Theorem 1.2.6. ([2]) *The selection principle* $\mathsf{S}_1^{(d)}(c_{2,+}^{0,P}, \mathrm{Tr}(R_{s_2},-\infty))$ *is satisfied.*

Proof. Let $(\mathbf{X}^{j,k} = (x_{m,n}^{j,k})_{m,n \in \mathbb{N}} : j,k \in \mathbb{N})$ be a double sequence of double sequences in $c_{2,+}^{0,P}$. We construct a new double sequence $\mathbf{Y} = (y_{j,k})_{j,k \in \mathbb{N}}$ as follows.
1. $y_{1,1} = x_{m,n}^{1,1}$ for arbitrary (and fixed) $m, n \in \mathbb{N}$.
2. Let $i \geq 2$.

(i) Choose $y_{i,1} = x_{m,n}^{i,1} \in \mathbf{X}^{i,1}$ so that $y_{i,1} < \left(\frac{1}{2}\right)^i y_{i-1,1}$. For $p \in \{2, 3, \cdots, i-1\}$ pick $y_{i,p} = x_{m,n}^{i,p}$ such that $y_{i,p} < \left(\frac{1}{2}\right)^i y_{i,p-1}$ and $y_{i,p} < \left(\frac{1}{2}\right)^i y_{i-1,p}$.

(ii) Similarly, $y_{1,i} = x_{m,n}^{1,i} \in \mathbf{X}^{1,i}$ such that $y_{1,i} < \left(\frac{1}{2}\right)^i y_{1,i-1}$. Select also $y_{p,i} = x_{m,n}^{p,i}$ such that $y_{p,i} < \left(\frac{1}{2}\right)^i y_{p-1,i}$ and $y_{p.i} < \left(\frac{1}{2}\right)^i y_{p,i-1}$.

(iii) Finally, choose $y_{i,i}$ to be some $x_{m,n}^{i,i} \in \mathbf{X}^{i,i}$ such that $y_{i,i} < \left(\frac{1}{2}\right)^i \min\{y_{i,i-1}, y_{i-1,i}\}$.

It remains to prove that $\mathbf{Y} \in \mathsf{Tr}(\mathsf{R}_{\mathsf{s}_2,-\infty})$. Let $\varepsilon > 0$ and $\alpha, \beta \geq 0$ with $\max\{\alpha, \beta\} \geq 1$ be given. Denote $h = h(\alpha, \beta) = [\alpha] + [\beta]$. Choose $r_0 \in \mathbb{N}$ such that $\left(\frac{1}{2}\right)^r \leq \varepsilon$ for all $r \geq r_0$. For $j \geq r_0$, $k \geq r_0$ we have

$$\frac{y_{j+1,k}}{y_{j,k}} \leq \left(\frac{1}{2}\right)^{r_0+1} \text{ and } \frac{y_{j,k+1}}{y_{j,k}} \leq \left(\frac{1}{2}\right)^{r_0+1},$$

and thus

$$\frac{y_{[j+\alpha],[k+\beta]}}{y_{j,k}} = \frac{y_{j+[\alpha],k+[\beta]}}{y_{j,k}} \leq \left(\frac{1}{2}\right)^{(r_0+1)h} \leq \left(\frac{1}{2}\right)^{r_0} \leq \varepsilon.$$

This means that $\mathbf{Y} \in \mathsf{Tr}(\mathsf{R}_{\mathsf{s}_2,-\infty})$. □

Theorem 1.2.7. ([8]) *Let $a \in \mathbb{R}$ and let $\leq_\varphi$ be as above. Then the selection hypothesis $\mathsf{S}_1^\varphi(c_2^{a,P}, \overline{c}_2^{a,P})$ is satisfied.*

Proof. Let $(\mathbf{X}^k : k \in \mathbb{N})$, $\mathbf{X}^k = (x_{m,n}^k)_{m,n \in \mathbb{N}}$, be a sequence in $c_2^{a,P}$. Construct a double sequence $\mathbf{Y} = (y_{s,t})_{s,t \in \mathbb{N}}$ as follows.

Fix $k \in \mathbb{N}$. Let $(s(k), t(k)) = \varphi^{-1}(k)$, and let $p(k) = \max\{s(k), t(k)\}$. There is $n_0(k) \in \mathbb{N}$ such that $|x_{m,n}^k - a| < 2^{-p(k)}$ for all $m, n \geq n_0(k)$. Set $y_{s(k),t(k)} = x_{n_0(k),n_0(k)}^k$ and $Y = (y_{s(k),t(k)})_{k \in \mathbb{N}}$. Then, by the construction, $\mathbf{Y} \in \overline{c}_2^a$ and $\mathbf{Y}$ has exactly one common element with $\mathbf{X}^k$ for each $k \in \mathbb{N}$, i.e., $\mathbf{Y}$ is the desired selector. □

Theorem 1.2.8. ([2]) *The selection principle $\mathsf{S}_1^\varphi(c_{2,+}^{0,P}, \mathsf{Tr}(\mathsf{R}_{\mathsf{s}_2,-\infty}))$ is satisfied.*

1.3 α_2-selection principle and double sequences

Now we give certain results about applications of α_2-type selection principles to double sequences.

Lemma 1.3.1. ([8]) *For $a \in \mathbb{R}$, the selection principle $\alpha_2(c_2^{u,\Gamma}, \overline{c}_2^{a,\Gamma})$ is satisfied.*

Proof. Let $(\mathbf{X}^k : k \in \mathbb{N})$ be a sequence of elements from $c_2^{a,P}$ and let for each $k \in \mathbb{N}$, $\mathbf{X}^k = (x_{m,n}^k)_{m,n \in \mathbb{N}}$.
1. Form first an increasing sequence $j_1 < j_2 < \cdots < j_i < \cdots$ in $\mathbb{N}$ so that:

1.a. $j_1 = \min\{n_0 \in \mathbb{N} : |x^1_{m,n} - a| \leq \frac{1}{2} \forall m \geq n_0 \text{ and } \forall n \geq n_0\}$;
1.b. Let $i \geq 2$. Find $p_i = \min\{n_0 \in \mathbb{N} : |x^i_{m,n} - a| \leq \frac{1}{2^i} \forall m, n \geq n_0\}$, and then define

$$j_i = \begin{cases} p_i, & \text{if } p_i > j_{i-1}; \\ j_{i-1} + 1, & \text{if } p_i \leq j_{i-1}. \end{cases}$$

2. Define now a double sequence $\mathbf{Y} = (y_{s,t})_{s,t \in \mathbb{N}}$ in this way:
2.a. $y_{s,t} = x^1_{s,t}$ for each $1 \leq s < j_2$, $t \in \mathbb{N}$, and each $1 \leq t < j_2$, $s \in \mathbb{N}$;
2.b. For $i \geq 2$, $y_{s,t} = x^i_{s,t}$, for $j_i \leq s < j_{i+1}$, $t \geq j_i$, and $j_i \leq t < j_{i+1}$, $s \geq j_i$.

By construction, $\mathbf{Y} \in c_2^a$ and $\mathbf{Y}$ has infinitely many common elements with each $\mathbf{X}^k$, $k \in \mathbb{N}$, i.e., the selection principle $\alpha_2(c_2^a, c_2^a)$ is satisfied. □

Remark 1.3.2. Using the technique from [4], we can prove that the double sequence $\mathbf{Y}$ in the proof of the previous lemma can be chosen in such a way that Y has infinitely many common elements with each $\mathbf{X}^k$, $k \in \mathbb{N}$, but on the same (corresponding) positions.

Let for each $k \in \mathbb{N}$, $\mathbf{x}^k$ denote the sequence $(x^k_{m,m})_{m \in \mathbb{N}}$. Then each $\mathbf{x}^k$ converges to a, so that we have the sequence $(\mathbf{x}^k : k \in \mathbb{N})$ of sequences converging to a. Let $2 = p_1 < p_2 < p_3 < \cdots$ be a sequence of prime natural numbers. Take sequence $\mathbf{x}^1 = (x^1_{m,m})_{m \in \mathbb{N}}$. For each $i \in \mathbb{N}$, replace the elements of $\mathbf{x}^1$ on the positions p_i^h, $h \in \mathbb{N}$, by the corresponding elements of the sequence $\mathbf{x}^{i+1}$. One obtains the sequence $(z_m)_{m \in \mathbb{N}}$ converging to a which has infinitely many common elements with each $\mathbf{x}^k$ on the same positions as in $\mathbf{x}^k$. Define now the double sequence $\mathbf{Y} = (y_{s,t})_{s,t \in \mathbb{N}}$ so that $y_{s,s} = z_s$, $s \in \mathbb{N}$, and $y_{s,t} = a$ whenever $s \neq t$. By construction, $\mathbf{Y} \in \overline{c}_2^{a,P}$ and Y have infinitely many common positions with each $\mathbf{X}^k$.

Theorem 1.3.3. *Let $a \in \mathbb{R}$ be given. The selection principle $\alpha_2^{(d)}(c_2^{a,P}, \overline{c}_2^{a,P})$ is true.*

Proof. Let $(\mathbf{X}^{j,k} : j, k \in \mathbb{N})$ be a double sequence of elements in $c_2^{a,P}$ and let $\mathbf{X}^{j,k} = (x^{j,k}_{m,n})_{m,n \in \mathbb{N}}$. In a standard way form from this double sequence, a sequence can be arranged $(x^i_{m,n})_{m,n \in \mathbb{N}}$. Apply now Lemma 1.3.1 to this sequence and find a double sequence $\mathbf{Y} \in c_2^{a,P}$ such that $\mathbf{Y} \cap \mathbf{X}^i$ is infinite for each $i \in \mathbb{N}$. But then $\mathbf{Y} \cap \mathbf{X}^{j,k}$ is infinite for all $j, k \in \mathbb{N}$. □

Remark 1.3.4. Notice that the double sequence $\mathbf{Y}$ from the proofs of Lemma 1.3.1 and Theorem 1.3.3 satisfies: (a) Y is bounded; (b) Y is regular and $\lim_{n \to \infty} y_{m,n} = \lim_{m \to \infty} y_{m,n} = a$ for each $m \in \mathbb{N}$ and each $n \in \mathbb{N}$.

From Theorem 1.3.3 we have the following corollary.

Corollary 1.3.5. *Let $a \in \mathbb{R}$ be given. The selection principle $\alpha_2^{(d)}(c_2^{a,P}, c_2^{a,P})$ is true.*

This corollary can be improved by replacing the second coordinate in it with a smaller class.

Theorem 1.3.6. ([2]) *The selection principle* $\alpha_2^{(d)}(c_{2,+}^{0,P}, \mathsf{Tr}(\mathsf{R}_{\mathsf{s}_2,-\infty}))$ *is satisfied.*

Proof. Let $(\mathbf{X}^{j,k} = (x_{m,n}^{j,k})_{m,n\in\mathbb{N}} : j,k \in \mathbb{N})$ be a double sequence of double sequences belonging to $c_{2,+}^{0,P}$. We are going to create a new double sequence $\mathbf{Y} = (y_{p,q})_{p,q\in\mathbb{N}}$ in the following way.
Step 1. Using some standard method, arrange the given double sequence $(\mathbf{X}^{j,k} : j,k \in \mathbb{N})$ of double sequences into a sequence $(\mathbf{x}^r = (x_{n,m}^r)_{m,n\in\mathbb{N}} : r \in \mathbb{N})$ from $c_{2,+}^{0,P}$.
Step 2. Consider the sequence of sequences $(x_{n,n}^r)_{n,r\in\mathbb{N}}$. Notice that for each $r \in \mathbb{N}$, $(x_{n,n}^r)_{n,r\in\mathbb{N}} \in \mathbb{S}_0$, where $\mathbb{S}_0$ denotes the set of all sequences of positive real numbers converging to 0 (see, for example, [7]). Let $\mathbb{T}_0$ be the set $\{(a_n)_{n\in\mathbb{N}} : a_1 > 0, a_{n+1} \le \frac{a_n}{n+1}\}$ of sequences of positive real numbers. It holds $\mathbb{T}_0 \subsetneq \mathbb{S}_0$ and the selection principle $\mathsf{S}_1(\mathbb{S}_0, \mathbb{T}_0)$ is satisfied.
Step 3. (In this part of the proof we use some techniques from [4]) Take an increasing sequence $2 = p_1 < p_2 < p_3 < \dots$ of prime numbers and a fixed $r \in \mathbb{N}$. Consider subsequences $(x_{p_t^n, p_t^n}^r)_{t\in\mathbb{N}}$ of the sequence $(x_{n,n}^r)_{n\in\mathbb{N}}$. These subsequences are in the class $\mathbb{S}_0$. Varying t and r in $\mathbb{N}$, arrange those subsequences in a sequence of sequences from $\mathbb{S}_0$. Apply $\mathsf{S}_1(\mathbb{S}_0, \mathbb{T}_0)$ and find a sequence $\mathbf{z} = (z_q)_{q\in\mathbb{N}} \in \mathbb{T}_0$ such that $\mathbf{z}$ has infinitely many elements with the sequence $(x_{n,n}^r)_{n\in\mathbb{N}}$ for each $r \in \mathbb{N}$. In other words, we conclude that the selection principle $\alpha_2(\mathbb{S}_0, \mathbb{T}_0)$ is true.

Let now $y_{q,q} = z_q$, $q \in \mathbb{N}$. For $q \ge 2$ we choose $y_{u,q} = \sqrt{u+1} \cdot y_{u+1,q}$ for $u \in \{1, 2, \dots, q-1\}$, and $y_{q,u} = \sqrt{u+1} \cdot y_{q,u+1}$. It is easy to see that the double sequence $\mathbf{Y}$ constructed in this way has infinitely many common elements with each double sequence $\mathbf{X}^{j,k} = (x_{m,n}^{j,k})_{m,n\in\mathbb{N}}$ for arbitrary and fixed $(j,k) \in \mathbb{N} \times \mathbb{N}$.

It remains to prove $\mathbf{Y} \in \mathsf{Tr}(\mathsf{R}_{\mathsf{s}_2,-\infty})$. Let $\varepsilon > 0$ and $\alpha \ge 0$, $\beta \ge 0$ with $\max\{\alpha,\beta\} \ge 1$, be given. Set $h = [\alpha] + [\beta]$. There is $n_0 \in \mathbb{N}$ such that $\left(\frac{1}{\sqrt{N+1}}\right)^h \le \varepsilon$ for each $N \in \mathbb{N}$ with $N \ge n_0$ ($n_0 \ge \varepsilon^{-(2/h)} - 1$). For $p, t \ge n_0$ we have

$$\frac{y_{p+1,t}}{y_{p,t}} \le \frac{1}{\sqrt{n_0+1}} \quad \text{and} \quad \frac{y_{p,t+1}}{y_{p,t}} \le \frac{1}{\sqrt{n_0+1}}.$$

So

$$\frac{y_{[p+\alpha],[t+\beta]}}{y_{p,t}} = \frac{y_{p+[\alpha],t+[\beta]}}{y_{p,t}} \left(\frac{1}{\sqrt{n_0+1}}\right)^h \le \varepsilon,$$

i.e., $\mathbf{Y} \in \mathsf{Tr}(\mathsf{R}_{\mathsf{s}_2,-\infty})$. $\square$

Theorem 1.3.7. ([12]) *The selection principle* $\alpha_2^{(d)}(c_{2,+}^{0,P}, \ell_2^D)$ *is satisfied.*

Proof. Let $(\mathbf{X}^{k,l} : k, l \in \mathbb{N})$ be a double sequence of double sequences $\mathbf{X}^{k,l} = (x_{m,n}^{k,l})_{m,n\in\mathbb{N}}$ in $c_{2,+}^{0,P}$. In a standard way we reorganize $(\mathbf{X}^{k,l} : k, l \in \mathbb{N})$ in a sequence $(\mathbf{X}^t = (x_{m,n}^t)_{m,n\in\mathbb{N}} : t \in \mathbb{N})$ of double sequences such that for each

$t \in \mathbb{N}$, the double sequence $\mathbf{X}^t \in c_{2,+}^{0,P}$. We construct the double sequence $\mathbf{Y} = (y_{i,j})_{i,j \in \mathbb{N}}$ as follows.
[Step 1: $i = 1$]: Let $\mathbf{Y}_1 = (y_{s,t}^1)_{s,t \in \mathbb{N}}$ be a double sequence such that for all $s, t \in \mathbb{N}$ it holds $0 < y_{s,t}^1 \leq \frac{1}{M^2(2M-1)}$, where $M = \max\{s, t\}$.
[Step 2: $i \geq 2$] Suppose that double sequences $\mathbf{Y}_1, \mathbf{Y}_2, \ldots, \mathbf{Y}_{i-1}$ be defined. We construct the sequence $\mathbf{Y}_i$.

Take an increasing sequence $(p_i)_{i \in \mathbb{N}}$ of prime numbers with $p_1 = 2$ and a bijection $\varphi_i : \mathbb{N} \to \mathbb{N}$ such that the series $\sum_{q=1}^{\infty} x_{p_i^{\varphi_i(q)}}^i$ converges and $\leq \frac{1}{i^2}$. We replace now elements $y_{p_i^{\varphi_i(q)}, p_i^{\varphi_i(q)}} \in \mathbf{Y}_{i-1}$ with elements $x_{p_i^{\varphi_i(q)}, p_i^{\varphi_i(q)}}^i \in \mathbf{X}^i$. Proceed with this procedure as $i \to \infty$. We obtain the double sequence $\mathbf{Y}$ as required. Indeed, evidently $\mathbf{Y} \in c_{2,+}^{0,P}$ and $\mathbf{Y} \in \ell_2^D$, and, by construction, $\mathbf{Y}$ has infinitely many common elements with each sequence $\mathbf{X}^{k,l}$. □

Theorem 1.3.8. ([9]) *For a given $a \in \mathbb{R}$ the selection principle $\alpha_2^{(d)}(c_2^{a,P}, c_2^{a,\text{sum}})$ holds.*

Proof. Let $(\mathbf{X}^{j,k} = (x_{m,n}^{j,k})_{m,n \in \mathbb{N}} : j, k \in \mathbb{N})$ be a double sequence of double sequences in $c_{2,+}^{0,P}$. We are going to form a new double sequence $\mathbf{Y} = (y_{j,k})_{j,k \in \mathbb{N}}$ as follows.

For $j, k \in \mathbb{N}$ with $j \neq k$ we take $y_{j,k} = a$. Let $j = k$. Take first an increasing sequence $(p_i)_{i \in \mathbb{N}}$, $2 = p_1 < p_2 < p_3 < \ldots$, of prime numbers. Then, the initial double sequence of double sequences organize as a sequence $\mathbf{x} = (x_{m,n}^i)_{i \in \mathbb{N}}$. For $i \geq 2$ we take $y_{j,j} = x_{p_i^s, p_i^s}^i$ if $j = p_i^s$ for some $s \in \mathbb{N}$. If $\left| x_{p_i^s, p_i^s}^i - a \right| > \frac{1}{2^i}$, then already defined $y_{j,j}$ replace with $y_{j,j} = a$. Put $j_i = \min\{j \in \mathbb{N} : y_{j,j} = x_{p_i^s, p_i^s}^i\}$. If $j_{i+1} < j_i$, then already defined $y_{j,j}$, $j \in \{j_{i+1}, j_{i+1}+1, \ldots, j_i - 1\}$, replace by putting $y_{j,j} = a$. If p_i and s with $j = p_i^s$ do not exist, then take $y_{j,j} = a$. According to the construction of $\mathbf{Y}$ we have:

(1) $\mathbf{Y} = (y_{j,k})_{j,k \in \mathbb{N}} \in c_2^{a,\text{sum}}$, and

(2) $\mathbf{Y}$ has infinitely many common elements (at the same positions) with every double sequence $\mathbf{X}^{j,k}$.

This completes the proof. □

Double sequences of positive real numbers $\mathbf{X} = (x_{m,n})_{m,n \in \mathbb{N}}$ and $\mathbf{Y} = (y_{m,n})_{m,n \in \mathbb{N}}$ are said to be *P-strongly asymptotically equivalent* (or *P-asymptotically equal*), denoted $\mathbf{X} \overset{P}{\sim} \mathbf{Y}$, if for every $\varepsilon > 1$ there is $n_0 = n_0(\varepsilon)$ such that $\frac{1}{\varepsilon} \leq \frac{x_{m,n}}{y_{m,n}} \leq \varepsilon$ for all $m \geq n_0$, $n \geq n_0$. The relation $\overset{P}{\sim}$ is an equivalence relation of the set of double sequences of positive real numbers, and $[\mathbf{X}]_P$ denotes the equivalence class of $\mathbf{X}$.

In a similar way we define *max-strong asymptotic equivalence* relation $\overset{\max}{\sim}$ by

$$\mathbf{X} \overset{\max}{\sim} \mathbf{Y} \Leftrightarrow \max\text{-}\lim \frac{x_{m,n}}{y_{m,n}} = 1$$

Some applications of double sequences 13

and *sum-strong asymptotic equivalence* $\stackrel{\text{sum}}{\sim}$ by

$$\mathbf{X} \stackrel{\text{sum}}{\sim} \mathbf{Y} \Leftrightarrow \text{sum} - \lim \frac{x_{m,n}}{y_{m,n}} = 1.$$

The corresponding equivalence classes of $\mathbf{X}$ are denoted by $[\mathbf{X}]_{\max}$ and $[\mathbf{X}]_{\text{sum}}$, respectively. Evidently,

$$[\mathbf{X}]_{\text{sum}} \subsetneq [\mathbf{X}]_{\max} \subsetneq [\mathbf{X}]_{\text{P}}.$$

Theorem 1.3.9. *Let* $\mathbf{X} = (x_{m,n})_{m,n \in \mathbb{N}}$ *be a given double sequence of positive real numbers. Then the selection principle* $\alpha_2^{(d)}([\mathbf{X}]_{\text{P}}, [\mathbf{X}]_{\text{sum}})$ *holds.*

Proof. Let $(\mathbf{Y}^{j,k} = (y_{m,n}^{j,k})_{m,n \in \mathbb{N}} : j, k \in \mathbb{N})$ be a double sequence of double sequences in $[\mathbf{X}]_{\text{P}}$. Take an increasing sequence $(p_i)_{i \in \mathbb{N}}$ of prime numbers with $p_1 = 2$, and arrange the double sequence $(\mathbf{Y}^{j,k} : j, k \in \mathbb{N})$ into a sequence $(\mathbf{Y}^i = (y_{m,n}^i)_{m,n \in \mathbb{N}} : i \in \mathbb{N})$ of double sequences. Then we define the double sequence $\mathbf{Z} = (z_{j,k})_{j,k \in \mathbb{N}}$ as follows.

Inductively we construct a sequence $(\mathbf{Z}_i : i \in \mathbb{N})$ of double sequences. Let $\mathbf{Z}_1 = \mathbf{X}$. Then for $i \geq 2$ we construct the double sequence $\mathbf{Z}_i$ by replacing elements at positions (p_i^s, p_i^s), $s \in \mathbb{N}$, in the double sequence $\mathbf{Z}_{i-1}$ by elements $y_{p_i^s, p_i^s}^i$, $s \in \mathbb{N}$, if

$$\frac{2^i - 1}{2^i} \leq \frac{y_{p_i^s, p_i^s}^i}{x_{p_i^s, p_i^s}} \leq \frac{2^i}{2^i - 1}.$$

Continuing this procedure as $i \to \infty$ we get the required double sequence $\mathbf{Z}$. Indeed, by construction, Z has infinitely many common elements (at the same positions) with double sequences $\mathbf{Y}^i$, $i \in \mathbb{N}$, and $\mathbf{Z} \in [\mathbf{X}]_{\text{sum}}$. □

The following result is given without proof.

Theorem 1.3.10. *Let* $a \in \mathbb{R}$ *be given. Then the selection principle* $\mathsf{S}_1^{\varphi}(c_2^{a,P}, c_2^{a,\text{sum}})$ *holds.*

Theorem 1.3.11. *Let* $a \in \mathbb{R}$ *and let* $(\mathbf{X}^k : k \in \mathbb{N})$ *be a sequence of double sequences in* $c_2^{a,P}$, $\mathbf{X}^k = (x_{m,n}^k)_{m,n \in \mathbb{N}}$. *Then there is a double sequence* $\mathbf{Y} = (y_{s,t})_{s,t \in \mathbb{N}}$ *in* $c_2^{a,P}$ *such that for each* $k \in \mathbb{N}$ *the set* $\{(s,t) \in \mathbb{N} \times \mathbb{N} : y_{s,t} = x_{m,n}^k \text{ for some } (m,n) \in \mathbb{N} \times \mathbb{N}\}$ *is infinite.*

Proof. The double sequence $\mathbf{Y}$ is defined in the following way:
Let $k \in \mathbb{N}$. There is $i_k \in \mathbb{N}$ such that $|x_{m,n}^k - a| < 2^{-k}$ for all $m, n \geq i_k$. Let

$$s^* = \begin{cases} i_k, & \text{for } s = k; \\ i_k + p, & \text{for } s = k + p, p \in \mathbb{N}. \end{cases}$$

and

$$t^* = \begin{cases} i_k, & \text{for } t = k; \\ i_k + p, & \text{for } t = k + p, p \in \mathbb{N}. \end{cases}$$

For $t \geq k$ let $y_{k,t} = x_{i_k,t^*}^k$, and for $s \geq k$ let $y_{s,k} = x_{s^*,i_k}^k$. The double sequence $\mathbf{Y} = (y_{s,t})_{s,t \in \mathbb{N}}$ constructed in this way is as required, because Y has the following properties:

(1) $\mathbf{Y} \in c_2^{a,P}$;

(2) The set $B^k = \{y_{k,t} : t \geq k\} \cup \{y_{s,k} : s \geq k\}$ is a subset of $A^k = \{x_{m,n}^k : m,n \in \mathbb{N}\}$;

(3) For each $k \in \mathbb{N}$, B^k is countable;

(4) $\bigcup_{k \in \mathbb{N}} B^k = \{y_{s,t} : s,t \in \mathbb{N}\}$.

$\square$

Another similar result is given in the next theorem.

Theorem 1.3.12. *Let $a \in \mathbb{R}$ and let $(\mathbf{X}^k : k \in \mathbb{N})$ be a sequence of double sequences in $c_2^{a,P}$, $\mathbf{X}^k = (x_{m,n}^k)_{m,n \in \mathbb{N}}$. Then there is a double sequence $\mathbf{Y} = (y_{s,t})_{s,t \in \mathbb{N}}$ in $c_2^{a,P}$ which has one common row with $\mathbf{X}^k$ for each $k \in \mathbb{N}$.*

Proof. For each $k \in \mathbb{N}$ there is $n_0(k) \in \mathbb{N}$ such that $|x_{m,n}^k - a| < 2^{-k}$ for all $m, n \geq n_0(k)$, $n_0(k_1) > n_0(k_2)$ whenever $k_1 > k_2$, and $n_0(k) \geq \min\{i(k) \in \mathbb{N} : |x_{m,n}^{k+1} - a| < 2^{-k}$ for all $m, n \geq i(k)\}$. Then the desired double sequence $\mathbf{Y}$ is defined in such a way that its $n_0(k)$th row is the $n_0(k)$th row of $\mathbf{X}^k$, i.e., $y_{n_0(k),n} = x_{n_0(k),n}^k$ ($n \in \mathbb{N}$), and $y_{s,t} = a$ otherwise. Let us prove that $\mathbf{Y} \in c_2^{a,P}$. Indeed, if $\varepsilon > 0$ is given, then choose $p \in \mathbb{N}$ such that $2^{-p} < \varepsilon$. Then for each $k \in \mathbb{N}$ we have $|x_{m,n}^k - a| < \varepsilon$ for all $m, n \geq p$. By construction of $\mathbf{Y}$ we have actually that $|y_{m,n} - a| < \varepsilon$ for all $m, n \geq p$, i.e., $\mathbf{Y} \in c_2^{a,P}$. $\square$

1.4 Double sequences and the exponent of convergence

The notion of exponent of convergence of single real sequences play an important role in the theory of convergence/divergence of sequences. This notion was implicitly defined by Pringsheim [30]. In 1931, Serbian mathematician M. Petrović [28] introduced the notion of sequence of exponents of convergence and gave an important contribution to this field.

The first application of the exponent of convergence for single sequences in the theory of selection was presented in [6]. The authors of [3] defined the exponent of convergence for double sequences and studied its applications in selection principles theory.

Definition 1.4.1. ([3]) *A real number λ is said to be the exponent of convergence (in the Prinsheim sense) of a double sequence $\mathbf{X} = (x_{m,n})_{m,n \in \mathbb{N}} \in c_{2,+}^{0,P}$ if for every $\varepsilon > 0$, the double sequence $\mathbf{X}_\varepsilon^+ = (x_{m,n}^{\lambda+\varepsilon})_{m,n \in \mathbb{N}}$ has a finite P-sum, while the double sequence $\mathbf{X}_\varepsilon^- = (x_{m,n}^{\lambda-\varepsilon})_{m,n \in \mathbb{N}}$ does not have.*

If for every $\varepsilon > 0$, the double sequence $\mathbf{X}_\varepsilon = (x_{m,n}^\varepsilon)_{m,n \in \mathbb{N}}$ does not have a finite P-sum, then we say that $\lambda = \infty$ is the exponent of convergence of $\mathbf{X}$.

Let $\lambda \in [0, \infty]$ and let $c_{2,+}^{0,P}(\lambda)$ denote the set of all double sequences from $c_{2,+}^{0,P}$ which P-converge to zero and whose exponent of convergence is λ.

Theorem 1.4.2. ([3]) *The selection principle* $S_1^{(d)}(c_{2,+}^{0,P}, c_{2,+}^{0,P}(\lambda))$ *is satisfied for* $\lambda = 0$.

Proof. Let $(\mathbf{X}^{k,l} = (x_{m,n}^{k,l})_{m,n \in \mathbb{N}} : k, l \in \mathbb{N})$ be a double sequence of double sequences from the class $c_{2,+}^{0,P}$. We will form a double sequence $\mathbf{Y} = (y_{k,l})_{k,l \in \mathbb{N}}$ in the following way:
[Step 1: $n=1$] Choose an element $y_{1,1}$ from double sequence $\mathbf{X}^{1,1}$ such that $y_{1,1} \leq \frac{1}{2}$.
[Step 2: $n \geq 2$] For $(k, l) \in \{(i, n), (n, i) : i \leq n\}$, choose $y_{k,l}$ from the double sequence $\mathbf{X}^{(k,l)}$ such that $y_{k,l} \leq \frac{1}{2^n}$.
Claim 1. $\mathbf{Y} \in c_{2,+}^{0,P}$.
For any $n \in \mathbb{N}$ we have
$$v_n(\mathbf{Y}) \leq \frac{2n-1}{2^n}$$
and thus
$$0 < \sum_{n=1}^{\infty} v_n(\mathbf{Y}) \leq \sum_{n=1}^{\infty} \frac{2n-1}{2^n}.$$
The series $\sum_{n=1}^{\infty} \frac{2n-1}{2^n}$ is convergent in $\mathbb{R}$, hence the series $\sum_{n=1}^{\infty} v_n(y)$ is convergent in $\mathbb{R}$. Thus, we conclude that the double sequence $\mathbf{Y}$ has the finite diagonal sum
$$D\text{-}\Sigma \mathbf{Y} = \lim_{n \to \infty} d_n(\mathbf{Y}) = \lim_{n \to \infty} \left(\sum_{k=1}^{n} \sum_{l=1}^{n} y_{k,l} \right)$$
in $\mathbb{R}$ (more about diagonal sums can be seen in [12]). By results obtained in [12, Proposition 1.2], we conclude that $\mathbf{Y} \in c_{2,+}^{0,P}$.
Claim 2. For any $\varepsilon > 0$, $\mathbf{Y}_\varepsilon^+ = (x_{m,n}^{0+\varepsilon})_{m,n \in \mathbb{N}}$ has finite P-sum.
For $n, k, l \in \mathbb{N}$ we have $y_{k,l}^\varepsilon \leq \frac{1}{2^{\varepsilon n}}$. Therefore, for the double sequence $\mathbf{Y}^\varepsilon = (y_{k,l}^\varepsilon)_{k,l \in \mathbb{N}}$ it holds $v_n(\mathbf{Y}^\varepsilon) \leq \frac{2n-1}{2^{\varepsilon n}}$, and thus
$$D\text{-}\Sigma \mathbf{Y}^\varepsilon \leq \sum_{n=1}^{\infty} \frac{2n-1}{2^{\varepsilon n}}.$$
Since
$$\lim_{n \to \infty} \frac{(2n+1)2^{\varepsilon n}}{(2n-1)2^{\varepsilon(n+1)}} = \frac{1}{2^\varepsilon} < 1,$$
the series $\sum_{n=1}^{\infty} \frac{2n-1}{2^{\varepsilon n}}$ is convergent. Again by results from [12, Proposition 1.2] we obtain that the double sequence Y^ε has finite P-sum $P\text{-}\Sigma \mathbf{Y}_\varepsilon^+$.
Claim 3. For any $\varepsilon > 0$, $\mathbf{Y}_\varepsilon^- = (x_{m,n}^{0-\varepsilon})_{m,n \in \mathbb{N}}|$ does not have finite P-sum.
By construction of the double sequence $\mathbf{Y}$, we have that $\lim_{n \to \infty} y_{n,n} = 0$, which implies $\lim_{n \to \infty} y_{n,n}^{0-\varepsilon} = \infty$, and thus the double sequence $\mathbf{Y}$ does not

P-converge to zero. By results from [12, Proposition 1.3] the double sequence $\mathbf{Y}_\varepsilon^-$ does not have finite P-sum. □

The following theorem we give without proof.

Theorem 1.4.3. ([3]) *The selection principle* $S_1^\varphi(c_{2,+}^{0,P}, c_{2,+}^{0,P}(\lambda))$ *is satisfied for* $\lambda = 0$.

Theorem 1.4.4. ([3]) *The selection principle* $\alpha_2^{(d)}(c_{2,+}^{0,P}, c_{2,+}^{0,P}(\lambda))$ *is satisfied for* $\lambda = 0$.

Proof. Let $(\mathbf{X}^{k,l} = (x_{m,n}^{k,l})_{m,n \in \mathbb{N}} : k, l \in \mathbb{N})$ be a double sequence of double sequences from the class $c_{2,+}^{0,P}$. As in the proof of Theorem 1.4.2 create the double sequence $\mathbf{Y} = (y_{k,l})_{k,l \in \mathbb{N}}$; therefore, for a fixed $n \in \mathbb{N}$ we have that for $(k, l) \in \{(i, n), (n, i) : i \leq n\}$, $z_{k,l} \leq \frac{1}{2^n}$. By the standard method, arrange the given double sequence of double sequences in a sequence $(\mathbf{X}_t = (x_t^{k,l})_{k,l \in \mathbb{N}} : t \in \mathbb{N})$ of double sequences belonging to $c_{2,+}^{0,P}$.

Take a sequence $2 = p_1 < p_2 < p_< \ldots$ of prime numbers. For a fixed $t \in \mathbb{N}$, consider a sequence $(x_{p_t^s, p_t^s}^t)_{s \in \mathbb{N}}$. Clearly, this sequence converges to zero when $s \to \infty$. There exist $s_{p_t} \in \mathbb{N}$ and a subsequence $(x_{p_t^{h(s)}, p_t^{h(s)}}^t)$, such that

$$\sum_{s=s_{p_t}}^\infty x_{p_t^{h(s)}, p_t^{h(s)}}^t \leq \frac{1}{2^t}.$$

For each $s \geq s_{p_t}$ it holds $x_{p_t^{h(s)}, p_t^{h(s)}}^t \leq \frac{1}{2^t}$, which implies $x_{p_t^{h(s)}, p_t^{h(s)}}^{\varepsilon t} \leq \frac{1}{2^{\varepsilon t}}$ for the same s.

In the double sequence $\mathbf{Y}$, replace elements $y_{k,k}$, where $k = p_t^{h(s)}$ for $s \geq s_{p_t}$, with the elements $x_{p_t^{h(s)}, p_t^{h(s)}}^t$. This will be done for every $t \in \mathbb{N}$. In this way, we obtain the double sequence $\overline{\mathbf{Y}} = (\overline{y}_{k,l})$. Then we have

$$0 < D\text{-}\sum \overline{\mathbf{Y}} \leq D\text{-}\sum \mathbf{Y} + \sum_{t=1}^\infty \frac{1}{2^t} < \infty,$$

and therefore $P\text{-}\lim \overline{\mathbf{Y}} = 0$. Moreover, the following hold:

(1) $\overline{y} \cap (x_{k,l}^t)$ is an infinite set, for every $t \in \mathbb{N}$;

(2) $\overline{\mathbf{Y}} \in c_{2,+}^{0,P}$;

(3) $P\text{-}\lim \overline{\mathbf{Y}} \in \mathbb{R}$.

Let an arbitrary $\varepsilon > 0$ be given. Then

$$0 < D\text{-}\sum \overline{\mathbf{Y}}_+^\varepsilon \leq D\text{-}\sum \mathbf{Y}_+^\varepsilon + \sum_{t=1}^\infty \frac{1}{2^{\varepsilon t}} < \infty,$$

since $\lim_{t\to\infty} \frac{2^{\varepsilon t}}{2^{\varepsilon(t+1)}} = 2^{-\varepsilon} < 1$. Thus, $D-\sum \overline{\mathbf{Y}}_+^\varepsilon$ is finite, so the double sequence $\overline{\mathbf{Y}}_+^\varepsilon$ has a finite P-sum according to [12, Proposition 1.2].

Let now $\varepsilon < 0$ be arbitrary. Consider the double sequence $\overline{\mathbf{Y}}^\varepsilon = (\overline{y}^\varepsilon)_{k,l}$. Since $\lim_{k\to\infty} \overline{y}_{k,k} = 0$ implies $\lim_{k\to\infty} \overline{y}_{k,k}^\varepsilon = \infty$, we can conclude that double sequence $\overline{\mathbf{Y}}^\varepsilon$ does not have finite P-sum. This completes the proof of the theorem. □

Bibliography

[1] Çakalli, H. and Savaş, E. 2010. Statistical convergence of double sequences in topological groups, *J. Comput. Anal. Appl.* 12:2: 421–426.

[2] Caserta, A., Djurčić, D. and Mitrović, M. 2014. Selection principles and double sequences-II, *Hacettepe J. Math. Stat.* 43:5: 725–730.

[3] Damljanović, N., Djurčić, D. and Žizović, M.R. 2017. Exponent of convergence for double sequences and selection principles, *Filomat* 31:9: 2821–2825.

[4] Kočinac, Lj. D.R. and Žizović, M.R. 2007. Relations between sequences and selection properties, *Abst. Appl. Anal.* 2007: Article ID 43081, 8 pages.

[5] Djurčić, D., Kočinac, Lj. D.R. Žizović, M.R. 2008. Classes of sequences of real numbers, games and selection properties, *Topol. Appl.* 156:1 46–55.

[6] Djurčić, D., Kočinac, Lj. D.R. and Žizović, M.R. 2011. Exponents of convergence and games, *Advances Dyn. Syst. Appl.* 6:1: 41–48.

[7] Djurčić, D., Kočinac, Lj. D.R. and Žizović, M.R. 2012. On the class $\mathbb{S}_0$ of real sequences, *Appl. Math. Letters* 25:10: 1296–1298.

[8] Djurčić, Kočinac, Lj. D.R. Žizović, M.R. 2012. Double sequences and selections, *Abst. Appl. Anal.* 2012: Article ID 497594, 6 pages.

[9] Djurčić, D., Žizović, M.R. and Petojević, A. 2012. Note on selection principles of Kočinac, *Filomat* 26:6: 1291–1295.

[10] Dündar, E. 2016. On rough $\mathcal{I}_2$-convergence of double sequences, *Numer. Funct. Anal. Optim.* 37:4: 480–491.

[11] Dutta, H. and Kočinac, Lj. D.R. 2015. On difference sequence spaces defined by Orlicz functions without convexity, *Bull. Iranian Math. Soc.* 41:2: 477–489.

[12] Elez, N. Djurčić, D. and Sebeković, A. 2015. Summation of double sequences and selection principles, *Filomat* 29:4: 781–785.

[13] Hardy, G.H. 1917. On the convergence of certain multiple series, *Math. Proc. Cambridge Phil. Soc.* 19: 86–95.

[14] Hobson, E.W. 1926. *The Theory of Functions of a Real Variable*, Vol. II (2nd edition), Cambridge University Press, Cambridge.

[15] Kočinac, Lj. D.R. 2004. Selected results on selection principles, In: Proceedings of the Third Seminar on Geometry and Topology (July 15–17, 2004, Tabriz, Iran) pp. 71–104.

[16] Kočinac, Lj. D.R. 2008. Selection principles related to α_i-properties, *Taiwanese J. Math.* 12:3: 561–571.

[17] Kočinac, Lj. D.R. 2011. On the α_i-selection principles and games, *Contemporary Math.* 533: 107–124.

[18] Kočinac, Lj. D.R., Djurčić, D. and Manojlović, J.V. 2018. Regular and Rapid Variations and Some Applications, In: Michael Ruzhansky, Hemen Dutta, Ravi P. Agarwal (eds.), Mathematical Analysis and Applications: Selected Topics, John Wiley & Sons, Inc. pp. 414–474.

[19] Kočinac, Lj. D.R. and Rashid, M.H.M 2017. On ideal convergence of double sequences in the topology induced by a fuzzy 2-norm, *TWMS J. Pure Appl. Math.* 8:1: 97–111.

[20] Kostyrko, P. and Šalát, T. 1982. On the exponent of convergence, *Rendi. Circ. Mat. Palermo II* 31:2: 187–194.

[21] Mursaleen, M. and Edely, Osama H.H. 2003. Statistical convergence of double sequences, *J. Math. Anal. Appl.* 288:1: 223–231.

[22] Mursaleen, M. and Mohiuddine, S.A. 2014. *Convergence Methods for Double Sequences and Applications*, Springer, New Delhi, Heidelberg, New York, Dordrecht, London (Springer India) pp. ix + 171.

[23] Nayak, L. and Baliarsingh, P. 2019. On difference double sequences and their applications, In: H. Dutta, Lj.D.R. Kočinac, H.M. Srivastava (eds.), Current Trends in Mathematical Analysis and its Interdisciplinary Applications, Birkhäuser, pp. 809–829.

[24] Patterson, R.F. and Çakalli, H. 2015. Quasi Cauchy double sequences, *Tbilisi Math. J.* 8:2: 211–219.

[25] Patterson, R.F. and Çakalli, H. 2016. Functions preserving slowly oscillating double sequences, *An. Stiint, Univ. Al. I. Cuza Ia.si Mat. (N.S.)* 72:2: 531–536.

[26] Patterson, R.F. and Savaş, E. 2011. A category theorem for double sequences, *Appl. Math. Letters* 24:10: 1680–1684.

[27] Patterson, R.F. and Savaş, E. 2013. On double sequences of continuous functions having continuous P-limits II, *Filomat* 27:5: 931–935.

[28] Petrović, M. 1931. On the exponent of convergence, Glas SKAN CXLIII, 149–167 (In Serbian).

[29] Pringsheim, A. 1990. Zur Theorie der zweifach unendlichen Zahlenfolgen, *Math. Annalen* 53:3: 289–321.

[30] Pringsheim, A. 1904. Elementare Theorie der ganzen transcedenten Funkcionen von endlicher Ordnung, *Math. Annalen* 58:3: 257–342.

[31] Robison, G.M. 1926. Divergent double sequences and series, *Trans. Amer. Math. Soc.* 28:1: 50–73.

[32] Šalát, T. 1984. Exponents of convergence of subsequences, Czechoslovak Math. J. 34:3: 362–370.

[33] Savaş, E. and Patterson, R.F. 2011. Double sequence spaces defined by a modulus, *Math. Slovaca.* 61:2: 245—256.

[34] Zeltser, M. 2001. *Investigation of Double Sequence Spaces by Soft and Hard Analytical Methods*, Tartu University Press, Tartu.

Chapter 2

Convergent triple sequences and statistical cluster points

Mehmet Gürdal

Mualla Birgül Huban

2.1	Introduction	21
2.2	Known definitions and properties	22
2.3	$\mathcal{I}_3$-statistical cluster points of triple sequences	23
	2.3.1 $\Gamma^{\mathcal{I}_3}$-statistical convergence	28
2.4	Lacunary $\mathcal{I}_3$-statistical cluster points	31
	Bibliography	35

2.1 Introduction

Statistical convergence theory is a hotly debated issue in various fields. Statistical convergence has been researched under several titles trigonometric series and summability theory [8, 10], in number theory [28] over the years. Fast [6] suggested statistical convergence as an expansion of the conventional idea of sequential limit. Some key features of statistical convergence established by Šalát [34]. Şahiner et al. [32] and Mursaleen and Edely [25] recently introduced the concept of statistical convergence for multiple sequences and there are various studies on the ideal and statistical convergence of these multiple sequences (look into the literature [4, 5, 15, 17–19, 23, 27]). On the other hand, the lacunary statistical convergence notation was studied by Fridy and Orhan [11]. This principle has a number of uses in [16, 30]. In [20], Kostyrko et al. proposed the idea of ideal convergence, which generalized the concept of statistical convergence. More research in this area, as well as applications of ideals, may be found in [12, 13, 15, 21, 26]. In a different path, [3] presented a different kind of convergence known as $\mathcal{I}$-statistical convergence.

Because sequence convergence is so important, there are numerous convergence notions in approximation theory, classical measure theory,

DOI: 10.1201/9781003330868-2

summability theory, probability theory and turnpike theory, and their connections are examined. Mohiuddine and Alamri [22], Hazarika et al. [16], Gürdal and Huban [14], Das et al. [2], Pehlivan et al. [29], and Pehlivan and Mamedov [31], the monograph of Mursaleen and Başar [24] are available to interested readers for the aforementioned theory. As a result, this chapter will do more research on the properties of triple sequences. Section 1.2 summarizes several well-known notions and theorems from summability theory. Section 1.3 examines and discusses the set of $\mathcal{I}_3$-statistical cluster points of triple sequences ($\mathcal{I}_3$-stat.c.p.t.s.) in finite dimensional spaces (f.d.s.). We also establish the concept of $\Gamma^{\mathcal{I}_3}$-statistical convergence ($\Gamma^{\mathcal{I}_3}$stat.c.). Finally, we extend those ideas by employing a lacunary triple sequence. Natural inclusion theorems will be taught in addition to these definitions.

2.2 Known definitions and properties

To begin, we will discuss some fundamental concepts linked to $\mathcal{I}$-statistically convergent, statistically convergent, lacunary convergent sequences, and convergent triple sequences.

The concept of statistical convergence is based on the asymptotic density of the subsets of the set $\mathbb{N}$ of positive integers. The idea of density of sets $K \subseteq \mathbb{N}$ is introduced axiomatically in [7].

Assume that T is a subset of the natural number set $\mathbb{N}$. T_n is the number of elements in the set T that are less or equal to $n \in \mathbb{N}$. Also denoted as $|T_n|$ is the cardinality of the set T_n. The asymptotic (natural) density of T is defined as

$$\delta(T) = \lim_{n \to \infty} \frac{1}{n} |T_n|$$

whenever a limit exists. We should similarly remember that $\delta(\mathbb{N} \setminus T) = 1 - \delta(T)$. A sequence $x = (x_k)_{k \in \mathbb{N}}$ is said to be statistically convergent to L if for every $\varepsilon > 0$,

$$\delta(\{k \in \mathbb{N} : |x_k - L| \geq \varepsilon\}) = 0.$$

Here, we use $st\text{-}\lim x_k = L$.

A statistical limit point (s.l.p.) of a sequence $x = (x_k)$ is a number $L \in \mathbb{R}$ if there is a set $\{k_1 < k_2 < ...k_n < ...\} \subseteq \mathbb{N}$ with an asymptotic density that is not zero such that $\lim_{n \to \infty} x_{k_n} = L$.

Let X be a finite dimensional Banach space, $x = (x_k)$ be an X-valued sequence, and $\mu \in X$. The sequence (x_k) is norm statistically convergent to μ provided that $\delta(\{k : \|x_k - \mu\| \geq \varepsilon\}) = 0$ for all $\varepsilon > 0$ (see [1]). Let C be any closed subset of X. Let $d(C, \mu)$ stand for the distance from a point μ to the closed set C, where $d(C, \mu) = \min_{c \in C} \|c - \mu\|$. We formulate the definition of statistical cluster point (s.c.p.) and some of its properties proved in [9]. Let X

be a finite dimensional Banach space and $x = (x_k)$ be an X-valued sequence. A point $L \in X$ is called a s.c.p. if for every $\varepsilon > 0$

$$\limsup_{n\to\infty} n^{-1} |\{k \in \mathbb{N} : \|x_k - L\| < \varepsilon\}| > 0.$$

The set of ordinary limit points, s.l.p., and s.c.p. of x will be denoted by symbols L_x, Λ_x and Γ_x, respectively. It is clear that $\Gamma_x \subseteq L_x$.

The definitions and features of the ideal symbolized by the $\mathcal{I}$ needed in this article are shown in [20]. We should also mention that [3] investigates the idea of $\mathcal{I}$-statistically convergent. In this context, we use $\mathcal{I}\text{-st} - \lim_{k\to\infty} x_k = L$.

We now recollect the basic principles from [32, 33, 35] that will be required throughout the article.

Definition 2.2.1. *Let $\mathcal{I}_3$ be an admissible ideal on $2^{\mathbb{N}^3}$. A triple sequence (x_{def}) is said to be $\mathcal{I}_3$-convergent to L if for each $\varepsilon > 0$,*

$$\{(d,e,f) \in \mathbb{N}^3 : |x_{def} - L| \geq \varepsilon\} \in \mathcal{I}_3.$$

In this context, one writes $\mathcal{I}_3\text{-}\lim x_{def} = L$.

Remark 2.2.2. *(i) $\mathcal{I}_3(f)$ represents the family of all finite subsets of $\mathbb{N}^3$. Afterward the convergence of $\mathcal{I}_3(f)$ corresponds to the convergence of triple sequences in [32].*
(ii) Let $\mathcal{I}_3(\delta) = \{A \subset \mathbb{N}^3 : \delta(A) = 0\}$. Then, in [32], $\mathcal{I}_3(\delta)$ convergence is related with statistical convergence.

Example 2.2.3. *Let $\mathcal{I} = \mathcal{I}_3(\delta)$ and*

$$(x_{def}) = \begin{cases} 1, & d,e,f \text{ are cubes} \\ 4, & \text{otherwise.} \end{cases}$$

After that, for each $\varepsilon > 0$

$$\delta\left(\{(d,e,f) \in \mathbb{N}^3 : |x_{def} - 4| \geq \varepsilon\}\right) \leq \lim_{d,e,f} \frac{\sqrt{d}\sqrt{e}\sqrt{f}}{def} = 0.$$

It means that $\mathcal{I}_3\text{-}\lim x_{def} = 4$. However, the triple sequence (x_{def}) is not convergent to 4.

2.3 $\mathcal{I}_3$-statistical cluster points of triple sequences

We will discuss some of the features of the set of $\mathcal{I}_3$stat.c.p.t.s. in f.d.s. in this part. We also make a contribution to the investigation of $\Gamma^{\mathcal{I}_3}$-stat.c. It is demonstrated that the triple sequence x is $\Gamma^{\mathcal{I}_3}$stat.c. iff the set $\{k : d(G, x_k) \geq \varepsilon\}$ has $\mathcal{I}_3$-density zero for all $\varepsilon > 0$.

Firstly, we will discuss some of the features of the set of $\mathcal{I}_3$stat.c.p. in f.d.s. Assume $\mathbb{R}^m$ is a finite dimensional space with norm $\|.\|$. Take a triple sequence $x = (x_{def})$, $x_{def} \in \mathbb{R}^m$, $d,e,f \in \mathbb{N}$, and a point $L \in \mathbb{R}^m$.

Definition 2.3.1. *A sequence* $x = (x_{def})$ *is* $\mathcal{I}_3$-*statistically convergent* ($\mathcal{I}_3$ *stat.c.*) *triple sequence to* L *if for each* $\gamma > 0$ *and* $\varepsilon > 0$,

$$\left\{ (u,v,w) \in \mathbb{N}^3 : \frac{1}{uvw} |\{d \leq u, e \leq v, f \leq w : \|x_{def} - L\| \geq \varepsilon\}| \geq \gamma \right\} \in \mathcal{I}_3.$$

or equivalently if for each $\gamma > 0$ *and* $\varepsilon > 0$,

$$\delta_{\mathcal{I}_3} (A(\varepsilon, \gamma)) = \mathcal{I}_3 - \lim_{d,e,f} \frac{|A_{def}(\varepsilon, \gamma)|}{def} = 0,$$

where $A_{def}(\varepsilon, \gamma) = \{d \leq u, \ e \leq v, \ f \leq w : \|x_{def} - L\| \geq \varepsilon\}$.

Definition 2.3.2. *For triple sequence* $x = (x_{def})$, *the point* L *is said to be an* $\mathcal{I}_3$-*statistical cluster points* ($\mathcal{I}_3$-*stat.c.p.*) *if and only if for every* $\varepsilon > 0$,

$$\delta_{\mathcal{I}_3} (\{d \leq u, \ e \leq v, \ f \leq w : \|x_{def} - L\| < \varepsilon\}) \neq 0.$$

Here

$$\mathcal{I}_3 - \lim_{n \to \infty} \sup \frac{1}{uvw} |\{d \leq u, \ e \leq v, \ f \leq w : \|x_{def} - L\| < \varepsilon\}| > 0.$$

By the symbol $\Gamma_x^{\mathcal{I}_3}$, we define the set of $\mathcal{I}_3$-stat.c.p. of triple sequence.

Remark 2.3.3. *Let* $\mathcal{I} = \mathcal{I}_3(f)$. *Then* $\mathcal{I}_3$ *stat.c. corresponds to statistical convergence and thus the concepts of* $\mathcal{I}_3$ *stat.l.p. and* $\mathcal{I}_3$ *stat.c.p. coincide with the concepts of s.l.p. and s.c.p. respectively for triple sequences in* [19].

Let $d(A, \xi)$ represent the distance between a point and the closed set A: $d(A, \xi) = \min_{y \in A} \|y - \xi\|$. Let $N_\varepsilon(A) = \{y \in \mathbb{R}^m : d(A, y) < \varepsilon\}$ be A's open ε-neighborhood.

Lemma 2.3.4. *Assume* $A \subset \mathbb{R}^m$ *is a compact set and* $A \cap \Gamma_x^{\mathcal{I}_3} = \emptyset$. *Then*

$$\delta_{\mathcal{I}_3} \left(\{(d,e,f) \in \mathbb{N}^3 : x_{def} \in A\} \right) = 0.$$

Proof. Since $A \cap \Gamma_x^{\mathcal{I}_3} = \emptyset$, thus $L \in A$, there is a positive number $\varepsilon = \varepsilon(L) > 0$ such that

$$\delta_{\mathcal{I}_3} \{(d,e,f) \in \mathbb{N}^3 : \|x_{def} - L\| < \varepsilon\} = 0.$$

Let $N_\varepsilon(L) = \{y \in \mathbb{R}^m : \|y - L_i\| < \varepsilon\}$. The open sets $\{N_\varepsilon(L) : L_i \in A\}$. therefore, forms an open cover of A. Because A is a compact set, a finite subcover of A exists, say $\{N_i = N_{\varepsilon_i}(L_i) : i = 1, 2, ..., z\}$. As a result $A \subset \cup_i N_i$ and

$$\delta_{\mathcal{I}_3} \{(d,e,f) \in \mathbb{N}^3 : \|x_{def} - L_i\| < \varepsilon_i\} = 0$$

for every i. We are able to write

$$\mathcal{I}_3 - \lim_{u,v,w} \frac{|\{(d,e,f) : d \leq u, \ e \leq v, \ f \leq w, \ x_{def} \in A\}|}{uvw}$$

$$\leq \sum_{i=1}^{z} \mathcal{I}_3 - \lim_{u,v,w} \frac{|\{d \leq u, \ e \leq v, \ f \leq w : \|x_{def} - L_i\| < \varepsilon\}|}{uvw} = 0.$$

This means that $\delta_{\mathcal{I}_3} \left(\{(d,e,f) \in \mathbb{N}^3 : x_{def} \in A\} \right) = 0$. $\square$

Remark 2.3.5. *Lemma 2.3.4 may not be valid if the set A is not compact. Consider the triple sequence $x = (x_{def})$ in $\mathbb{R}$, which is defined as $x_{def} = \{0, 1, 0, 2, 0, 3, ...\}$ with suitable index selection. Then $\Gamma_x^{\mathcal{I}_3} = \{0\}$. Now if we select $A = [1, \infty)$, then $A \cap \Gamma_x^{\mathcal{I}_3} = \emptyset$, but $\delta_{\Gamma_x^{\mathcal{I}_3}}\{(n, k, l) \in \mathbb{N}^3 : x_{nkl} \in A\} = 1/2 \neq 0$.*

Definition 2.3.6. *Let $K = \{(d_m e_n f_o) : d, e, f \in \mathbb{N}\}$ and $\{x\}_K = \{x_{d_m e_n f_o}\}$ be a subsequence of $x = (x_{def})$. If $\delta_{\mathcal{I}_3}(K) = 0$, then $\{x\}_K$ is an $\mathcal{I}_3$-thin subsequence ($\mathcal{I}_3$t.s.) of the sequence $x = (x_{def})$. If the set K does not have $\mathcal{I}_3$-density zero, the subsequence $\{x\}_K$ is known an $\mathcal{I}_3$-nonthin subsequence ($\mathcal{I}_3$nont.ss.) of x.*

Theorem 2.3.7. *If a triple sequence $x = (x_{def})$ has a bounded $\mathcal{I}_3$nont.ss., then the set $\Gamma_x^{\mathcal{I}_3}$ is a non-empty closed set.*

Proof. Let $\{x_{d_m e_n f_o}\}_{m,n,o \in \mathbb{N}}$ be a bounded $\mathcal{I}_3$nont.ss. of x and there is a compact set A such that $x_{def} \in A$ for each $(d, e, f) \in K$, where $K = \{(d_m e_n f_o) : m, n, o \in \mathbb{N}\}$. If $\Gamma_x^{\mathcal{I}_3}$ is empty, then $A \cap \Gamma_x^{\mathcal{I}_3} = \emptyset$. As a result of Lemma 2.3.4, we have

$$\delta_{\mathcal{I}_3}\left(\{(d, e, f) \in \mathbb{N}^3 : x_{def} \in A\}\right) = 0.$$

However,

$$\left|\{(d, e, f) \in \mathbb{N}^3 : d \leq u, \ e \leq v, \ f \leq w, \ (d, e, f) \in K\}\right|$$
$$\leq \left|\{(d, e, f) \in \mathbb{N}^3 : d \leq u, \ e \leq v, \ f \leq w, \ x_{def} \in A\}\right|$$

and as a result, $\delta_{\mathcal{I}_3}(K) = 0$. Thus $\Gamma_x^{\mathcal{I}_3} \neq \emptyset$. We demonstrated that $\Gamma_x^{\mathcal{I}_3}$ is nonempty. Now to explain $\Gamma_x^{\mathcal{I}_3}$ is closed, let L be a limit point of $\Gamma_x^{\mathcal{I}_3}$. Then we have $N_\varepsilon(\xi) \cap \Gamma_x^{\mathcal{I}_3} \neq \emptyset$ for every $\varepsilon > 0$. Let $\alpha \in N_\varepsilon(\xi) \cap \Gamma_x^{\mathcal{I}_3}$. Now we may select $\varepsilon' > 0$ such that $N_{\varepsilon'}(\alpha) \subset S_\varepsilon(L)$. With $\alpha \in \Gamma_x^{\mathcal{I}_3}$ so

$$\delta_{\mathcal{I}_3}\left(\{(d, e, f) \in \mathbb{N}^3 : \|x_{def} - \alpha\| < \varepsilon'\}\right) \neq \emptyset$$

as a result

$$\delta_{\mathcal{I}_3}\left(\{(d, e, f) \in \mathbb{N}^3 : \|x_{def} - L\| < \varepsilon\}\right) \neq \emptyset.$$

So $L \in \Gamma_x^{\mathcal{I}_3}$. □

And now we'll define $\mathcal{I}_3$-statistically bounded sequences.

Definition 2.3.8. *A triple sequence $x = (x_{def})$ is called to be $\mathcal{I}_3$-statistically bounded ($\mathcal{I}_3$statb) if there exists a compact set B so that for each $\gamma > 0$, the set*

$$\left\{(u, v, w) \in \mathbb{N}^3 : \frac{1}{uvw}|\{d \leq u, \ e \leq v, \ f \leq w, \ x_{def} \notin B\}| \geq \gamma\right\}$$

belongs to $\mathcal{I}_3$.

Corollary 2.3.9. Let $x = (x_{def})$ be $\mathcal{I}_3 statb$. Then the set $\Gamma_x^{\mathcal{I}_3}$ is non-empty and compact.

Proof. Assume that B is a compact set with
$$\delta_{\mathcal{I}_3}\left(\{(d,e,f) \in \mathbb{N}^3 : x_{def} \notin B\}\right) = 0.$$
Then $\delta_{\mathcal{I}_3}\left(\{(d,e,f) \in \mathbb{N}^3 : x_{def} \in B\}\right) = 1 > 0$, which implying that B includes an $\mathcal{I}_3$nont.ss. of x. As a result of Theorem 2.3.7, $\Gamma_x^{\mathcal{I}_3}$ is non-empty and closed.

To demonstrate that $\Gamma_x^{\mathcal{I}_3}$ is compact, it is sufficient to show that $\Gamma_x^{\mathcal{I}_3} \subset B$. Assume, if feasible that $L \in \Gamma_x^{\mathcal{I}_3}$ but $L \notin B$. Because B is compact, there is a number $\varepsilon > 0$ such that $N_\varepsilon(L) \cap B = \emptyset$. In this situation we have
$$\{(d,e,f) \in \mathbb{N}^3 : \|x_{def} - L\| < \varepsilon\} \subset \{(d,e,f) \in \mathbb{N}^3 : x_{def} \notin B\}$$
as well as
$$\delta_{\mathcal{I}_3}\{(d,e,f) \in \mathbb{N}^3 : \|x_{def} - L\| < \varepsilon\} = 0.$$
This contradicts the fact that $L \in \Gamma_x^{\mathcal{I}_3}$. □

Theorem 2.3.10. Let $x = (x_{def})$ be an $\mathcal{I}_3 stat.b.t.s.$ Then for every $\varepsilon > 0$
$$\delta_{\mathcal{I}_3}\left\{(d,e,f) \in \mathbb{N}^3 : d\left(\Gamma_x^{\mathcal{I}_3}, x_{def}\right) \geq \varepsilon\right\} = 0, \qquad (2.1)$$
where $d\left(\Gamma_x^{\mathcal{I}_3}, x_{def}\right) = \inf_{y \in \Gamma_x^{\mathcal{I}_3}} \|y - x_{def}\|$ is the distance between x_{def} and the set $\Gamma_x^{\mathcal{I}_3}$.

Proof. Assume B be a compact set such that $\delta_{\mathcal{I}_3}\left(\{(d,e,f) \in \mathbb{N}^3 : x_{def} \notin B\}\right) = 0$. Then, according to Corollary 2.3.9, the set $\Gamma_x^{\mathcal{I}_3}$ is non-empty and clearly $\Gamma_x^{\mathcal{I}_3} \subset B$.

Assume the equality (2.1) does not hold. In this example, there is a number $\varepsilon > 0$ such that
$$\mathcal{I}_3 - \lim_{u,v,w \to \infty} \sup \frac{1}{uvw} \left|\{d \leq u,\ e \leq v,\ f \leq w : d\left(\Gamma_x^{\mathcal{I}_3}, x_{def}\right) \geq \varepsilon\}\right| > 0.$$

We may now define $N_\varepsilon\left(\Gamma_x^{\mathcal{I}_3}\right) = \{y : d\left(\Gamma_x^{\mathcal{I}_3}, y\right) < \varepsilon\}$ and let $A = B \setminus N_\varepsilon\left(\Gamma_x^{\mathcal{I}_3}\right)$. Then A is a compact set that contains $\mathcal{I}_3$-nonthin subsequence of x. According to Lemma 2.3.4, then $A \cap \Gamma_x^{\mathcal{I}_3} \neq \emptyset$; i.e., A includes an $\mathcal{I}_3 stat.c.p.$ This is a contradiction. Consequently
$$\delta_{\mathcal{I}_3}\left\{(d,e,f) \in \mathbb{N}^3 : d\left(\Gamma_x^{\mathcal{I}_3}, x_{def}\right) \geq \varepsilon\right\} = 0. \qquad □$$

Remark 2.3.11. *Theorem 2.3.10 does not have to be true if the triple sequence x is not $\mathcal{I}_3$-bounded. For example, if $x_{def} = \{1, 0, 2, 0, 3, 0, 4, 0, ...\}$ we have $\Gamma_x^{\mathcal{I}_3} = \{0\}$, but*
$$\delta_{\mathcal{I}_3}\left(\{(d,e,f) \in \mathbb{N}^3 : d\left(\Gamma_x^{\mathcal{I}_3}, x_{def}\right) \geq \varepsilon\}\right) = 1/2 > 0.$$

Huban [17] recently defined the idea of $\mathcal{I}_3$-statistically pre-Cauchy ($\mathcal{I}_3$stat.pC.) as follows:

For every $\delta > 0$ and $\varepsilon > 0$, a triple sequence $x = (x_{def})$ is called to be $\mathcal{I}_3$stat.pC., if

$$\{(m,n,o) \in \mathbb{N}^3 : \tfrac{1}{m^2 n^2 o^2} |\{(d,e,f) : d \leq u,\ e \leq v,\ f \leq w,\ |x_{def} - x_{pqr}| \geq \varepsilon\}| \geq \delta\} \in \mathcal{I}_3.$$

Condition 2.3.12. If $x = (x_{def})$ is an $\mathcal{I}_3$stat.pC. triple sequence and $x_{def} \notin (\alpha, \beta)$ for all $(d,e,f) \in \mathbb{N}^3$, where (α, β) is an open interval in $\mathbb{R}$, then

$$\delta_{\mathcal{I}_3}\left(\{(d,e,f) \in \mathbb{N}^3 : x_{def} \leq \alpha\}\right) = 0$$

or

$$\delta_{\mathcal{I}_3}\left(\{(d,e,f) \in \mathbb{N}^3 : x_{def} \geq \beta\}\right) = 0$$

must be true.

Theorem 2.3.13. Let $x = (x_{def})$ be an $\mathcal{I}_3$stat.pC. triple sequence. If x's limit points set is no-where dense and x has an $\mathcal{I}_3$stat.c.p., then x is $\mathcal{I}_3$stat.c. under Condition 2.3.12.

Proof. Assume that x has an $\mathcal{I}_3$stat.c.p. $L \in \mathbb{R}$. So we have for any $\varepsilon > 0$

$$\delta_{\mathcal{I}_3}\left(\{(d,e,f) \in \mathbb{N}^3 : |x_{def} - L| < \varepsilon\}\right) \neq 0.$$

Assume that x is $\mathcal{I}_3$stat.pC. meets Condition 2.3.12 but is not $\mathcal{I}_3$stat.c. Then there is an $\varepsilon_0 > 0$ such that

$$\delta_{\mathcal{I}_3}\left(\{(d,e,f) \in \mathbb{N}^3 : |x_{def} - L| \geq \varepsilon_0\}\right) \neq 0.$$

We assume, without sacrificing generality,

$$\delta_{\mathcal{I}_3}\left(\{(d,e,f) \in \mathbb{N}^3 : x_{def} \leq L - \varepsilon_0\}\right) \neq 0.$$

We assert that every point of $(L - \varepsilon_0, L)$ is a l.p. of x. If not, then we can identify an interval $(\alpha, \beta) \subset (L - \varepsilon_0, L)$ such that $x_{def} \notin (\alpha, \beta)$ for every $(d,e,f) \in \mathbb{N}^3$. As a result,

$$\delta_{\mathcal{I}_3}\left(\{(d,e,f) \in \mathbb{N}^3 : x_{def} \leq \alpha\}\right) \neq 0.$$

Furthermore, since L is a $\mathcal{I}_3$-statistical cluster point, we have

$$\delta_{\mathcal{I}_3}\left(\{(d,e,f) \in \mathbb{N}^3 : x_{def} \geq \beta\}\right) \neq 0.$$

However, this opposes the Condition 2.3.12. So, every point of $(L - \varepsilon_0, L)$ is a l.p. of x, contradicting the statement that the set of l.p. of x is a nowhere dense set. Therefore, x is $\mathcal{I}_3$-statistically convergent. □

We now establish an extended form of a result in [9]. This theorem and proof are applicable to the finite dimensional situation.

Theorem 2.3.14. *There exists a triple sequence* $y = (y_{def})$ *in which*
(i) $L_y^{\mathcal{I}_3} = \Gamma_x^{\mathcal{I}_3}$; *where* $L_y^{\mathcal{I}_3}$ *denotes a collection of l.p. of the triple sequence* y;
(ii) $y_{def} = x_{def}$ *for almost all* d, e, f; *that is,*

$$\delta_{\mathcal{I}_3}\left(\{(d,e,f) \in \mathbb{N}^3 : y_{def} \neq x_{def}\}\right) = 0.$$

2.3.1 $\Gamma^{\mathcal{I}_3}$-statistical convergence

Definition 2.3.15. *If a closed set* $G \subset \mathbb{R}^m$, *then it satisfies*

$$\{(d,e,f) \in \mathbb{N}^3 : d(G, x_{def}) \geq \varepsilon\} \in \mathcal{I}_3 \qquad (2.2)$$

for every $\varepsilon > 0$. *Then set* G *is called to be an* $\mathcal{I}_3$-*minimal closed set* ($\mathcal{I}_3$m.c.s.) *if for every closed set* $G' \subset G$, *where* $G \setminus G' \neq \emptyset$ *seems a number* $\varepsilon' > 0$ *such that*

$$\{(d,e,f) \in \mathbb{N}^3 : d(G', x_{def}) \geq \varepsilon'\} \notin \mathcal{I}_3. \qquad (2.3)$$

Definition 2.3.16. *If* G *is a non-empty* $\mathcal{I}_3$m.c.s., *a triple sequence* $x = (x_{def})$ *is called to be* $\Gamma^{\mathcal{I}_3}$ stat.c. *to the set* G.

We now provide the following lemma.

Lemma 2.3.17. *If* $x = (x_{def})$ *is* $\Gamma^{\mathcal{I}_3}$ stat.c., *then the* $\mathcal{I}_3$-*limit set is unique.*

Proof. Let x be $\Gamma^{\mathcal{I}_3}$stat.c. to the sets G_1 and G_2. Then $\delta_{\mathcal{I}_3}(K_1) = \delta_{\mathcal{I}_3}(K_2) = 0$, where

$$K_1 = \{(d,e,f) \in \mathbb{N}^3 : d(G_1, x_{def}) \geq \varepsilon\}$$

and

$$K_2 = \{(d,e,f) \in \mathbb{N}^3 : d(G_2, x_{def}) \geq \varepsilon\}$$

for every $\varepsilon > 0$. Assume $d(G_1, G_2) \geq 2\varepsilon$ with $\varepsilon > 0$. Now

$$2\varepsilon \leq d(G_1, x_{def}) + d(G_2, x_{def}).$$

Consequently

$$\{(d,e,f) \in \mathbb{N}^3 : d(G_1, x_{def}) < \varepsilon\} \subseteq \{(d,e,f) \in \mathbb{N}^3 : d(G_2, x_{def}) \geq \varepsilon\}.$$

But $\delta_{\mathcal{I}_3}\left(\{(d,e,f) \in \mathbb{N}^3 : \rho(G_1, x_{def}) < \varepsilon\}\right) = 0$ thereby contradicting that fact
$\delta_{\mathcal{I}_3}\left(\{(d,e,f) \in \mathbb{N}^3 : d(G_1, x_{def}) \geq \varepsilon\}\right) = 0$. □

Lemma 2.3.18. *If the* $\mathcal{I}_3$-*limit in Lemma 2.3.17 given above is a single point, then the triple sequence is* $\mathcal{I}_3$stat.c. *to that point.*

Example 2.3.19. *When a triple sequence* $x = (x_{def})$ *is* $\Gamma^{\mathcal{I}_3}$ stat.c.; *that is, when an* $\mathcal{I}_3$m.c.s. *existing between all closed sets* G *fulfilling* (2.2).

It is worth noting that triple sequence $x = (x_{def})$ does not have to be $\Gamma^{\mathcal{I}_3}$stat.c. This is how we explain it. G_α denotes the system of all closed sets fulfilling (2.2). This system is clearly non-empty; for instance, the condition holds if $G = \mathbb{R}^m$. Consider the intersection $G = \bigcap_\alpha G_\alpha$. Is this set G a satisfactory $\mathcal{I}_3$m.c.s. (2.2)? This is not generally true. Furthermore, the set G may be empty. For example if $x_{def} = \{0, 1, -1, 2, -2, 3, -3, ...\}$, we may take $G_\alpha = (-\infty, -\alpha] \cup [\alpha; \infty)$ (with $\alpha > 0$) and obviously $G = \emptyset$, and the triple sequence $x = (x_{def})$ is not $\Gamma^{\mathcal{I}_3}$stat.c.

The following conclusion demonstrates that if $x = (x_{def})$ is $\mathcal{I}_3$-statistically bounded then $G = \bigcap_\alpha G_\alpha = \Gamma^{\mathcal{I}_3}$ and thereby G is non-empty.

Theorem 2.3.20. *If $x = (x_{def})$ is an $\mathcal{I}_3$stat.b.t.s., then it is $\Gamma^{\mathcal{I}_3}$stat.c. to the set $\Gamma_x^{\mathcal{I}_3}$.*

Proof. According to Theorem 2.3.10, $\Gamma_x^{\mathcal{I}_3}$ is a non-empty compact set, and the condition (2.2) holds for this set. Consider that is not $\mathcal{I}_3$m.c.s., that is, there exists a closed set G fulfilling (2.2) such that $G \subset \Gamma_x^{\mathcal{I}_3}$ and $\Gamma_x^{\mathcal{I}_3} \setminus G \neq \emptyset$. In this context, there is a point $L \in \Gamma_x^{\mathcal{I}_3}$ such that $L \notin G$. So, there exists a number $\varepsilon > 0$ such that $S_\varepsilon(L) \cap S_\varepsilon(G) = \emptyset$. Because L is an $\mathcal{I}_3$stat.c.p. according to Definition 2.3.8,

$$\mathcal{I}_3 - \lim_{u,v,w \to \infty} \sup \frac{1}{uvw} |\{d \leq u, \ e \leq v, \ f \leq w : x_{def} \in S_\varepsilon(L)\}| > 0.$$

Then from $\{(d, e, f) : x_{def} \in S_\varepsilon(L)\} \subset \{(d, e, f) : x_{def} \notin S_\varepsilon(G)\}$, we have

$$\mathcal{I}_3 - \lim_{u,v,w \to \infty} \sup \frac{1}{uvw} |\{d \leq u, \ e \leq v, \ f \leq w : x_{def} \notin S_\varepsilon(G)\}| > 0.$$

This is contrary to (2.2). $\square$

Let $x = (x_{def})$ be $\Gamma^{\mathcal{I}_3}$stat.c. to the set G. Theorem 2.3.20 indicates that if this sequence is $\mathcal{I}_3$stat.b., then $G = \Gamma_x^{\mathcal{I}_3}$. We are now looking into triple sequences that are not $\mathcal{I}_3$sb. This sequence may also be $\Gamma^{\mathcal{I}_3}$sc. Define the triple sequence $x = (x_{def})$ of real numbers by $(x_{d,e,f}) = pqr$, where $d = 2^{p-1}(2t+1)$, $e = 2^{q-1}(2s+1)$, $f = 2^{r-1}(2u+1)$; that is, $(p-1)$, $(q-1)$, and $(r-1)$ are the power 2 in the prime factorization of d, e, f respectively and $\mathcal{I}_3$ be a strongly admissible ideal. Then we may demonstrate that the triple sequence is not $\mathcal{I}_3$stat.b. Therefore, $\Gamma_x^{\mathcal{I}_3} = \{pqr\}_{p,q,r=1}^{\infty}$ and x is $\Gamma^{\mathcal{I}_3}$stat.c. to $\Gamma_x^{\mathcal{I}_3}$.

We may instantly derive the following from the evidence of Theorem 2.3.20.

Corollary 2.3.21. *If $\delta_{\mathcal{I}_3}\left(\{(d, e, f) : d\left(\Gamma_x^{\mathcal{I}_3}, x_{def}\right) \geq \varepsilon\}\right) = 0$ for every $\varepsilon > 0$, then the triple sequence x is $\Gamma^{\mathcal{I}_3}$stat.c. to $\Gamma_x^{\mathcal{I}_3}$.*

The following conclusion demonstrates that if x is $\Gamma^{\mathcal{I}_3}$stat.c., then this limit set may be only the set of $\mathcal{I}_3$stat.c.p. $\Gamma_x^{\mathcal{I}_3}$.

Theorem 2.3.22. If $x = (x_{def})$ be $\Gamma^{\mathcal{I}_3}$ stat.c. to the set G, then $G = \Gamma_x^{\mathcal{I}_3}$.

Proof. We explain that $\Gamma_x^{\mathcal{I}_3} \subset G$. However, assume there is a point $L \in \Gamma_x^{\mathcal{I}_3}$ such that $L \notin G$. Now closedness of G means that there exists a number $\varepsilon > 0$ for which $N_\varepsilon(L) \cap N_\varepsilon(G) = \emptyset$, which consequently entails that

$$\{(d,e,f) \in \mathbb{N}^3 : x_{def} \notin N_\varepsilon(G)\} \supset \{(d,e,f) \in \mathbb{N}^3 : x_{def} \in N_\varepsilon(L)\}.$$

From

$$\delta_{\mathcal{I}_3}\left(\{(d,e,f) \in \mathbb{N}^3 : x_{def} \notin N_\varepsilon(G)\}\right) = 0,$$

we have

$$\delta_{\mathcal{I}_3}\left(\{(d,e,f) \in \mathbb{N}^3 : x_{def} \in N_\varepsilon(L)\}\right) = 0.$$

This indicates that $L \in \Gamma_x^{\mathcal{I}_3}$. Therefore, $\Gamma_x^{\mathcal{I}_3} \subset G$.

Now we demonstrate that $G \subset \Gamma_x^{\mathcal{I}_3}$. Consider $L \in G$, but $L \notin \Gamma_x^{\mathcal{I}_3}$. Then there is a number $\varepsilon' > 0$ such that

$$\delta_{\mathcal{I}_3}\left(\{(d,e,f) \in \mathbb{N}^3 : x_{def} \in N_\varepsilon(L)\}\right) = 0$$

for every $\varepsilon \leq \varepsilon'$. Point L might now be an isolated point or a limit point of G. So we consider the following cases.

Case 1. Assume L is an isolated point of G. So, there is a number $\varepsilon \leq \varepsilon'$ such that $N_\varepsilon(L) \cap N_\varepsilon(G \setminus \{L\}) = \emptyset$. This implies that $N_\varepsilon(G) = N_\varepsilon(L) \cup N_\varepsilon(G \setminus \{L\})$. Thereby,

$$\left|\{(d,e,f) \in \mathbb{N}^3 : x_{def} \notin N_\varepsilon(G \setminus \{L\})\}\right|$$
$$= \left|\{(d,e,f) \in \mathbb{N}^3 : x_{def} \in N_\varepsilon(L)\}\right| + \left|\{(d,e,f) \in \mathbb{N}^3 : x_{def} \notin N_\varepsilon(G)\}\right|.$$

Then, using (2.2), we have

$$\delta_{\mathcal{I}_3}\left(\{(d,e,f) \in \mathbb{N}^3 : x_{def} \notin N_\varepsilon(G \setminus \{L\})\}\right) = 0$$

for every sufficiently small $\varepsilon > 0$. This suggests that the set $G \setminus \{L\}$ also fulfills (2.2). This demonstrates that G is not an $\mathcal{I}_3$ m.s.

Case 2. Let L be a l.p. of the set G. Hence, there is a sequence (L_m) in G such that (L_m) converges to L and $L_j \neq L_i$ if $j \neq i$. Let $\varepsilon > 0$ be given. $L' = L_m$ should be chosen so that $\|L - L'\| = 2\delta$ with $4\delta < \varepsilon$.

We conclude that $N_\delta(G) \subset N_\varepsilon(G \setminus N_\delta(L))$. Let $x \in N_\delta(G)$ and $x' \in G$ be such that $\|x - x'\| < \delta$. If $x' \notin N_\delta(L)$ then $x' \in G \setminus N_\delta(L)$ and so $x \in N_\delta(G \setminus N_\delta(L)) \subset N_\varepsilon(G \setminus N_\delta(L))$. If $x' \in N_\delta(L)$ is again present then

$$\|x - L'\| \leq \|x - x'\| + \|x' - L\| + \|L - L'\| < 4\delta < \varepsilon.$$

But $\|L - L'\| = 2\delta$ and thus $L' \in G \setminus N_\delta(L)$. Therefore, $x \in N_\varepsilon(G \setminus N_\delta(L))$ and then $N_\delta(G) \subset N_\varepsilon(G \setminus N_\delta(L))$ which indicates

$$\left|\{(d,e,f) \in \mathbb{N}^3 : x_{def} \notin N_\varepsilon(G \setminus N_\delta(L))\}\right| \leq \left|\{(d,e,f) \in \mathbb{N}^3 : x_{def} \notin N_\delta(G)\}\right|$$

and

$$\delta_{\mathcal{I}_3}\{(d,e,f) \in \mathbb{N}^3 : x_{def} \notin N_\varepsilon(G \setminus N_\delta(L))\} = 0$$

As a result, G is not an $\mathcal{I}_3$-minimal set. $\square$

The ultimate outcome of this part is as follows:

Corollary 2.3.23. *A triple sequence* $x = (x_{def})$ *is* $\Gamma^{\mathcal{I}_3}$ *stat.c. iff*
$$\delta_{\mathcal{I}_3}\left(\{(d,e,f) \in \mathbb{N}^3 : d\left(\Gamma^{\mathcal{I}_3}_x, x_{def}\right) \geq \varepsilon\}\right) = 0 \text{ for every } \varepsilon > 0.$$

2.4 Lacunary $\mathcal{I}_3$-statistical cluster points

The goal of this section is to present the idea of a lacunary $\mathcal{I}_3$stat.c.p. and to show various features of the set of a lacunary $\mathcal{I}_3$stat.c.p in f.d.s. To begin, we will discuss some of the features of the set of s.c.p. in f.d.s.

An X-valued triple sequence (x_{def}) is called to be $\mathcal{I}_3$stat.c. to a closed subset Y if the set $\{(d, e, f) \in \mathbb{N}^3 : d(Y, x_{def}) \geq \alpha\}$ for every $\alpha > 0$, has zero $\mathcal{I}_3$-density.

The set Y is said to be a m.c.s. if it satisfies the following property

$$\mathcal{I}_3 - \lim_{u,v,w \to \infty} \frac{1}{uvw} |\{d \leq u,\ e \leq v,\ f \leq w : d(Y, x_{def}) \geq \alpha\}| = 0 \quad (2.4)$$

for every $\alpha > 0$.

$Z = \cap_\sigma Y_\sigma$ is the smallest nonempty closed set (s.none.c.s.) for which (2.4) holds.

The triple sequence $x = (x_{def})$ is said to be $\mathcal{I}_3$stat.b. if

$$\delta_{\mathcal{I}_3}\left(\{(d,e,f) \in \mathbb{N}^3 : \|x_{def}\| > M\}\right) = 0 \text{ for some } M > 0.$$

The theorem that follows creates a connection between $\Gamma^{\mathcal{I}_3}_x$ and the s.none.c.s. that satisfies (2.4).

Theorem 2.4.1. *Consider* $x = (x_{def})$ *to be an* $\mathcal{I}_3$*stat.b. Then s.none.c.s.* Z *is* $\Gamma^{\mathcal{I}_3}_x$.

Proof. Suppose that $a \in Z$ and $a \notin \Gamma^{\mathcal{I}_3}_x$ then given $\varepsilon > 0$

$$\mathcal{I}_3 - \limsup_{u,v,w \to \infty} \frac{1}{uvw} |\{d \leq u,\ e \leq v,\ f \leq w : \|x_{def} - a\| < \varepsilon\}| > 0.$$

Define $\overset{\circ}{N}_{\varepsilon/2}(a) = \{y : \|y - a\| < \frac{\varepsilon}{2}\}$ for $0 < \alpha < \frac{\varepsilon}{2}$. Take this into consideration: $\widetilde{Z} = Z \backslash \overset{\circ}{N}_{\varepsilon/2}(a)$. $\widetilde{Z}$ is clearly a compact set.

$$\left\{d \leq u,\ e \leq v,\ f \leq w : d\left(\widetilde{Z}, x_{def}\right) \geq \alpha\right\}$$
$$\subset \{d \leq u,\ e \leq v,\ f \leq w : d(Z, x_{def}) \geq \alpha\}$$
$$\cup \{d \leq u,\ e \leq v,\ f \leq w : \|x_{def} - a\| < \varepsilon\}.$$

Hence,

$$\frac{1}{uvw}\left|\left\{d \leq u,\ e \leq v,\ f \leq w : d\left(\widetilde{Z}, x_{def}\right) \geq \alpha\right\}\right|$$
$$\leq \frac{1}{uvw}\left|\{d \leq u,\ e \leq v,\ f \leq w : d(Z, x_{def}) \geq \alpha\}\right|$$
$$+ \frac{1}{uvw}\left|\{d \leq u,\ e \leq v,\ f \leq w : \|x_{def} - a\| < \varepsilon\}\right|.$$

for infinitely many u, v, w. Then

$$\mathcal{I}_3 - \lim_{u,v,w\to\infty} \frac{1}{uvw}\left|\left\{d \leq u,\ e \leq v,\ f \leq w : d\left(\widetilde{Z}, x_{def}\right) \geq \alpha\right\}\right| = 0.$$

Then we get $\widetilde{Z} \subset Z$ for this $\widetilde{Z}$. We have $a \in \Gamma_x^{\mathcal{I}_3}$ as a result of this contradiction. This suggests that $Z \subset \Gamma_x^{\mathcal{I}_3}$. Assume, on the other hand, that $a \in \Gamma_x^{\mathcal{I}_3}$ and $a \notin Z$. We shall define a closed ε-neighborhood of the closed set Z by $N_\varepsilon(Z) = \{y \in X : d(Z, y) \leq \varepsilon\}$. There exists $\varepsilon > 0$ such that $N_\varepsilon(a) \cap N_\varepsilon(Z) = \emptyset$. Then

$$\{d \leq u,\ e \leq v,\ f \leq w : \|x_{def} - a\| < \varepsilon\}$$
$$\subset \{d \leq u,\ e \leq v,\ f \leq w : d(Z, x_{def}) \geq \varepsilon\}$$

since the right-hand set's $\mathcal{I}_3$-density is zero,

$$\mathcal{I}_3 - \lim_{u,v,w\to\infty} \frac{1}{uvw}|\{d \leq u,\ e \leq v,\ f \leq w : \|x_{def} - a\| < \varepsilon\}| = 0.$$

As a result, a is not in $\Gamma_x^{\mathcal{I}_3}$, and we have $a \in Z$. So, $\Gamma_x^{\mathcal{I}_3} \subset Z$ and $\Gamma_x^{\mathcal{I}_3} = Z$ as desired. □

The lacunary triple sequence (l.t.s.) $\theta_3 = \theta_{p,q,r} = \{(d_p, e_q, f_r)\}$ to be used in the following definitions and results are given by Esi and Savaş [5].

Definition 2.4.2. *Let $\theta_3 = \theta_{u,v,w} = \{(d_u, e_v, f_w)\}$ be a l.t.s. A point γ is a lacunary $\mathcal{I}_3$-statistical cluster points (lac$\mathcal{I}_3$stat.c.p.) of the X-valued triple sequence (Xv.t.s.) $x = (x_{def})$, If for every $\varepsilon > 0$*

$$\mathcal{I}_3 - \limsup_{p,q,r\to\infty} \frac{1}{h_{p,q,r}}|\{(d, e, f) \in I_{p,q,r} : \|x_{def} - \gamma\| < \varepsilon\}| > 0.$$

The set of all $\mathcal{I}_3$stat.c.p. of the Xv.t.s. x is denoted by $\Gamma_x^{\mathcal{I}_3}(\theta_3)$.

Lemma 2.4.3. *Let $\theta_3 = \theta_{u,v,w} = \{(d_u, e_v, f_w)\}$ be a l.t.s. and $\Gamma_x^{\mathcal{I}_3}(\theta_3) \neq \emptyset$. Then $\Gamma_x^{\mathcal{I}_3}(\theta_3)$ is a closed set.*

Proof. Let $\gamma_{m,n,o} \in \Gamma_x^{\mathcal{I}_3}(\theta_3)$ and $\gamma_{m,n,o} \to \gamma$. We will demonstrate that $\gamma \in \Gamma_x^{\mathcal{I}_3}(\theta_3)$. Let $\varepsilon > 0$ be any positive number. There are numbers m', n', o' such that $\|\gamma_{m',n',o'} - \gamma\| < \frac{\varepsilon}{2}$. Since $\gamma_{m',n',o'} \in \Gamma_x^{\mathcal{I}_3}(\theta_3)$ according to Definition 2.4.2, we have

$$\mathcal{I}_3 - \lim\sup_{p,q,r\to\infty} \frac{1}{h_{p,q,r}} \left|\left\{(d,e,f) \in I_{p,q,r} : \|x_{def} - \gamma_{m',n',o'}\| < \frac{\varepsilon}{2}\right\}\right| > 0$$

and hence

$$\left\{(d,e,f) \in I_{p,q,r} : \|x_{def} - \gamma_{m',n',o'}\| < \frac{\varepsilon}{2}\right\}$$
$$\subset \{(d,e,f) \in I_{p,q,r} : \|x_{def} - \gamma\| < \varepsilon\}.$$

Thereby

$$\mathcal{I}_3 - \lim\sup_{p,q,r\to\infty} \frac{1}{h_{p,q,r}} |\{(d,e,f) \in I_{p,q,r} : \|x_{def} - \gamma\| < \varepsilon\}| > 0$$

it suggests $\gamma \in \Gamma_x^{\mathcal{I}_3}(\theta_3)$. $\square$

Lemma 2.4.4. *Let $\theta_3 = \theta_{u,v,w} = \{(d_u, e_v, f_w)\}$ be a l.t.s. and $\Gamma_x^{\mathcal{I}_3}(\theta_3)$ be set of $\mathcal{I}_3$stat.c.p. of the X v.t.s. $x = (x_{def})$ and $M \subset X$ be an $\mathcal{I}_3$b.c.s. If*

$$\mathcal{I}_3 - \lim\sup_{p,q,r\to\infty} \frac{1}{h_{p,q,r}} |\{(d,e,f) \in I_{p,q,r} : x_{def} \in M\}| > 0, \quad (2.5)$$

then $\Gamma_x^{\mathcal{I}_3}(\theta_3) \cap M \neq \emptyset$.

Proof. Let $\Gamma_x^{\mathcal{I}_3}(\theta_3) \cap M = \emptyset$. Then there exists a positive number $\alpha > 0$ such that $N_\alpha(\Gamma_x^{\mathcal{I}_3}(\theta_3)) \cap N_\alpha(M) = \emptyset$. Let $\gamma \in M$. Since $\gamma \notin S_\alpha(\Gamma_x^{\mathcal{I}_3}(\theta_3))$ by Definition 2.4.2, there is a number $\varepsilon(\gamma) < \alpha$ such that

$$\mathcal{I}_3 - \lim\sup_{p,q,r\to\infty} \frac{1}{h_{p,q,r}} |\{(d,e,f) \in I_{p,q,r} : \|x_{def} - \gamma\| < \varepsilon(\gamma)\}| = 0. \quad (2.6)$$

Obviously, $M \subset \cup_{\gamma \in M} N_{\varepsilon(\gamma)}(\gamma)$. Because M is a compact set, we may select a finite subcovering of sets $N_{m,n,o} = N_{\varepsilon(\gamma_{mno})}(\gamma_{mno})$, $m, n, o = 1, 2, ..., D$ such that $M \subset \cup_{m,n,o=1}^{D} N_{m,n,o}$. So, for every

$$|\{(d,e,f) \in I_{u,v,w} : x_{def} \in M\}|$$
$$\leq \sum_{m=1}^{D}\sum_{n=1}^{D}\sum_{o=1}^{D} |\{(d,e,f) \in I_{p,q,r} : \|x_{def} - \gamma_{mno}\| < \varepsilon(\gamma)\}|.$$

Then, given (2.4),

$$\mathcal{I}_3 - \lim\sup_{p,q,r\to\infty} \frac{1}{h_{p,q,r}} |\{(d,e,f) \in I_{p,q,r} : x_{def} \in M\}| = 0,$$

which contradicts itself (2.4). $\square$

Theorem 2.4.5. Let $\theta_3 = \theta_{u,v,w} = \{(d_u, e_v, f_w)\}$ be a l.t.s. and $\{x_{def} : x_{def} \in X\}$ an $\mathcal{I}_3$ b.s. Then,

(i) $\Gamma_x^{\mathcal{I}_3}(\theta_3)$ is a non-empty compact set.

(ii) $\mathcal{I}_3 - \lim_{p,q,r \to \infty} \frac{1}{h_{p,q,r}} \left| \{(d,e,f) \in I_{u,v,w} : d(\Gamma_x^{\mathcal{I}_3}(\theta_3), x_{def}) \geq \varepsilon \} \right| = 0$ for every $\varepsilon > 0$.

Proof. Let x_{def} be an $\mathcal{I}_3$ b.t.s. and $M \subset X$ a compact set so that $x_{def} \in M$ for every d, e, f. We already have

$$\mathcal{I}_3 - \lim\sup_{p,q,r \to \infty} \frac{1}{h_{p,q,r}} \left| \{(d,e,f) \in I_{p,q,r} : x_{def} \in M\} \right| = 1 > 0.$$

Then, according to Lemma 5, $\Gamma_x^{\mathcal{I}_3}(\theta_3) \cap M \neq \emptyset$, i.e., the set $\Gamma_x^{\mathcal{I}_3}(\theta_3)$ is non-empty. In this situation, Lemma 2.4.3 shows that $\Gamma_x^{\mathcal{I}_3}(\theta_3)$ is similarly a compact set. We can now demonstrate (ii). Let

$$\mathcal{I}_3 - \lim\sup_{p,q,r \to \infty} \frac{1}{h_{p,q,r}} \left| \{(d,e,f) \in I_{p,q,r} : \rho(\Gamma_x^{\mathcal{I}_3}(\theta_3), x_{def}) \geq \varepsilon \} \right| > 0.$$

Then, given the set $\widetilde{M} = M \setminus \overset{\circ}{N}_\varepsilon \left(\Gamma_x^{\mathcal{I}_3}(\theta_3) \right)$ we obtain

$$\mathcal{I}_3 - \lim\sup_{p,q,r \to \infty} \frac{1}{h_{p,q,r}} \left| \{(d,e,f) \in I_{p,q,r} : x_{def} \in \widetilde{M} \subset X\} \right| > 0.$$

In this situation, Lemma 2.4.4 says $\Gamma_x^{\mathcal{I}_3}(\theta_3) \cap \widetilde{M} \neq \emptyset$. This completes the proof of the theorem as desired. □

Theorem 2.4.6. Let $\theta_3 = \theta_{u,v,w} = \{(d_u, e_v, f_w)\}$ be a l.t.s. and let $M \subset X$ be a compact set and $\Gamma_x^{\mathcal{I}_3}(\theta_3) \cap M = \emptyset$. So,

$$\delta_{\mathcal{I}_3}^{\theta_3} \left(\{(d,e,f) \in I_{p,q,r} : x_{def} \in M\} \right) = 0.$$

Proof. There is a positive number $\varepsilon = \varepsilon(\gamma) > 0$ for any point $\gamma \in M$ such that

$$\delta_{\mathcal{I}_3}^{\theta_3} \left(\{(d,e,f) \in I_{p,q,r} : x_{def} \in M\} \right) = 0.$$

Let $N_\varepsilon(\gamma) = \{y \in X : \|y - \gamma\| < \varepsilon\}$. The open sets $N_\varepsilon(\gamma), \gamma \in M$, are made up of an open covering of M. However, M is a compact set, there exists a finite subcover of M, say $N_{m,n,o} = N_{\varepsilon_{m,n,o}}(\gamma_{mno})$, $m, n, o = 1, 2, ..., D$. Clearly, $M \subset \cup_{m,n,o} N_{m,n,o}$ and

$$\mathcal{I}_3 - \lim_{p,q,r} \frac{1}{h_{p,q,r}} \left| \{(d,e,f) \in I_{p,q,r} : \|x_{def} - \gamma_{mno}\| < \varepsilon_{m,n,o}\} \right| = 0$$

for every m, n, o. We are able to write

$$\left| \{(d,e,f) \in I_{p,q,r} : x_{def} \in M\} \right|$$
$$\leq \sum_{m=1}^{D} \sum_{n=1}^{D} \sum_{o=1}^{D} \left| \{(d,e,f) \in I_{p,q,r} : \|x_{def} - \gamma_{mno}\| < \varepsilon_{m,n,o}\} \right|$$

and hence

$$\mathcal{I}_3 - \lim_{p,q,r} \frac{1}{h_{p,q,r}} |\{(d,e,f) \in I_{p,q,r} : x_{def} \in M\}|$$
$$\leq \sum_{m=1}^{D}\sum_{n=1}^{D}\sum_{o=1}^{D} \mathcal{I}_3 - \lim_{p,q,r} \frac{1}{h_{p,q,r}} |\{(d,e,f) \in I_{p,q,r} : \|x_{def} - \gamma_{mno}\| < \varepsilon_{m,n,o}\}|$$
$$= 0.$$

This means that

$$\delta_{\mathcal{I}_3}^{\theta_3}\left(\{(d,e,f) \in I_{p,q,r} : x_{def} \in M\}\right) = 0.$$

As a consequence, the theorem is established. □

Bibliography

[1] Connor, J., Ganichev, M. and Kadets, V. 2000. A Characterization of Banach Spaces with separable duals via weak statistical convergence. *J. Math. Anal. Appl.* 244:251–261.

[2] Das, P., Dutta, S., Mohiuddine, S.A. and Alotaibi. A. 2014. A-statistical cluster points in finite dimensional spaces and application to turnpike theorem. *Abst. Appl. Anal.* 2014:7 pages.

[3] Das, P., Savaş, E. and Ghosal, S.K. 2011. On generalizations of certain summability methods using ideals. *Appl. Math. Letters* 24(9):1509–1514.

[4] Demirci, I.A. and Gürdal, M. 2021. Lacunary statistical convergence for sets of triple sequences via Orlicz function. *Theory Appl. Math. Comput. Sci.*, 11(1):1–13.

[5] Esi, A. and Savaş, E. 2015. On lacunary statistically convergent triple sequences in probabilistic normed space. *Appl. Math. Inform. Sci.* 9(5):2529–2534.

[6] Fast, H. 1951. Sur la convergence statistique. *Colloq. Math.* 2:241–244.

[7] Freedman, A.R. and Sember, J.J. 1981. Densities and summability. *Pacific J. Math.* 95:293–305.

[8] Fridy, J.A. 1985. On statistical convergence. *Analysis* 5:301–313.

[9] Fridy, J.A. 1993. Statistical limit points. *Proc. Amer. Math. Soc.* 118(4):1187–1192.

[10] Fridy, J.A. and Miller, H.I. 1991. A matrix characterization of statistical convergence. *Analysis* 11:55–66.

[11] Fridy, J.A. and Orhan, C. 1993. Lacunary statistical convergence. *Pacific J. Math.* 160(1):43–51.

[12] Gürdal, M. 2004. *Some type of convergence*, Doctoral Diss., S. Demirel Univ., Isparta.

[13] Gürdal, M. and Huban, M.B. 2012. $\mathcal{I}$-limit points in random 2-normed spaces. *Theory Appl. Math. Comput. Sci.*, 2(1):15–22.

[14] Gürdal, M. and Huban, M.B. 2012. On $\mathcal{I}$-convergence of double sequences in the topology induced by random 2-norms. *Matematicki Vesnik*, 66(1):73–83, 2012.

[15] Gürdal, M. and Şahiner, A. 2008. Extremal $\mathcal{I}$-limit points of double sequences. *Appl. Mathematics E-Notes* 8:131–137.

[16] Hazarika, B., Alotaibi, A. and Mohiudine, S.A. 2020. Statistical convergence in measure for double sequences of fuzzy-valued functions. *Soft Computing*, 24(9):6613–6622.

[17] Huban, M.B. 2021. Generalized statistically pre-Cauchy triple sequences via Orlicz functions. *J. Nonlinear Sci. Appl.* 14(6):414–422.

[18] Huban, M.B. and Gürdal, M. 2021. Wijsman lacunary invariant statistical convergence for triple sequences via Orlicz function. *J. Classical Anal.* 17(2):119–128.

[19] Huban, M.B., Gürdal, M. and Savaş, E. 2020. I-statistical limit superior and I-statistical limit inferior of triple sequences. *Proceeding Book of ICRAPAM 2021: 7th International Conference on Recent Advances in Pure and Applied Mathematics*, pages 42–49.

[20] Kostyrko, P., Macaj, M. and Šalát, T. 2000. $\mathcal{I}$-Convergence. *Real Anal. Exchange*, 26(2):669–686.

[21] Malik, P., Das, S. and Ghosh, A. 2020. $\mathcal{I}$-statistical limit points and $\mathcal{I}$-statistical cluster points of double sequences. *J. Anal.*, 28: 753–768.

[22] Mohiuddine S.A. and Alamri, B.A.S. 2019. Generalization of equi-statistical convergence via weighted lacunary sequence with associated Korovkin and Voronovskaya type approximation theorems. *Rev. Real Acad. Cienc. ExactasFis. Nat.-A: Mat.* 113(3): 1955–1973.

[23] Mohiuddine, S.A., Hazarika, B. and Alotaibi, A. 2017. On statistical convergence of double sequences of fuzzy valued functions. *J. Intell. Fuzzy Systems* 32(6): 4331–4342.

[24] Mursaleen, M. and Başar, F. 2020. *Sequence Spaces: Topics in Modern Summability Theory.* CRC Press, Taylor & Francis Group, Series: Mathematics and Its Applications.

[25] Mursaleen, M. and Edely, O.H.H. 2003. Statistical convergence of double sequences. *J. Math. Anal. Appl.* 288:223–231.

[26] Nabiev, A.A., Pehlivan, S. and Gürdal, M. 2007. On I-Cauchy sequences. *Taiwan J. Math.* 11(2): 569–576.

[27] Nabiev, A.A., Savaş, E. and Gürdal, M. 2019. Statistically localized sequences in metric spaces. *J. Appl. Anal. Comput.* 9(2):739–746.

[28] Niven, I., Zuckerman, H.S. and Montgomery, H. 1991. *An Introduction to the Theory of Numbers.* Fifth Ed., Wiley, New York.

[29] Pehlivan, S., Güncan, A. and Mamedov, M.A. 2004. Statistical cluster points of sequences in finite dimensional spaces. *Czechoslovak Math. J.* 54(1): 95–102.

[30] Pehlivan, S., Gürdal, M. and Fisher, B. 2006. Lacunary statistical cluster points of sequences. *Math. Commun.* 11: 39–46.

[31] Pehlivan, S. and Mamedov, M.A. 2000. Statistical cluster points and turnpike. *Optimization,* 48(1): 91–106.

[32] Sahiner, A., Gürdal, M. and Düden, F.K. 2007. Triple sequences and their statistical convergence. *Selçuk J. Appl. Math.* 8(2):49–55.

[33] Sahiner A. and Tripathy, B.C. 2008. Some $\mathcal{I}$-related properties of triple sequences. *Selçuk J. Appl. Math.* 9(2):9–18.

[34] Šalát, T. 1980. On statistically convergent sequences of real numbers. *Math. Slovaca,* 30(2):139–150.

[35] Subramanian, N. and Esi. A. 2018. Wijsman rough lacunary statistical convergence on $\mathcal{I}$ Cesaro triple sequences. *Int. J. Anal. Appl.* 16:5:643–653.

Chapter 3

Relative uniform convergence of sequence of positive linear Functions

Kshetrimayum Renubebeta Devi

Binod Chandra Tripathy

3.1	Introduction	39
3.2	Preliminaries and definitions	40
3.3	Relative uniform convergence of single sequence of functions	42
3.4	Statistical convergence of sequence	44
3.5	Double sequences	48
3.6	Relative uniform convergence of difference double sequence of positive linear functions	52
	Bibliography	54

3.1 Introduction

A sequence of functions (f_n) defined on a compact domain D is said to be pointwise convergent to a limit function f if for every $\varepsilon > 0$ and for each $x \in D$, there exists an integer $n_0 = n_0(x, \varepsilon)$ such that

$$|f_n(x) - f(x)| < \varepsilon, \text{ for all } n \geq n_0.$$

A sequence of functions (f_n) is said to converge uniformly on compact domain D to limit function f if for every $\varepsilon > 0$ and for all $x \in D$, there exists an integer $n_0 = n_0(\varepsilon)$ such that

$$|f_n(x) - f(x)| < \varepsilon, \text{ for all } n \geq n_0.$$

Example 3.1.1. *Let us consider a sequence of functions, $(f_n(x))$, $f_n : [0,1] \to R$ defined by $f_n(x) = x^n$. We have,*

$$f(x) = \begin{cases} 0, & \text{for } 0 \leq x < 1; \\ 1, & \text{for } x = 1. \end{cases}$$

DOI: 10.1201/9781003330868-3

$(f_n(x))$ converges pointwise to discontinuous function $f(x)$. Hence, $(f_n(x))$ does not converge uniformly on $[0,1]$ but converges uniformly on $[0,0.9]$.

Moore [18] introduced the notion of relative uniform convergence of sequence of functions with respect to a scale function. Uniform convergence is the special case of relative uniform convergent in which the scale function is a non-zero constant which will be discussed in the following section.

In order to extend the notion of convergence of sequences, statistical convergence of sequences was introduced by Fast [10], Buck [2], and Schoenberg [25] independently. It is also found in Zygmund [30]. Later on it was studied from sequence space point of view and linked with summability theory by Šalát [24], Fridy [11], Connor [6], and many others [9,13,14,19,22,26,27]. The basic concept of statistical convergence depends on the notion of asymptotic density of subsets of the set N of natural numbers.

The notion of double sequence and its convergence was introduced by Pringsheim [21]. Some earlier works on double sequences are found in Bromwich [1]. The notion of regular convergence of double sequences was introduced by Hardy [15]. Double sequence was studied from different aspects by using different notion by many others. Different classes of double sequences have been introduced and their different algebraic and topological properties have been investigated by [28].

Korovkin theorem is the study of sequence of positive linear operator on the Banach space of all continuous real valued functions on a compact metric space to provide uniform approximations of continuous functions. For details on this theory, one may refer to [12,17,20].

Kizmaz [16] defined the difference sequence spaces $\ell_\infty(\Delta), c(\Delta), c_0(\Delta)$ as follows:
$$Z(\Delta) = \{x = (x_k) : (\Delta x_k) \in Z\},$$
for $Z = \ell_\infty,\ c,\ c_0$ where $\Delta x_k = x_k - x_{k+1}, k \in N$.

These sequence spaces are Banach space under the norm
$$\|(x_k)\|_\Delta = |x_1| + \sup_{k \in N} |\Delta x_k|.$$

The notion was further investigated by [8,29].

3.2 Preliminaries and definitions

Moore [18] introduced the notion of uniform convergence of a sequence of functions relative to a scale function. One may refer to Chittenden [3] for the definition of relative uniform convergence of sequence of functions, which is defined as follows.

Chittenden [3–5] published 3 articles almost one hundred years back in which he studied different properties of relative uniform convergence of singles sequences of positive linear functions. The works in this direction was not done. In the beginning of twenty first century, the potential of the notion was realized by the researcher on sequence spaces and summability theory. Different classes of sequences of positive linear functions have been introduced and investigated some of their properties, which is evidenced from the list of references.

Definition 3.2.1. *A real single-valued functions $(f_n(x))$ of a real variable x, ranging over a compact subset D of real numbers, converges relatively uniformly on D in case there exist functions g and σ, defined on D, and for every $\varepsilon > 0$, there exists an integer n_o (dependent on ε) such that for every $n \geq n_o$, the inequality*

$$\mid g(x) - f_n(x) \mid < \varepsilon \mid \sigma(x) \mid,$$

holds for every element x of D.

The function σ of the above definition is called a scale function. The sequence (f_n) is said to converge uniformly relative to the scale function σ.

Example 3.2.2. *Let us consider the sequence of function $(f_n(x))$, $f_n : [0,1] \to R$ be defined by*

$$f_n(x) = \begin{cases} \frac{1}{nx}, & \text{for } x \in (0,1]; \\ 0, & \text{for } x = 0. \end{cases}$$

$(f_n(x))$ does not converge uniformly but converges uniformly to zero function w.r.t. a scale function $\sigma(x)$ defined by

$$\sigma(x) = \begin{cases} \frac{1}{x}, & \text{for } x \in (0,1]; \\ 1, & \text{for } x = 0. \end{cases}$$

Equivalently Definition 3.2.1 can be defined as for every m, there exists an integer n_m (depending on m), such that for every $n \geq n_m$, the inequality

$$m \mid f_n(x) - g(x) \mid \leq \mid \sigma(x) \mid,$$

holds for every element x of D.

The following results on relative uniform convergence of double sequences of real functions are straightforward from the Definition 3.2.1.

Result 3.2.3. *Uniform convergence of double sequence relative to a constant scale function different from zero is equivalent to uniform convergence of double sequence.*

Result 3.2.4. *If a double sequence is not convergent uniformly, but converges relative to a scale function, then the scale function σ is not bounded.*

Result 3.2.5. *Uniform convergence of double sequence relative to σ implies uniform convergence relative to every function τ such that $\mid \tau(x) \mid > \mid \sigma(x) \mid$, for all $x \in D$.*

Result 3.2.6. *Uniform convergence of double sequence relative to a scale function σ such that $A \leq |\sigma| \leq B$, where A and B are positive implies uniform convergence of double sequence.*

Result 3.2.7. *If a sequence of functions (f_n) is defined on a domain D and if D may be divided into a finite number of parts such that the sequence converges relatively uniformly on each part, then the sequence convergence relatively uniformly on D.*

Result 3.2.8. *If a sequence of functions (f_n) converges relatively uniformly on D, D may be divided into a sequence (D_n) such that no two sets $D_{n_1}, D_{n_2} (n_1 \neq n_2)$ have a common element and such that on each D_n, the sequence converges uniformly.*

Result 3.2.9. *If a sequence of functions (f_n) is defined on an enumerable set D and converges on D, it converges relatively uniformly on D.*

Result 3.2.10. *If a sequence of functions (f_n) is divided and converges relatively uniformly on each of a sequence of classes D_i, it converges relatively uniformly on the class D of all elements which are in some class D_i, the least common super class of the classes D_i.*

3.3 Relative uniform convergence of single sequence of functions

Theorem 3.3.1. *(Chittenden [3], Theorem 1) A necessary and sufficient condition that a sequence (f_n) of functions f_n converges relatively uniformly on D to a limit function $g(x)$ is that there exists a sequence (n_m) of positive integers such that the sequence $m\phi_{n_m}(x)$ has an upper bound $B(x)$ for every x, where $\phi_n(x)$ is the least upper bound of $|g(x) - f_m(x)|$, for all $m \geq n$.*

Proof. Let (f_n) be uniformly convergent to $g(x)$ relative to the scale function $\sigma(x)$, then for every m, there exists an integer n_m such that

$$m|g(x) - f_m(x)| \leq |\sigma(x)|,$$

is satisfied by $m \geq n_m$.
Hence, it can be written as

$$m\phi_{n_m}(x) \leq |\sigma(x)|,$$

holds for every $x \in D$. □

Theorem 3.3.2. *(Chittenden [3], Theorem 2) A necessary and sufficient condition that a sequence of functions (f_n), converges relatively uniformly on an interval D is that it converges relatively uniformly on the derived set E' of the set E of the points of D, which are points of non-uniformity.*

Corollary 3.3.3. *(Chittenden [3], Corollary I) If the derive set E' is enumerable, the sequences converges relatively uniformly on D.*

Result 3.3.4. *(Chittenden [3], Corollary II) If a sequence of functions does not converge relatively uniformly on D, the corresponding set E' is not enumerable, i.e., E is dense on a perfect set.*

The converse of Result 3.2.10 is not true is shown in Example 3.3.6.

Definition 3.3.5. *A point x is a point of non-uniform convergence in case the measure of non-uniformity of convergence of the sequence is greater than zero at x.*

Example 3.3.6. *Let us consider a sequence of functions (f_n) defined on $(0, 1)$ as follows:*

$$f_n(x) = \begin{cases} 0, & \text{if } x \text{ is irrational;} \\ 0, & \text{if } x = 0; \\ 0, & \text{if } x = \frac{m}{k} (m \text{ and } k \text{ are relatively prime, and } k \neq n); \\ 1, & \text{if } x = \frac{m}{n} (m \text{ and } n \text{ relatively prime)}. \end{cases}$$

We have seen that every x is a point of non-uniformity of convergence for sequence of functions (f_n). However, (f_n) is zero on the irrational points. Hence by Result 3.2.6 and 3.2.8, (f_n) converges relatively uniformly.

He studied that a sequence of continuous functions may converge to a continuous limit without converging relatively uniformly. This is shown in the following theorem:

Theorem 3.3.7. *(Chittenden [3], Theorem 4) If a sequence of continuous functions converges on an interval D to a continuous limit in such a way that the set E of all points of non-uniformity of convergence is dense on D, the sequence does not converge relatively uniformly.*

Theorem 3.3.8. *(Chittenden [4], Lemma) If $(f_n(x))$ is a sequence of continuous functions converging uniformly on an open set Q, the limit function $f(x)$ of the sequence is continuous, not only on Q, but also on $Q^0 = Q + Q'$.*

Proof. Since the convergence is uniform on Q, for every $\varepsilon > 0$, there exists an integer n_o (dependent on ε) such that for every $n \geq n_o$ and for every point $x \in Q$, the inequality

$$|f_n(x) - f_{n+p}(x)| \leq \varepsilon,$$

where $p > 1$ is arbitrary.

The same inequality holds on Q^0, since the functions $f_n(x)$ are continuous. Hence, the sequence converges uniformly on Q^0. Therefore, $f(x)$ is continuous on Q^0. □

He also studied that if $f(x)$ is a limit function of sequence of continuous functions then the discontinuities of $f(x)$ relative to any perfect set P are non-dense on P as stated in the following theorem:

Theorem 3.3.9. *(Chittenden [4], Theorem 3) A necessary and sufficient condition that a function $f(x)$ defined on an interval D be the limit of a sequence of continuous functions converging relatively uniformly on D is that the discontinuities of $f(x)$ relative to every perfect set P of D be non-dense on P.*

3.4 Statistical convergence of sequence

For the following result, we consider the double sequences (f_{nk}) with elements chosen from the complete subset $B(D)$ of $C(D)$, which is a Banach space with respect to the norm of $C(D)$.

In this section we discuss about statistical convergence and its aspects using the notion of relative uniform convergence.

Definition 3.4.1. *Let A be a subset of N, we say that A is said to possesses asymptotic density $\delta(A)$ if*

$$\delta(A) = \lim_{n \to \infty} \frac{1}{n} \sum_{k=1}^{n} \chi_A(k)$$

exists, where χ_A is the characteristic function of A.

Clearly, all finite subsets of N have zero natural density and

$$\delta(A^c) = \delta(N - A) = 1 - \delta(A).$$

Properties of density of subsets of N

1. All finite subsets of N have zero natural density.

2. $\delta(A^c) = \delta(N - A) = 1 - \delta(A)$, provided $\delta(A)$ exists.

3. If $A \subseteq B$ then, $\delta(A) \leq \delta(B)$.

4. Let $A, B \subseteq N$ then, $\delta(A \cup B) = \delta(A) + \delta(B) - \delta(A \cap B)$.

Definition 3.4.2. *(Fridy [11]) The sequence x is statistically convergent to L if for each $\varepsilon > 0$,*

$$\lim_{n \to \infty} \frac{1}{n} |\{k \leq n : |x_k - L| \geq \varepsilon\}| = 0.$$

It is denoted by $st - \lim x_k = L$.

Example 3.4.3. *(Fridy [11]) Let us define a sequence* (x_k) *as*

$$x_k = \begin{cases} 1, & \text{if } k = i^2, i \in N; \\ 0, & \text{otherwise.} \end{cases}$$

Then,

$$\delta(x_k) = \delta(\{k \leq n : x_k = i^2 : i \in N\}) \leq \lim_{n \to \infty} \frac{\sqrt{n}}{n} = 0.$$

This implies that $st - \lim x_k = 0$.

Definition 3.4.4. *(Fridy [11]) The sequence x is statistically Cauchy to L if for each $\varepsilon > 0$, there exists a positive integer $N = N(\varepsilon)$ such that*

$$\lim_{n \to \infty} \frac{1}{n} |\{k \leq n : |x_n - x_k| \geq \varepsilon\}| = 0.$$

Definition 3.4.5. *Two sequences (x_k) and (y_k) are said to be equal almost all k(a.a.k) if*

$$\delta(\{k \leq n : x_k \neq y_k\}) = 0.$$

Theorem 3.4.6. *(Fridy [11], Theorem 1) The following statements are equivalent:*

1. (x_k) *is a statistically convergent sequence.*

2. (x_k) *is a statistically Cauchy sequence.*

3. (x_k) *is a sequence for which there is a convergent sequence (y_k) such that $x_k = y_k$ a.a.k.*

Theorem 3.4.7. *(Šalát [24], Lemma 1.1) The sequence (x_k) of real numbers is said to converge statistically to the real number L if and only if there exists such a set $K = \{k_1 < k_2 < k_3 < k_4 < ...k_n < ...\} \subset N$ that $\delta(K) = 1$ and $\lim_{n \to \infty} x_{k_n} = L$.*

Theorem 3.4.8. *(Connor [6], Theorem 2.3.) If $x \in \omega$ is strongly p-Cesáro summable or statistically convergent to L, then there is a convergent sequence y and a statistically null sequence z such that y is convergent to L, $x = y + z$ and $\lim_{n \to \infty} \frac{1}{n} |\{k \leq n; z_k \neq 0\}| = 0$. Moreover, if x is bounded then, z is bounded and $||z||_\infty \leq ||x||_\infty + |L|$.*

Demirci and Orhan [20] studied statistical uniform convergence of sequence of functions relative to a scale function and applied this notion to prove a Korovkin type approximation theorem.

Let $f, (f_n) \in C(D)$, which is the space of all continuous real valued functions on a compact subset D of the real numbers and $||f||_{C(D)}$ denotes the usual supremum norm of f in $C(D)$.

Definition 3.4.9. *(Duman and Orhan [9])* A sequence of functions $(f_n(x))$ is said to be statistically pointwise convergent to f on D if $st - \lim_n f_n(x) = f(x)$, for each $x \in D$. i.e., for every $\varepsilon > 0$ and for each $x \in D$,

$$\delta(\{n : |f_n(x) - f(x)| \geq \varepsilon\}) = 0.$$

Then it is denoted by $f_n \to f(stat)$ on D.

Definition 3.4.10. *(Duman and Orhan [9])* A sequence of functions $(f_n(x))$ is said to be statistically uniform convergent to f on D if

$$st - \lim_n \|f_n(x) - f(x)\|_{C(D)} = 0$$

or $\delta(\{n : \|f_n(x) - f(x)\|_{C(D)} \geq \varepsilon\}) = 0$,

for every $\varepsilon > 0$. This limit is denoted by $f_n \rightrightarrows f(stat)$ on D.

Definition 3.4.11. *(Demirci and Orhan [7], Definition 3)* A sequence of functions (f_n) is said to be statistically relatively uniform convergent to f on D if there exists a function $\sigma(x)$, such that for every $\varepsilon > 0$,

$$\delta\left(\left\{n : \sup_{x \in D} \left|\frac{|f_n(x) - f(x)|}{\sigma(x)}\right| \geq \varepsilon\right\}\right) = 0,$$

where $\sigma(x)$ is the scale function. This limit is denoted by $(st) - f_n \rightrightarrows f(D; \sigma)$.

The following result is straightforward from the above definition.

Result 3.4.12. *(Demirci and Orhan [7], Lemma 1)* Let (f_n) converges uniformly to f on D then, (f_n) converges statistically uniformly to f on D which also implies that (f_n) converges statistically relatively uniformly to f on D w.r.t. scale function $\sigma(x)$.

The following example shows that the converse of the above result is not always true.

Example 3.4.13. *(Demirci and Orhan [7], Example 2)* Let $g_n : [0, 1] \to R$, for each $n \in N$ define by

$$g_n(x) = \frac{2nx}{1 + n^2 x^2}.$$

We have seen that, $(g_n(x))$ converges statistically relatively uniformly to zero function on $[0, 1]$ w.r.t. the scale function $\sigma(x)$ defined by

$$\sigma(x) = \begin{cases} \frac{1}{x}, & 0 < x < 1; \\ 1, & x = 0 \end{cases}$$

but does not converge uniformly and hence is not statistical uniform convergent.

Let T be a linear operator from $C(D)$ into itself. Then, T is a linear operator provided that $f \geq 0 \Rightarrow T(f) \geq 0$.

Theorem 3.4.14. *(Demirci and Orhan [7], Theorem 2) Let (T_n) be a sequence of positive linear operators acting from $C(D)$ into itself. Then, for all $f \in C(D)$,*

$$(st) - T_n(f) \rightrightarrows f(D; \sigma)$$

if and only if

$$(st) - T_n(e_i) \rightrightarrows e_i(D; \sigma_i), i = 0, 1, 2,$$

where $\sigma(x) = \max\{|\sigma_i(x)|; i = 0, 1, 2\}$, $|\sigma_i(x)|$ is unbounded, $i = 0, 1, 2$.

They also studied the rates of statistically relatively uniformly convergence of a sequence of positive linear operators defined on $C(D)$ with the help of modulus continuity.

Definition 3.4.15. *The modulus of continuity of a function $f \in C(D)$ is defined by*

$$\omega(f, \delta) = \sup_{|y-x| \leq \delta; x, y \in D} |f(y) - f(x)| (\delta > 0).$$

Definition 3.4.16. *(Demirci and Orhan [7], Definition 4) A sequence (f_n) is said to converge statistically uniformly relative to the scale function $\sigma(x)$, $|\sigma(x)| > 0$, to f on X with the rate of $\beta \in (0,1)$ if for every $\varepsilon > 0$,*

$$\lim_{n \to \infty} \frac{\left| \left\{ k \leq n : \sup_{x \in D} \left| \frac{f_k(x) - f(x)}{\sigma(x)} \right| \geq \varepsilon \right\} \right|}{n^{1-\beta}} = 0.$$

It is denoted by $(st) - (f_n - f) = o(n^{-\beta})(D; \sigma)$.

They obtained the rates of convergence in Theorem 9 by using Definition 11 as follows:

Theorem 3.4.17. *(Demirci and Orhan [7], Theorem 3) Let D be a compact subset of the real numbers and let (T_n) be a sequence of positive linear operators acting from $C(D)$ into itself. Assume that the following conditions hold:*

1. $(st) - (T_n(e_0) - e_0) = o(n^{-\beta_0})(D; \sigma_0)$.

2. $(st) - \omega(f, \delta_n) = o(n^{-\beta_0})(D; \sigma_1)$, where $\delta_n(x) = \sqrt{T_n(\phi^2; x)}$ with $\phi(y) = (y - x)$.

Then, we have for all $f \in C(D)$,

$$(st) - (T_n(f) - f) = o(n^{(-\beta)})(D; \sigma),$$

where $\beta = \min\{\beta_0, \beta_1\}$, $\sigma(x) = \max\{|\sigma_0(x)|, |\sigma_1(x)|, |\sigma_0(x)\sigma_1(x)|\}$, $|\sigma_i(x)| > 0$, and $\sigma_i(x)$ is unbounded, $i = 0, 1$.

Note: Korovkin theorem does not work for the sequence of functions that does not converge uniformly.

3.5 Double sequences

Definition 3.5.1. *A double sequence $x = (x_{nk})$ is said to be convergent to L in Pringsheim's sense if for every $\varepsilon > 0$, there exists $n_0 \in N$ such that*

$$|x_{nk} - L| < \varepsilon,$$

for all $n, k \geq n_0$, which is denoted by

$$P - \lim_{n,k \to \infty} x_{nk} = L.$$

Definition 3.5.2. *A double sequence (x_{nk}) is said to converge regularly if it converges in the Pringsheim's sense and the following limit exists:*

$$\lim_{n,k \to \infty} x_{nk} = L_k, \text{ exists, for each } k \in N;$$

and

$$\lim_{n,k \to \infty} x_{nk} = M_n, \text{ exists, for each } n \in N.$$

Remark 3.5.3. *If $L = L_k = M_n = 0$, for all $n, k \in N$, we get the definition of regular null sequences.*

The definition is equivalent to the following statement:

$$\lim_{min(n,k) \to \infty} x_{nk} = 0.$$

Remark 3.5.4. *A double sequence which converges regularly is also convergent in Pringsheim's sense but not conversely.*

Example 3.5.5. *Let us consider the double sequence*

$$x_{nk} = \begin{cases} 1, & \text{for } n = 1, k \text{ is odd;} \\ 0, & \text{otherwise.} \end{cases}$$

We have seen that double sequence (x_{nk}) converges to 0 in Pringsheim's sense but not converges regularly.

Remark 3.5.6. *A double sequence which converges regularly is always bounded but Pringsheim's sense convergent is not necessarily bounded.*

Tripathy [28], F. Moricz and M. Mursaleen independently introduced the notion of statistical convergence of double sequence based on the natural density of the natural numbers $N \times N$ in the year 2003.

A subset E of $N \times N$ is said to have density $\delta(E)$ if,

$$\delta(E) = \lim_{p,q \to \infty} \frac{1}{pq} \sum\sum_{n \leq p k \leq q} \chi_E(n,k) \text{ exists},$$

where χ_E is the characteristics function of E.

Clearly, finite subsets of N have zero natural density and $\delta(E^c) = \delta(N - E) = 1 - \delta(E)$.

The following results are due to Tripathy [28].

Definition 3.5.7. *A double sequence (x_{nk}) is said to be statistically convergent to L if for any given $\varepsilon > 0$,*
$$\delta(\{(n,k) : |x_{nk} - L| \geq \varepsilon\}) = 0.$$

Definition 3.5.8. *A double sequence (x_{nk}) is said to be statistically null if it is statistically convergent to 0.*

Definition 3.5.9. *Let (x_{nk}) and (y_{nk}) be two double sequences, then we say that $x_{nk} = y_{nk}$ for almost all n and k (in short a.a.n and k) if,*
$$\delta(\{(n,k) : x_{nk} \neq y_{nk}\}) = 0.$$

Theorem 3.5.10. *(Tripathy [28], Theorem 1) The following statements are equivalent:*

1. *The double sequence (x_{nk}) is statistically convergent to L.*

2. *The double sequence $(x_{nk} - L)$ is statistically convergent to 0.*

3. *There exists a sequence $(y_{nk}) \in c_2^P$ such that $x_{nk} = y_{nk}$ for a.a.n and k, where c_2^P denote statistically convergent in Pringsheim's sense.*

4. *There exists a subset $M = \{(n_i, k_j) \in N \times N : i, j \in N\}$ such that $\delta(M) = 1$ and $(x_{n_i k_j}) \in c_2^P$.*

5. *There exist two sequences (a_{nk}) and (b_{nk}) such that $x_{nk} = a_{nk} + b_{nk}$, for all $n, k \in N$, where (a_{nk}) converges to L and $(b_{nk}) \in (0^c)_2^P$, where $(0^c)_2^P$ denote the statistically null in Pringsheim's sense.*

Theorem 3.5.11. *(Tripathy [28], Theorem 2) Let $A = (x_{nk})$ be a double sequence of real numbers. Then A is statistically convergent if and only if A is statistically Cauchy.*

Sahin and Dirik [23] introduced the concept of A-statistical relative uniform convergence of double sequences of functions defined on a compact subset D of the real two dimensional space. This concept is dependent on a non-negative RH-regular summability matrix A−density. They also proved Korovkin type approximation theorem. The definition and characterization of regularity for four dimensional matrices is known as Robison-Hamilton conditions or RH-regularity. They explained the concept of A−density as follows. Let $A = (x_{ijnk})$ be a non-negative RH-regular summability matrix and let $K \subset N^2 = N \times N$. Then A−density of K which is denoted by $\delta_{(A)}^2(K)$ is given by
$$\delta_{(A)}^2(K) = P - \lim_{i,j} \sum_{(n,k) \in K} x_{ijnk},$$

provided that the limit on the right-hand side exists in the Pringsheim's sense. A real two dimensional matrix transformation is said to be regular if it maps every convergent sequence into a convergent sequence with the same limit. A four dimensional matrix $A = (x_{ijnk})$ is said to be RH-regular if it maps every bounded Pringsheim's sense convergent sequence into a Pringsheim's sense convergent sequence with the same Pringsheim limit.

A real double sequence $x = (x_{nk})$ is said to be A-statistically convergent to L if for every $\varepsilon > 0$,

$$\delta^2_{(A)}(\{(n,k) \in N : |x_{nk} - L| \geq \varepsilon\}) = 0.$$

It is denoted by $st^2_{(A)} - \lim x_{nk} = L$.

Definition 3.5.12. *(Sahin and Dirik [23], Definition 3) A double sequence of function $(f_{nk}(x))$ is said to be statistically relatively uniform convergent to f on D if there exists a function $\sigma(x,y)$, $|\sigma(x,y)| > 0$, called a scale function $\sigma(x,y)$ such that for every $\varepsilon > 0$,*

$$\delta^2_{(A)}\left(\left\{(n,k): \sup_{(x,y) \in D}\left|\frac{f_{nk}(x) - f(x,y)}{\sigma(x,y)}\right| \geq \varepsilon\right\}\right) = 0.$$

This limit is denoted by $(st)^2_{(A)} - f_{nk} \rightrightarrows f(D; \sigma)$.

Result 3.5.13. *(Sahin and Dirik [23], Lemma 1) $f_{nk} \rightrightarrows f$ on D (in the ordinary sense) implies $f_{nk} \rightrightarrows f(stat)$ on D, which also implies $(st)^2_{(A)} - f_{nk} \rightrightarrows f(D; \sigma)$.*

The converse of the above result is not true is shown in the following example.

Example 3.5.14. *(Sahin and Dirik [23], Example 2) For each $(n,k) \in N^2$, define $g_{nk} : [0,1] \times [0,1] \to R$ by*

$$g_{nk}(x,y) = \frac{2n^2k^2xy}{1 + n^3k^3x^2y^2}.$$

(g_{nk}) is statistically relatively uniform convergent w.r.t. the following scale function $\sigma(x)$ defined by

$$\sigma(x,y) = \begin{cases} \frac{1}{xy}, & \text{for } (x,y) \in (0,1] \times (0,1]; \\ 1, & \text{for } x = 0 \text{ or } y = 0. \end{cases}$$

They studied Korovkin theory using the notion of statistical relative uniform convergence w.r.t. scale function and formulate the following theorem. Let T be a linear operator from $C(D)$ into itself. Then, we say that T is positive provided that $f \geq 0$ implies $T(f) \geq 0$. The value of $T(f)$ at a point $(x,y) \in D$ is denoted by $T(f(u,v); x, y)$ or $T(f; x, y)$.

Theorem 3.5.15. *(Sahin and Dirik [23], Theorem 3) Let $A = (x_{ijnk})$ be a non-negative RH-regular summability matrix method. Let (T_{nk}) be a double sequence of positive linear operators acting from $C(D)$ into itself. Then, for all $f \in C(D)$,*

$$(st)^2_{(A)} - T_{nk}(f) \rightrightarrows f(D;\sigma)$$

if and only if

$$(st)^2_{(A)} - T_{nk}(e_m) \rightrightarrows e_m(D;\sigma_m), m = 0,1,2,3,$$

where $\sigma(x,y) = \max\{|\sigma_m(x,y)| : m = 0,1,2,4\}, |\sigma_m(x,y)| > 0$ and $\sigma_m(x,y)$ is unbounded for $m = 0,1,2,3$.

They also studied the rates of statistical relative uniform convergence of a sequence of positive linear operators defined on $C(D)$ with the help of modulus of continuity.

Definition 3.5.16. *(Sahin and Dirik [23], Definition 4) Let $A = (x_{ijnk})$ be a non-negative RH-regular summability matrix and let (α_{nk}) be a positive non-increasing double sequence. A double sequence (f_{nk}) is said to converge statistically relatively uniform to the scale function $\sigma(x,y), |\sigma(x,y)| > 0$, to f on D with the rate of $o(\alpha_{nk})$ if for every $\varepsilon > 0$,*

$$P - \lim_{i,j} \frac{1}{\alpha_{i,j}} \sum_{(n,k) \in K(\varepsilon)} x_{ijnk} = 0,$$

where $K(\varepsilon) = \left\{ (n,k) : \sup_{(x,y) \in D} \left| \frac{f_{nk}(x,y) - f(x,y)}{\sigma(x,y)} \right| \geq \varepsilon \right\}.$

It is denoted by $(st)^2_{(A)} - (f_{nk} - f) = o(\alpha_{nk})(D;\sigma)$.

Definition 3.5.17. *(Sahin and Dirik [23], Definition 5) Let $A = (x_{ijnk})$ be a non-negative RH-regular summability matrix and let (α_{nk}) be a positive non-increasing double sequence. A double sequence (f_{nk}) is said to converge statistically relatively uniform to the scale function $\sigma(x,y), |\sigma(x,y)| > 0$, to f on D with the rate of $o_{nk}(\alpha_{nk})$ if for every $\varepsilon > 0$,*

$$P - \lim_{i,j} \sum_{(n,k) \in M(\varepsilon)} x_{ijnk} = 0,$$

where $M(\varepsilon) = \left\{ (n,k) : \sup_{(x,y) \in D} \left| \frac{f_{nk}(x,y) - f(x,y)}{\sigma(x,y)} \right| \geq \varepsilon \alpha_{nk} \right\}.$

It is denoted by $(st)^2_{(A)} - (f_{nk} - f) = o_{nk}(\alpha_{nk})(D;\sigma)$.

They obtained the rate of convergence in Theorem 13 by using the Definition 18 as follows:

Theorem 3.5.18. *(Sahin and Dirik [23], Theorem 4) Let (T_{nk}) be a double sequence of positive linear operators acting from $C(D)$ into itself. Assume that the following conditions hold:*

1. $(st)^2_{(A)} - (T_{nk}(e_0) - e_0) = o(\alpha_{nk})(D;\sigma)$,

2. $(st)^2_{(A)} - \omega(f, \delta_{nk}) = o(\beta_{nk})(D;\sigma_1)$, where $\delta_{nk} = \sqrt{||T_{nk}(\phi)||_{C(D)}}$ with $\phi(u,v) = \phi_{x,y}(u,v) = (u-x)^2 + (v-y)^2$.

3.6 Relative uniform convergence of difference double sequence of positive linear functions

Devi and Tripathy [8] introduced the notion of relative uniform convergence of difference double sequence of positive linear functions.

Definition 3.6.1. *A subset E of the set of all double sequence $_2w$ is said to be* solid *or* normal *if $(f_{nk}(x)) \in E \Rightarrow (\alpha_{nk} f_{nk}(x)) \in E$, for all (α_{nk}) of sequence of scalars with $|\alpha_{nk}| \leq 1$, for all $n, k \in N$.*

Definition 3.6.2. *Let*

$$K = \{(n_i, k_j) : i, j \in N; n_1 < n_2 < n_3 < \ldots \text{ and } k_1 < k_2 < k_3 < \ldots\} \subseteq N \times N$$

and E be a subset of the set of all double sequence $_2w$. A K-step space of E is a sequence space

$$\lambda_K^E = \{(f_{n_i k_j}(x)) \in_2 \omega : (f_{nk}(x)) \in E\}.$$

A canonical pre-image of a sequence of functions $(f_{n_i k_j}(x)) \in E$ is a sequence of functions $(g_{nk}(x)) \in E$ defined by

$$g_{nk}(x) = \begin{cases} f_{nk}(x), & \text{if } (n,k) \in K; \\ 0, & \text{otherwise}. \end{cases}$$

Definition 3.6.3. *A double sequence space E is said to be* monotone *if it contains the canonical pre-images of all its step spaces.*

Remark 3.6.4. *From the above notions, it follows that if a sequence space E is solid then, E is monotone.*

Definition 3.6.5. *A double sequence space E is said to be* symmetric *if $(f_{nk}(x)) \in E \Rightarrow (f_{\pi(n,k)}(x)) \in E$, where π is a permutation of N.*

Definition 3.6.6. *A difference double sequence of functions $(\Delta f_{nk}(x))$ defined on a compact domain D is said to be* relatively uniformly convergent *if there exists a function $\sigma(x)$ defined on D and for every $\varepsilon > 0$, there exists an integer $n_0 = n_0(\varepsilon)$ such that*

$$|\Delta f_{nk}(x) - f(x)| < \varepsilon |\sigma(x)|,$$

for all $n, k \geq n_0$ holds for every element x of D. The difference operator Δ is defined by $\Delta f_{nk}(x) = \Delta f_{nk}(x) - \Delta f_{nk}(x) - \Delta f_{nk}(x)$, for all $n, k \in N$.

Relative Uniform Convergence of Sequence

Definition 3.6.7. *A difference double sequence of functions $(\Delta f_{nk}(x))$ defined on a compact domain D is said to be regular relative uniform convergent if there exist functions $g(x), g_k(x), f_n(x), \sigma(x), \xi_n(x), \eta_k(x)$ defined on D, for every $\varepsilon > 0$, there exists an integer $n_0 = n_0(\varepsilon)$ such that for all $x \in D$,*

$|\Delta f_{nk}(x) - g(x)| < \varepsilon |\sigma(x)|$, *for all $n, k \geq n_0$*;

$|\Delta f_{nk}(x) - g_k(x)| < \varepsilon |\eta_k(x)|$, *for each $k \in N$ and for all $n \geq n_0$*;

$|\Delta f_{nk}(x) - f_n(x)| < \varepsilon |\xi_n(x)|$, *for each $n \in N$ and for all $k \geq n_0$.*

Devi and Tripathy [8] introduced the following difference double sequence spaces defined over the normed space $(D, \|\cdot\|_{(\Delta, \sigma)})$.

$$Z(\Delta, ru) = \{(f_{nk}(x)) : (\Delta f_{nk}(x)) \in Z \text{ relative uniformly w.r.t. } \sigma(x)\},$$

where $Z =_2 \ell_\infty$, $_2c_0{}^B$, $_2c^B$, $_2c^R$ $_2c_0{}^R$, $_2c$, and $_2c_0$.

The above sequence spaces are normed by the norm defined by

$$\|f(x)\|_{(\Delta, \sigma)} = \sup_{n \geq 1} \sup_{\|x\| \leq 1} \frac{\|f_{n1}(x)\| \|\sigma(x)\|}{\|x\|} +$$

$$\sup_{k \geq 1} \sup_{\|x\| \leq 1} \frac{\|f_{1k}(x)\| \|\sigma(x)\|}{\|x\|} + \sup_{n \geq 1; k \geq 1} \sup_{\|x\| \leq 1} \frac{\|\Delta f_{nk}(x)\| \|\sigma(x)\|}{\|x\|}$$

Theorem 3.6.8. *(Devi and Tripathy [8], Theorem 3.1.) The sequence spaces $Z(\Delta, ru)$ where, $Z =_2 \ell_\infty$, $_2c_0{}^B$, $_2c^B$, $_2c^R$, and $_2c_0{}^R$ are normed linear spaces.*

Theorem 3.6.9. *(Devi and Tripathy [8], Theorem 3.2.) Let $(D, \|\cdot\|_{(\Delta, \sigma)})$ be a complete normed space. The sequence spaces $Z(\Delta, ru)$ where, $Z =_2 \ell_\infty$, $_2c_0{}^B$, $_2c^B$, $_2c^R$, and $_2c_0{}^R$ are complete.*

Theorem 3.6.10. *(Devi and Tripathy [8], Result 3.1.) The sequence spaces $Z(\Delta, ru)$ where $Z =_2 \ell_\infty$, $_2c_0{}^B$, $_2c^B$, $_2c^R$, $_2c_0{}^R$, $_2c$, and $_2c_0$ are not monotone.*

Remark 3.6.11. *Since soild implies monotone and from the Result 3.1, it follows that the spaces $_2\ell_\infty(\Delta, ru)$, $_2c_0{}^B(\Delta, ru)$, $_2c^B(\Delta, ru)$, $_2c^R(\Delta, ru)$, $_2c(\Delta, ru)$, $_2c_0(\Delta, ru)$, $_2c_0{}^R(\Delta, ru)$ are not solid in general.*

Result 3.6.12. *(Devi and Tripathy [8], Result 3.2.) The sequence spaces $Z(\Delta, ru)$ where, $Z =_2 \ell_\infty$, $_2c_0{}^B$, $_2c^B$, $_2c^R$, $_2c_0{}^R$, $_2c$, and $_2c_0$ are not symmetric.*

Theorem 3.6.13. *(Devi and Tripathy [8], Theorem 3.3.)*

(i) $Z(ru) \subset Z(\Delta, ru)$, *for $Z =_2 \ell_\infty$, $_2c_0{}^B$, $_2c^B$, $_2c^R$, $_2c_0{}^R$, $_2c$, and $_2c_0$ and the inclusions are strict.*

(ii) $Z_0(\Delta, ru) \subset Z(\Delta, ru)$, *for $Z =_2 \ell_\infty$, $_2c_0{}^B$, $_2c^B$, $_2c^R$, $_2c_0{}^R$, $_2c$, and $_2c_0$ and the inclusions are strict.*

Result 3.6.14. *(Devi and Tripathy [8], Result 3.3.) A difference double sequence of functions $(\Delta f_{nk}(x))$ is regular relative uniform convergent over a compact subset D w.r.t. a scale function $\sigma(x)$ then, $(\Delta f_{nk}(x))$ is also relative uniform convergent w.r.t. a scale function $\sigma(x)$ but not conversely.*

Example 3.6.15. *(Devi and Tripathy [8], Example 3.5.) Consider the sequence of functions $(f_{nk}(x))$, $f_{nk}(x) : [0,1] \to R$ defined by*

$$f_{nk}(x) = \begin{cases} x, & \text{when } n = 1 \text{ and } k \text{ is odd}, k \in N; \\ 0 & \text{otherwise}. \end{cases}$$

$(\Delta f_{nk}(x))$ *is given by*

$$\Delta f_{nk}(x) = \begin{cases} x, & \text{when } n = 1; \ k \text{ is odd}, k \in N; \\ -x, & \text{when } n = 1; \ k \text{ is even}, k \in N; \\ 0 & \text{otherwise}. \end{cases}$$

We have seen that $(\Delta f_{nk}(x))$ is relative uniform convergent to the zero function θ on $[0,1]$ w.r.t. the constant scale function 1. However, in case of regular relative uniform convergence, the first row of the sequence of functions $(\Delta f_{nk}(x))$ fails to converge relatively uniformly w.r.t. a scale function $\sigma(x)$.

Bibliography

[1] Bromwich, T.J.I. 1965. An *Introduction to the Theory of Infinite Series*, MacMillan and Co. Ltd. New York.

[2] Buck, R.C. 1953. Generalized asymptotic density, *Amer. J. Math.* 75: 335–346.

[3] Chittenden, E.W. 1914. Relatively uniform convergence of sequences of functions, *Trans. Amer. Math. Soc.* 15: 197–201.

[4] Chittenden, E.W. 1919. On the limit functions of sequences of continuous functions converging relatively uniformly, *Trans. Amer. Math. Soc.* 20: 179–184.

[5] Chittenden, E.W. 1922. Relatively uniform convergence and classification of functions, *Trans. Amer. Math. Soc.* 23: 1–15.

[6] Connor, J.S. 1988. The statistical and strong p-Cesàro convergence of sequences, *Analysis* 8: 47–63.

[7] Demirci, K. and Orhan, S. 2016. Statistically Relatively Uniform Convergence of Positive Linear Operators, *Results. Math.* 69: 359—367.

[8] Devi, K.R. and Tripathy, B.C. Relative uniform convergence of difference double sequence of positive linear functions: *Ricerche di Matematica* (2021): Article no 181, https://doi.org/10.1007/s11587-021-00613-0.

[9] Duman, O. and Orhan, C. 2004. μ-Statistically Convergent Function Sequences, *Czechoslov. Math. J.* 54: 413–422.

[10] Fast, H. 1951. Sur la convergence statistique, *Colloq. Math.* 2: 241–244.

[11] Fridy, J.A. 1985. On statistical convergence, *Analysis* 5: 301–313.

[12] Gadjiev, A.D. and Orhan, C. 2002. Some approximation theorems via statistical convergence, *Rocky Mount. J. Math.* 32: 129–138.

[13] Gökhan, A. and Güngör, M. 2002. On pointwise statistical convergence, *Indian J. Pure appl. Math.* 33:9: 1379–1384.

[14] Gökhan, A., Güngör, M. and Et, M. 2007. Statistical convergence of double sequences of real-valued functions, *Int. Math. Forum* 2:8: 365–374.

[15] Hardy, G.H. 1917. On the convergence of certain multiple series, *Proc. Camb. Phil. Soc.* 19: 86–95.

[16] Kizmaz, H. 1981. On certain sequence spaces, *Canad. Math. Bull.* 24: 169–176.

[17] Korovkin, P.P. 1960. Linear operator and approximation theory, Hindustan Publ. Co., Delhi.

[18] Moore, E.H. 1910. An *Introduction to a form of general analysis*, The New Haven Mathematical Colloquium, Yale University Press, New Haven, p. 30.

[19] Nath, P.K. and Tripathy, B.C. 2020. Statistical convergence of complex uncertain sequences defined by Orlicz function, *Proyecciones* 39:2: 301–315.

[20] Orhan, S. and Demirci, K. 2015. Statistical approximation by double sequences of positive linear operators on modular spaces, *Positivity* 19: 23–36.

[21] Pringsheim, A. 1900. Zur Ttheorie der zweifach unendlichen Zahlenfolgen, *Math. Ann.* 53: 289–321.

[22] Rath, D. and Tripathy, B.C. 1996. Matrix maps on sequence spaces associated with sets of integers, *Indian J. Pure Appl. Math.* 27:2: 197–206.

[23] Sahin, P.O. and Dirik, F. 2017. Statistical Relative Uniform Convergence of Double Sequences of Positive Linear Operators, *Appl. Math. E-Notes* 18: 207–220.

[24] Šalát, T.1980. On statistically convergent sequences of real numbers, *Math. Slovaca* 30: 139–150.

[25] Schoenberg, I.J. 1959. The integrability of certain functions and related summability methods, *Amer. Math. Monthly* 66: 361–375.

[26] Steinhaus, H. 1951. Sur la convergence ordinarie et la convergence asymptotique, *Colloq. Math.* 2: 73–74.

[27] Tripathy, B.C. 1998. On statistically convergent sequences, *Bull. Calcutta. Math. Soc.* 90: 259–262.

[28] Tripathy, B.C. 2003. Statistically convergent double sequences, *Tamkang Jour. Math.* 34:3: 231–237.

[29] Tripathy, B.C. and Sarma, B. 2008. Statistically convergent difference double sequence spaces, *Acta Math. Sinica (Eng. Ser.)* 24:5: 737–742.

[30] Zygmund. A. 1979. *Trignometric series*, 2nd ed., Vol. II, Cambridge Univ. Press, London and New York.

Chapter 4

Almost convergent sequence spaces defined by Nörlund matrix and generalized difference matrix

Kuldip Raj

Manisha Devi

4.1	Introduction and preliminaries	57
4.2	Main results	62
	Bibliography	68

4.1 Introduction and preliminaries

Lorentz [10] gave the concept of almost convergence, which motivated various authors to construct new classes of sequence spaces. Further the topological, geometric, and algebraic properties of newly formed Banach spaces played a very significant role. Basically, these classes are matrix domains of generalized difference matrices in sequence spaces. Various inclusion relations among almost convergent spaces and some other similar spaces have been studied by a number of authors in past time. In recent years, topological properties of different types sequences like almost convergent sequence spaces are studied by several authors (see [18, 19]). Başar and Kirişçi [1] considered the generalized difference matrix $B(u,v)$ in sets of almost null and almost convergent sequences and investigated its matrix domains. After that, Esi studied the concepts of almost convergence and almost Cauchy for triple sequences in [4]. For more results on almost convergence (see [2,3,25]). Sönmez [24] investigated the triple band matrix $B(u,v,t)$. Subsequently, Kayaduman and Şengönül have introduced matrix domains of the Cesàro matrix of order 1 and Riesz matrix, which are almost convergent sequence spaces in sets of almost convergent and almost null sequences (see [5,8]). In their study they introduced some classes of matrix mappings, topological properties, and finally gave some core results.

DOI: 10.1201/9781003330868-4

A sequence space is basically a subspace of w, where w is the vector space of all real sequences. By c, ℓ_∞, and c_0, we denote the class of all convergent, bounded, and null convergent sequences, respectively. Also, by $\mathbb{C}$, we denote the complex field. Moreover, the classes of all convergent and bounded series are respectively denoted by cs and bs.

Let X and Y be two sequence spaces and $A = (a_{ri})$ be an infinite matrix, where a_{ri} are real or complex numbers with $i, r \in \mathbb{N}$. Then $A : X \to Y$ is called as a matrix mapping, if for every sequence $x = (x_i) \in X$, the sequence $Ax = \{A_r(x)\}$ lies in Y, where

$$A_r(x) = \sum_i a_{ri} x_i \quad (r \in \mathbb{N}). \tag{4.1}$$

The collection of all the matrices A where $A : X \to Y$ is matrix mapping is denoted by (X, Y). So we have $A \in (X, Y)$ if and only if the series lying on the right side of (4.1) converges for all $x \in X$ and each $r \in \mathbb{N}$ and also for all $x \in X$, $Ax \in Y$ holds.

Take A to be an infinite matrix in a sequence space X, then the matrix domain X_A is defined by

$$X_A = \{x = (x_i) \in w : Ax \in X\}, \tag{4.2}$$

which forms a sequence space.

Misiak [12] gave the concept of n-normed spaces. A real valued function $\|\cdot, \cdots, \cdot\|$ on Z^n such that

1. $\|(x_1, x_2, \cdots, x_n)\| = 0$ iff $(x_1, x_2, \cdots, x_n)$ are linearly dependent on Z,
2. $\|(x_1, x_2, \cdots, x_n)\|$ is invariant under permutation,
3. $\|(\gamma x_1, x_2, \cdots, x_n)\| = |\gamma| \|(x_1, x_2, \cdots, x_n)\|$ for any $\gamma \in K$,
4. $\|(x + x', x_2, \cdots, x_n)\| \leq \|(x, x_2, \cdots, x_n)\| + \|(x', x_2, \cdots, x_n)\|$.

is said to be a n-norm on Z. For basics of n-normed space (see [6, 7, 14]). Further, a sequence (x_i) taken in an n-normed space Z is called to be bounded if there exists a positive constant M, such that $\|(x_i, z_1, \cdots, z_{n-1})\| \leq M$ for all $z_1, \cdots, z_{n-1} \in Z$.

Also, a sequence (x_i) in $(Z, \|\cdot, \cdots, \cdot\|)$ is called to be Cauchy under the n-norm if we have

$$\lim_{i,p \to \infty} \|(x_i - x_p, z_1, \cdots, z_{n-1})\| = 0, \forall z_1, \cdots, z_{n-1} \in Z.$$

Consider a sequence (t_i) of nonnegative real numbers with $t_0 > 0$ such that $T_m = \sum_{i=0}^{m} t_i$ for each $m \in \mathbb{N}$. Then, the Nörlund means $\mathcal{N}^t = (c_{mi}^t)$ is defined by

$$c_{mi}^t = \begin{cases} \frac{t_{m-i}}{T_m}, & \text{if } 0 \leq i \leq m, \\ 0, & \text{if } i > m \end{cases}$$

for every $i, m \in \mathbb{N}$. To know more about Nörlund spaces one may refer ([11, 20, 26]) and the reference therein. Let $t_0 = D_0 = 1$ and define D_m for $m \in \{1, 2, 3, ...\}$ by

$$D_m = \begin{vmatrix} t_1 & 1 & 0 & 0 & \cdots & 0 \\ t_2 & t_1 & 0 & 0 & \cdots & 0 \\ t_3 & t_2 & t_1 & 0 & \cdots & 0 \\ \vdots & \vdots & \vdots & \vdots & \ddots & \vdots \\ t_{m-1} & t_{m-2} & t_{m-3} & t_{m-4} & \ddots & 1 \\ t_m & t_{m-1} & t_{m-2} & t_{m-3} & \ddots & t_1 \end{vmatrix}.$$

The inverse matrix $V^t = (\nu^t_{mi})$ of the matrix $N^t = (c^t_{mi})$ (see [11]) as follows:

$$\nu^t_{mi} = \begin{cases} (-1)^{m-i} D_{m-i} T_i, & 0 \leq i \leq m \\ 0, & i > m, \end{cases}$$

for every $i, m \in \mathbb{N}$. Also, for $i \in \{1, 2, 3, ...\}$ we have

$$D_i = \sum_{j=1}^{i-1} (-1)^{j-1} D_{i-j} + (-1)^{i-1} t_i.$$

The generalized difference matrix $B(u, v) = (b_{mi}(u, v))$ for each $m, i \in \mathbb{N}$ and non-zero real numbers u, v is defined by

$$b_{mi}(u, v) = \begin{cases} u, & i = m, \\ v, & i = m-1, \\ 0, & 0 \leq i \leq m-1 \text{ or } i > m. \end{cases} \quad (4.3)$$

The inverse of the generalized difference matrix is denoted and defined as

$$B^{-1}(u, v) = \hat{b}_{mi}(u, v) = \begin{cases} \frac{1}{u}(-\frac{v}{u})^{m-i}, & 0 \leq i \leq m, \\ 0, & i > m. \end{cases}$$

for all $i, m \in \mathbb{N}$.

The sequence $z = (z_i)$ is said to the $B(u,v)$-transform of a sequence $x = (x_i)$ such that

$$z_i = ux_i + vx_{i-1} \quad (i \in \mathbb{N}). \quad (4.4)$$

Now we consider the sets of almost convergent sequences. A linear functional ψ continuous on ℓ_∞ is said to be a Banach limit if

(i) $\psi(x) \geq 0$ for $x = (x_i)$, $x_i \geq 0$ for every i,

(ii) $\psi(e) = 1$, where $e = (1, 1, 1, ...)$ and;

(iii) $\psi(x_{\delta(i)}) = \psi(x_i)$, where δ is an operator defined on w as $\delta(i) = i + 1$ and is called as the shift operator.

If all Banach limits of x are a, then we say that the sequence $x = (x_i)$ in ℓ_∞ is almost convergent to the generalized limit a ([10]) and one can denote it by $\mathfrak{f} - \lim x = a$. We can simply say that, $\lim\limits_{p} \dfrac{(x_m + x_{m+1} + \ldots + x_{m+p-1})}{p} = a$, uniformly in m if and only if $\mathfrak{f} - \lim x_i = a$ uniformly in m. The spaces of all almost null convergent and almost convergent sequences are respectively denoted by $\mathfrak{f}_0$ and $\mathfrak{f}$.

Definition 4.1.1. *A function p mapping from linear metric space X to $\mathbb{R}$ is said to be a paranorm, if it has the following properties*

(1) $p(x) \geq 0, \ \forall \, x \in X$,

(2) $p(x + y) \leq p(x) + p(y), \ \forall \, x, y \in X$,

(3) $p(-x) = p(x), \ \forall \, x \in X$,

(4) *if (γ_m) is a sequence of scalars with $\gamma_m \to \gamma$ as $m \to \infty$ and (x_m) is a sequence of vectors with $p(x_m - x) \to 0$ as $m \to \infty$, then $p(\gamma_m x_m - \gamma x) \to 0$ as $m \to \infty$.*

A paranorm p is said to be total paranorm if $p(x) = 0$ implies $x = 0$ and the total paranormed space is denoted by pair (X, p). Recall that for any linear metric space, the metric is given by the total paranorm (refer [27], Theorem 10.4.2, pp. 183).

A function O continuous from $[0, \infty)$ to $[0, \infty)$ is called an Orlicz function if it is non-decreasing and convex and satisfies the properties such as $O(0) = 0$; for $x > 0$, $O(x) > 0$ and $O(x)$ approaches ∞ as x goes to ∞. Lindenstrauss and Tzafriri [9] formed the new sequence space using Orlicz function which are as follows

$$\ell_O = \left\{ x = (x_i) \in w : \sum_{i=1}^{\infty} O\left(\frac{|x_i|}{\rho}\right) \text{ is finite, for some } \rho > 0 \right\}.$$

Such a sequence space is named as an Orlicz sequence space.

It can be easily seen in [9] that there is always a subspace of every Orlicz sequence space ℓ_O such that it is isomorphic to $\ell_p (p \geq 1)$. In integral form, an Orlicz function O can be expressed as $O(x) = \int_0^x \zeta(t) dt$, where ζ is the kernel of function O. For $t \geq 0$, ζ is right differentiable. At $t = 0$, ζ takes the value zero and for $t > 0$ we have $\zeta(t) > 0$. Further, ζ is non-decreasing and it approaches ∞ as t approaches ∞.

The sequence $\mathcal{O} = (O_i)$ of Orlicz functions is named as the Musielak-Orlicz function (refer [13,17]). The complementary function of function $\mathcal{O}$ is denoted by a sequence $\mathcal{P} = (P_i)$ and is defined by

$$P_i(\nu) = \sup\{|\nu|s - O_i(s) : s \geq 0\}, \text{ where } i = 1, 2, \ldots$$

For positive cone i^1_+ of i^1, if there exist a sequence $c = (c_i)_{i=1}^{\infty} \in i^1_+$ which satisfy the inequality

$$O_i(2s) \leq KO_i(s) + c_i, \quad a, K > 0 \text{ are constants,}$$

for all $s \in \mathbb{R}^+$ (the set of strictly positive real numbers) and $i \in \mathbb{N}$, whenever $O_i(s) \leq a$, then we can say that Musielak-Orlicz function satisfy Δ_2-condition. To get more information about these spaces one can refer [15, 16, 21–23] and references therein.

Consider a bounded sequence $p = (p_i)$ of positive real numbers, $s = (s_i)$ as a sequence formed by the strictly positive real numbers and $\mathcal{O} = (O_i)$ be a sequence of Orlicz functions. By considering the $B(u,v)$-transform of sequences $x = (x_i)$, some new sequences spaces are defined as follows:

$$[\mathcal{N}^t, \mathcal{O}, s, p] = \left\{ (x_i) \in w : \exists; a \in \mathbb{C} \ni \lim_m \frac{1}{T_m} \sum_{i=0}^{m} \left[s_i O_i \left(\left\| t_{m-i} \frac{|ux_{i+r} + vx_{i+r-1}|}{\rho}, z_1, z_2, \ldots, z_{n-1} \right\| \right) \right]^{p_i} = a, \text{ uniformily in } r, \text{ some } \rho > 0 \right\}$$

and

$$[\mathcal{N}^t, \mathcal{O}, s, p]_0 = \left\{ (x_i) \in w : \lim_m \frac{1}{T_m} \sum_{i=0}^{m} \left[s_i O_i \left(\left\| t_{m-i} \frac{|ux_{i+r} + vx_{i+r-1}|}{\rho}, z_1, z_2, \ldots, z_{n-1} \right\| \right) \right]^{p_i} = 0, \text{ uniformily in } r \text{ some } \rho > 0 \right\}.$$

For $O_i(x) = x$, where $i \in \mathbb{N}$, the above sequence spaces get reduced to $[\mathcal{N}^t, s, p]$ and $[\mathcal{N}^t, s, p]_0$, where

$$[\mathcal{N}^t, s, p] = \left\{ (x_i) \in w : \exists a \in \mathbb{C} \ni \lim_m \frac{1}{T_m} \sum_{i=0}^{m} \left[s_i \left(\left\| t_{m-i} \right\| |ux_{i+r} + vx_{i+r-1}|, z_1, z_2, \ldots, z_{n-1} \right\| \right) \right]^{p_i} = a, \text{ uniformily in } r \right\}$$

and

$$[\mathcal{N}^t, s, p]_0 = \left\{ (x_i) \in w : \lim_m \frac{1}{T_m} \sum_{i=0}^{m} \left[s_i \left(\left\| t_{m-i} \right\| |ux_{i+r} + vx_{i+r-1}|, z_1, z_2, \ldots, z_{n-1} \right\| \right) \right]^{p_i} = 0, \text{ uniformily in } r \text{ some} \rho > 0 \right\}.$$

If $(p_i) = 1$ and $(s_i) = 1$, for all $i \in \mathbb{N}$, then the sequence spaces get reduced to $[\mathcal{N}^t, \mathcal{O}]$ and $[\mathcal{N}^t, \mathcal{O}]_0$, where

$$[\mathcal{N}^t, \mathcal{O}] = \left\{ (x_i) \in w : \exists\, a \in \mathbb{C} \ni \lim_m \frac{1}{T_m} \sum_{i=0}^{m} \left[O_i \left(t_{m-i} \left\| \frac{|ux_{i+r} + vx_{i+r-1}|}{\rho}, z_1, z_2, \ldots, z_{n-1} \right\| \right) \right] = a, \text{ uniformily in } r \text{ some } \rho > 0 \right\}$$

and

$$[\mathcal{N}^t, \mathcal{O}]_0 = \left\{ (x_i) \in w : \lim_m \frac{1}{T_m} \sum_{i=0}^{m} \left[O_i \left(t_{m-i} \left\| \frac{|ux_{i+r} + vx_{i+r-1}|}{\rho}, z_1, z_2, \ldots, z_{n-1} \right\| \right) \right] = 0, \text{ uniformily in } r \text{ and some } \rho > 0 \right\}.$$

Throughout the chapter, for $0 \leq p_i \leq \sup p_i = E$ and $F = \max(1, 2^{E-1})$, we use the inequality

$$|a_i + b_i|^{p_i} \leq K\{|a_i|^{p_i} + |b_i|^{p_i}\} \tag{4.5}$$

for all i and $a_i, b_i \in \mathbb{C}$. Further, for all $a \in \mathbb{C}$, we have $|a|^{p_i} \leq \max(1, |a|^E)$.

4.2 Main results

Theorem 4.2.1. *The sequence spaces $[\mathcal{N}^t, \mathcal{O}, s, p]_0$ and $[\mathcal{N}^t, \mathcal{O}, s, p]$ forms a linear spaces over $\mathbb{C}$.*

Proof. Let $x = (x_i)$ and $y = (y_i)$ lies in $[\mathcal{N}^t, \mathcal{O}, s, p]_0$. Then we can find two positive numbers ρ_1 and ρ_2 such that

$$\frac{1}{T_m} \sum_{i=0}^{m} \left[s_i O_i \left(t_{m-i} \left\| \frac{|ux_{i+r} + vx_{i+r-1}|}{\rho_1}, z_1, z_2, \ldots, z_{n-1} \right\| \right) \right]^{p_i} \to 0, \text{ as } m \to \infty$$

and

$$\frac{1}{T_m} \sum_{i=0}^{m} \left[s_i O_i \left(t_{m-i} \left\| \frac{|uy_{i+r} + vy_{i+r-1}|}{\rho_2}, z_1, z_2, \ldots, z_{n-1} \right\| \right) \right]^{p_i} \to 0, \text{ as } m \to \infty.$$

Let $\rho_3 = \max(2|\alpha|\rho_1, 2|\beta|\rho_2)$, where $\alpha, \beta \in \mathbb{C}$. As we know that $\mathcal{O} = (O_i)$ is convex and nondecreasing so from inequality (4.5), we get

$$\frac{1}{T_m} \sum_{i=0}^{m} \left[s_i O_i \left(t_{m-i} \left\| \frac{|\alpha(ux_{i+r} + vx_{i+r-1}) + \beta(uy_{i+r} + vy_{i+r-1})|}{\rho_3}, \right.\right.\right.$$
$$\left.\left.\left. z_1, z_2, \ldots, z_{n-1} \right\| \right) \right]^{p_i}$$
$$= \frac{1}{T_m} \sum_{i=0}^{m} \left[s_i O_i \left(t_{m-i} \left\| \frac{|\alpha(ux_{i+r} + vx_{i+r-1})|}{\rho_3}, z_1, z_2, \ldots, z_{n-1} \right\| \right) \right.$$
$$\left. + s_i O_i \left(t_{m-i} \left\| \frac{|\beta(uy_{i+r} + vy_{i+r-1})|}{\rho_3}, z_1, z_2, \ldots, z_{n-1} \right\| \right) \right]^{p_i}$$
$$\leq K \frac{1}{T_m} \sum_{i=0}^{m} \left[s_i O_i \left(t_{m-i} \left\| \frac{|ux_{i+r} + vx_{i+r-1}|}{\rho_1}, z_1, z_2, \ldots, z_{n-1} \right\| \right) \right]^{p_i}$$
$$+ K \frac{1}{T_m} \sum_{i=0}^{m} \left[s_i O_i \left(t_{m-i} \left\| \frac{|uy_{i+r} + vy_{i+r-1}|}{\rho_2}, z_1, z_2, \ldots, z_{n-1} \right\| \right) \right]^{p_i} \to 0,$$

as $m \to \infty$.
Therefore, $\alpha x + \beta y \in [\mathcal{N}^t, \mathcal{O}, s, p]$. Hence, $[\mathcal{N}^t, \mathcal{O}, s, p]$ forms a linear space. Proof is similar for the space $[\mathcal{N}^t, \mathcal{O}, s, p]_0$, so we omit it. □

Theorem 4.2.2. *The sequence spaces $[\mathcal{N}^t, \mathcal{O}, s, p]_0$ and $[\mathcal{N}^t, \mathcal{O}, s, p]$ forms paranormed space under paranorm*

$$f(x) = \inf \left\{ \rho^{\frac{p_i}{M}} : \left(\frac{1}{T_m} \sum_{i=0}^{m} \left[s_i O_i \left(t_{m-i} \left\| \frac{|ux_{i+r} + vx_{i+r-1}|}{\rho}, \right.\right.\right.\right.\right.$$
$$\left.\left.\left.\left.\left. z_1, z_2, \ldots, z_{n-1} \right\| \right) \right]^{p_i} \right)^{\frac{1}{M}} \leq 1, uniformly\ in\ r > 0,\ \rho > 0 \right\},$$

where $0 \leq p_i \leq \sup p_i = E$ and $M = \max(1, E)$.

Proof. Since, the proof for both the spaces $[\mathcal{N}^t, \mathcal{O}, s, p]$ and $[\mathcal{N}^t, \mathcal{O}, s, p]_0$ is similar, so we take only the space $[\mathcal{N}^t, \mathcal{O}, s, p]_0$. It is Clear that $f(0) = 0$, $f(x) \geq 0$, and $f(x) = f(-x)$. Let $(x_i), (y_i)$ lies in $[\mathcal{N}^t, \mathcal{O}, s, p]_0$ and $\rho_1, \rho_2 > 0$ be such that

$$\left(\frac{1}{T_m} \sum_{i=0}^{m} \left[s_i O_i \left(t_{m-i} \left\| \frac{|ux_{i+r} + vx_{i+r-1}|}{\rho_1}, z_1, z_2, \ldots, z_{n-1} \right\| \right) \right]^{p_i} \right)^{\frac{1}{M}} \leq 1$$

and

$$\left(\frac{1}{T_m} \sum_{i=0}^{m} \left[s_i O_i \left(t_{m-i} \left\| \frac{|uy_{i+r} + vy_{i+r-1}|}{\rho_2}, z_1, z_2, \ldots, z_{n-1} \right\| \right) \right]^{p_i} \right)^{\frac{1}{M}} \leq 1.$$

Let $\rho = \rho_1 + \rho_2$. Then, using inequality (4.5), we get

$$\left(\frac{1}{T_m}\sum_{i=0}^{m}\left[s_iO_i\left(t_{m-i}\left\|\frac{|(ux_{i+r}+vx_{i+r-1})+(uy_{i+r}+vy_{i+r-1})|}{\rho}\right.\right.\right.\right.$$

$$\left.\left.\left.\left.z_1,z_2,\ldots,z_{n-1}\right\|\right)\right]^{p_i}\right)^{\frac{1}{M}}$$

$$\leq \left(\frac{\rho_1}{\rho_1+\rho_2}\right)\left(\frac{1}{T_m}\sum_{i=0}^{m}\left[s_iO_i\left(t_{m-i}\left\|\frac{|ux_{i+r}+vx_{i+r-1}|}{\rho_1}\right.\right.\right.\right.,$$

$$\left.\left.\left.\left.z_1,z_2,\ldots,z_{n-1}\right\|\right)\right]^{p_i}\right)^{\frac{1}{M}}$$

$$+\left(\frac{\rho_2}{\rho_1+\rho_2}\right)\left(\frac{1}{T_m}\sum_{i=0}^{m}\left[s_iO_i\left(t_{m-i}\left\|\frac{|uy_{i+r}+vy_{i+r-1}|}{\rho_2}\right.\right.\right.\right.,$$

$$\left.\left.\left.\left.z_1,z_2,\ldots,z_{n-1}\right\|\right)\right]^{p_i}\right)^{\frac{1}{M}}$$

$$\leq 1.$$

Thus,

$$f(x+y) = \inf\left\{(\rho)^{\frac{p_i}{M}} : \left(\frac{1}{T_m}\sum_{i=0}^{m}\left[s_iO_i\right.\right.\right.$$

$$\left(t_{m-i}\left\|\frac{|(ux_{i+r}+vx_{i+r-1})+(uy_{i+r}+vy_{i+r-1})|}{\rho_1+\rho_2}\right.\right.,$$

$$\left.\left.\left.\left.z_1,z_2,\ldots,z_{n-1}\right\|\right)\right]^{p_i}\right)^{\frac{1}{M}} \leq 1\right\}$$

$$\leq \inf\left\{(\rho_1)^{\frac{p_i}{M}} : \left(\frac{1}{T_m}\sum_{i=0}^{m}\left[s_iO_i\left(t_{m-i}\left\|\frac{|ux_{i+r}+vx_{i+r-1}|}{\rho_1}\right.\right.\right.\right.,\right.$$

$$\left.\left.\left.\left.z_1,z_2,\ldots,z_{n-1}\right\|\right)\right]^{p_i}\right)^{\frac{1}{M}} \leq 1\right\}$$

$$+\inf\left\{(\rho_2)^{\frac{p_i}{M}} : \left(\frac{1}{T_m}\sum_{i=0}^{m}\left[s_iO_i\left(t_{m-i}\left\|\frac{|uy_{i+r}+vy_{i+r-1}|}{\rho_2}\right.\right.\right.\right.,\right.$$

$$\left.\left.\left.\left.z_1,z_2,\ldots,z_{n-1}\right\|\right)\right]^{p_i}\right)^{\frac{1}{M}} \leq 1\right\}.$$

Therefore, $f(x+y) \leq f(x) + f(y)$. Here, we show the continuity of scalar multiplication. For this, take any complex number say γ. Using definition, we obtain

$$f(\gamma x) = \inf\left\{(\rho)^{\frac{p_i}{M}} : \left(\frac{1}{T_m}\sum_{i=0}^{m}\left[s_i O_i\left(t_{m-i}\left\|\frac{|\gamma(ux_{i+r} + vx_{i+r-1})|}{\rho}, z_1, z_2, \ldots, z_{n-1}\right\|\right)\right]^{p_i}\right)^{\frac{1}{M}} \leq 1\right\}$$

$$= \inf\left\{(|\gamma|t)^{\frac{p_i}{M}} : \left(\frac{1}{T_m}\sum_{i=0}^{m}\left[s_i O_i\left(t_{m-i}\left\|\frac{|ux_{i+r} + vx_{i+r-1}|}{t}, z_1, z_2, \ldots, z_{n-1}\right\|\right)\right]^{p_i}\right)^{\frac{1}{M}} \leq 1\right\},$$

where $t = \frac{\rho}{|\gamma|} > 0$. As $|\gamma|^{p_i} \leq \max(1, |\gamma|^{\sup p_i})$, thus we have

$$f(\gamma x) \leq \max(1, |\gamma|^{\sup p_i}) \inf\left\{t^{\frac{p_i}{M}} : \left(\frac{1}{T_m}\sum_{i=0}^{m}\left[s_i O_i\left(t_{m-i}\left\|\frac{|ux_{i+r} + vx_{i+r-1}|}{t}, z_1, z_2, \ldots, z_{n-1}\right\|\right)\right]^{p_i}\right)^{\frac{1}{M}} \leq 1\right\}.$$

The above inequality implies the continuity of the scalar multiplication. □

Theorem 4.2.3. Let $s = (s_i)$ be a sequence in $\mathbb{R}^+$ and $\mathcal{O} = (O_i)$ be a sequence of Orlicz functions. Consider two bounded sequences $p = (p_i)$ and $q = (q_i)$ of positive real numbers, where $0 \leq p_i \leq q_i < \infty$ for each i. Then, $[\mathcal{N}^t, \mathcal{O}, s, p]_0 \subseteq [\mathcal{N}^t, \mathcal{O}, s, q]$.

Proof. Let $x \in [\mathcal{N}^t, \mathcal{O}, s, p]_0$. Then

$$\frac{1}{T_m}\sum_{i=0}^{m}\left[s_i O_i\left(t_{m-i}\left\|\frac{|ux_{i+r} + vx_{i+r-1}|}{\rho}, z_1, z_2, \ldots, z_{n-1}\right\|\right)\right]^{p_i} \longrightarrow 0,$$

as $m \to \infty$.
Then for sufficiently large values of i, we have

$$\left[s_i O_i\left(t_{m-i}\left\|\frac{|ux_{i+r} + vx_{i+r-1}|}{\rho}, z_1, z_2, \ldots, z_{n-1}\right\|\right)\right]^{p_i} \leq 1.$$

Since $p_i \leq q_i$ and O_i is increasing, so we have

$$\frac{1}{T_m} \sum_{i=0}^{m} \left[s_i O_i \left(t_{m-i} \left\| \frac{|ux_{i+r} + vx_{i+r-1}|}{\rho}, z_1, z_2, \ldots, z_{n-1} \right\| \right) \right]^{q_i}$$

$$\leq \frac{1}{T_m} \sum_{i=0}^{m} \left[s_i O_i \left(t_{m-i} \left\| \frac{|ux_{i+r} + vx_{i+r-1}|}{\rho}, z_1, z_2, \ldots, z_{n-1} \right\| \right) \right]^{p_i}$$

$$\longrightarrow 0, \text{ as } m \to \infty.$$

Hence, $x \in [\mathcal{N}^t, \mathcal{O}, s, q]$. $\square$

Theorem 4.2.4. *Suppose the sequence of Orlicz functions $\mathcal{O} = (O_i)$ satisfies the Δ_2-condition and $\beta = \lim_{t \to \infty} \frac{O_i(t)}{t} > 0$. Then, we get that $[\mathcal{N}^t, \mathcal{O}, s, p]_0 \subseteq [\mathcal{N}^t, s, p]_0$.*

Proof. For $\beta > 0$, we get $O_i(t) \geq \beta(t)$, for every $t > 0$. As $\beta > 0$, we have $t \leq \frac{1}{\beta} O_i(t)$ for every $t > 0$.
Let $x = (x_i) \in [\mathcal{N}^t, \mathcal{O}, s, p]_0$. Thus, we have

$$\frac{1}{T_m} \sum_{i=0}^{m} \left[s_i \left(t_{m-i} \left\| \frac{|ux_{i+r} + vx_{i+r-1}|}{\rho}, z_1, z_2, \ldots, z_{n-1} \right\| \right) \right]^{p_i}$$

$$\leq \frac{1}{\beta T_m} \sum_{i=0}^{m} \left[s_i O_i \left(t_{m-i} \left\| \frac{|ux_{i+r} + vx_{i+r-1}|}{\rho}, z_1, z_2, \ldots, z_{n-1} \right\| \right) \right]^{p_i}$$

from which it follows that $x = (x_i) \in [\mathcal{N}^t, s, p]_0$. This completes the proof. $\square$

Theorem 4.2.5. *Consider two sequences $\mathcal{O}' = (O'_i)$ and $\mathcal{O}'' = (O''_i)$ of Orlicz functions satisfying the Δ_2-condition. Then, we have*

$$[\mathcal{N}^t, \mathcal{O}', s, p]_0 \cap [\mathcal{N}^t, \mathcal{O}'', s, p]_0 \subseteq [\mathcal{N}^t, \mathcal{O}' + \mathcal{O}'', s, p]_0.$$

Proof. Let $x = (x_i) \in [\mathcal{N}^t, \mathcal{O}', s, p]_0 \cap [\mathcal{N}^t, \mathcal{O}'', s, p]_0$. Therefore,

$$\frac{1}{T_m} \sum_{i=0}^{m} \left[s_i O'_i \left(t_{m-i} \left\| \frac{|ux_{i+r} + vx_{i+r-1}|}{\rho}, z_1, \ldots, z_{n-1} \right\| \right) \right]^{p_i} \longrightarrow 0, \text{ as } m \to \infty$$

and

$$\frac{1}{T_m} \sum_{i=0}^{m} \left[s_i O''_i \left(t_{m-i} \left\| \frac{|ux_{i+r} + vx_{i+r-1}|}{\rho}, z_1, z_2, \ldots, z_{n-1} \right\| \right) \right]^{p_i}$$

$$\longrightarrow 0, \text{ as } m \to \infty.$$

Thus, we have

$$\frac{1}{T_m} \sum_{i=0}^{m} \left[s_i(O'_i + O''_i) \left(t_{m-i} \left\| \frac{|ux_{i+r} + vx_{i+r-1}|}{\rho}, z_1, z_2, \ldots, z_{n-1} \right\| \right) \right]^{p_i}$$

$$\leq K \left\{ \frac{1}{T_m} \sum_{i=0}^{m} \left[s_i O'_i \left(t_{m-i} \left\| \frac{|ux_{i+r} + vx_{i+r-1}|}{\rho}, z_1, z_2, \ldots, z_{n-1} \right\| \right) \right]^{p_i} \right\}$$

$$+ K \left\{ \frac{1}{T_m} \sum_{i=0}^{m} \left[s_i O''_i \left(t_{m-i} \left\| \frac{|ux_{i+r} + vx_{i+r-1}|}{\rho}, z_1, z_2, \ldots, z_{n-1} \right\| \right) \right]^{p_i} \right\}$$

$$\longrightarrow 0, \text{ as } m \to \infty.$$

Therefore,

$$\frac{1}{T_m} \sum_{i=0}^{m} \left[s_i(O'_i + O''_i) \left(t_{m-i} \left\| \frac{|ux_{i+r} + vx_{i+r-1}|}{\rho}, z_1, z_2, \ldots, z_{n-1} \right\| \right) \right]^{p_i}$$

$$\longrightarrow 0, \text{ as } m \to \infty.$$

Hence, $x = (x_i) \in [\mathcal{N}^t, \mathcal{O}' + \mathcal{O}'', s, p]_0$. $\square$

Theorem 4.2.6. *Let $\mathcal{O} = (O_i)$ and $\mathcal{O}' = (O'_i)$ are two sequences of Orlicz functions. Then $[\mathcal{N}^t, \mathcal{O}', s, p]_0 \subseteq [\mathcal{N}^t, \mathcal{O} \circ \mathcal{O}', s, p]_0$.*

Proof. Let $x = (x_i) \in [\mathcal{N}^t, \mathcal{O}', s, p]_0$. Then we have

$$\lim_{m \to \infty} \frac{1}{T_m} \sum_{i=0}^{m} \left[s_i(O'_i) \left(t_{m-i} \left\| \frac{|ux_{i+r} + vx_{i+r-1}|}{\rho}, z_1, z_2, \ldots, z_{n-1} \right\| \right) \right]^{p_i}$$

$$= 0.$$

For $\epsilon > 0$, consider $\eta > 0$ with $0 < \eta < 1$ such that $O_i(t) < \epsilon$ for $0 \leq t \leq \eta$. Write $y_i = \left[s_i(O'_i) \left(t_{m-i} \left\| \frac{|ux_{i+r} + vx_{i+r-1}|}{\rho}, z_1, z_2, \ldots, z_{n-1} \right\| \right) \right]$ and consider

$$\frac{1}{T_m} \sum_{i=0}^{m} [O_i(y_i)]^{p_i} = \frac{1}{T_m} \sum_{\substack{i=0, \\ y_i \leq \eta}}^{m} [O_i(y_i)]^{p_i} + \frac{1}{T_m} \sum_{\substack{i=0, \\ y_i > \eta}}^{m} [O_i(y_i)]^{p_i},$$

where $\sum_{\substack{i=0, \\ y_i \leq \eta}}^{m} [O_i(y_i)]^{p_i}$ is represented by $\sum_{1} [O_i(y_i)]^{p_i}$ and $\sum_{\substack{i=0, \\ y_i > \eta}}^{m} [O_i(y_i)]^{p_i}$ is represented by $\sum_{2} [O_i(y_i)]^{p_i}$. So by the continuity of O_i, we have

$$\frac{1}{T_m} \sum_{1} [O_i(y_i)]^{p_i} < \epsilon^E. \tag{4.6}$$

Also, for $y_i > \eta$, we use the fact that

$$y_i < \frac{y_i}{\eta} \leq 1 + \frac{y_i}{\eta}.$$

Therefore, by the definition, for $y_i > \eta$ we have

$$O_i(y_i) < 2O_i(1)\frac{y_i}{\eta}.$$

Hence,

$$\frac{1}{T_m}\sum_2 [O_i(y_i)]^{p_i} \leq \max\left(1, (2O_i(1)\eta^{-1})^E\right)\frac{1}{T_m}\sum_i [y_i]^{p_i}. \tag{4.7}$$

Thus, equations (4.6) and (4.7) imply that

$$[\mathcal{N}^t, \mathcal{O}', s, p]_0 \subseteq [\mathcal{N}^t, \mathcal{O} \circ \mathcal{O}', s, p]_0.$$

Hence the theorem. □

Bibliography

[1] Başar, F. and Kirişçi, M. 2011. Almost convergence and generalized difference matrix, *Comput. Math. Appl.* 61: 602–611.

[2] Esi, A. 2009. Lacunary Strong Almost Convergence of Generalized Difference Sequences With Respect to a Sequence of Moduli, *J. Adv. Res. Math.* 1: 9–18.

[3] Esi, A. 2012. Strongly almost summable sequence spaces in 2-normed spaces defined by ideal convergence and an Orlicz function, *Stud. Univ. Babeş-Bolyai Math.* 57: 75–82.

[4] Esi, A. and Çatalbaş, M.N. 2014. Almost convergence of triple sequences, *Glob. J. Math. Anal.* 2: 6–10.

[5] Kayaduman, K. and Şengönül, M. 2012. The space of Cesàro almost convergent sequence and core theorems, *Acta Math. Sci.* 6: 2265–2278.

[6] Gunawan, H. 2001. On n-inner product, n-norms and the Cauchy-Schwartz inequality, *Sci. Math. Jpn.* 5: 47–54.

[7] Gunawan, H. 2001. The space of p-summable sequence and its natural n-norm, *Bull. Aust. Math. Soc.* 64: 137–147.

[8] Kayaduman, K. and Şengönül, M. 2012. On the Riesz almost convergent sequences space, *Abstr. Appl. Anal.* Article ID 691694: 18 pages.

[9] Lindenstrauss, J. and Tzafriri, L. 1971. On Orlicz sequence spaces, *Israel J. Math.* 10: 379–390.

[10] Lorentz, G.G. 1948. A contribution to the theory of divergent series, *Acta Math.* 80: 167–190.

[11] Mears, F.M. 1943. The inverse Nörlund mean, *Ann. Math.* 44: 401–409.

[12] Misiak, A. 1989. n-inner product spaces, *Math. Nachr.* 140: 299–319.

[13] Maligranda, L. 1989. *Orlicz spaces and interpolation*, Seminars in Mathematics 5: Polish Academy of Science.

[14] Mursaleen, M. and Raj, K. 2018. Some vector valued sequence spaces of Musielak-Orlicz functions and their operator ideals, *Math. Slovaca* 68: 115–134.

[15] Mursaleen, M., Sharma, S.K., Mohiuddine, S.A. and Kiliçman, A. 2014. New difference sequence spaces defined by Musielak-Orlicz function, *Abstr. Appl. Anal.* Art. ID 691632, pp. 9.

[16] Mursaleen, M. 1996. Generalized spaces of difference sequences, *J. Math. Anal. Appl.* 203: 738–745.

[17] Musielak, J. 1983. *Orlicz spaces and modular spaces*, Lecture Notes in Mathematics, Vol. 1034.

[18] Malkowsky, E. and Özger, F. 2012. A note on some sequence spaces of weighted means, *Filomat* 26: 511–518.

[19] Malkowsky, E. and Özger, F. 2012. Compact operators on spaces of sequences of weighted means, *AIP Conference Proceedings* 1470: 179–182.

[20] Peyerimhoff, A. 1969. *Lectures on Summability*, Lecture Notes in Mathematics, Springer, New York, NY, USA.

[21] Raj, K. and Kilicman, A. 2015. On certain generalized paranormed spaces, *J. Inequal. Appl.* 2015: 37.

[22] Raj, K. and Sharma, C. 2016. Applications of strongly convergent sequences to Fourier series by means of modulus functions, *Acta Math. Hungar.* 150: 396–411.

[23] Raj, K. and Sharma, S.K. 2014. Some multiplier generalized difference sequence spaces over n-normed spaces defined by a Musielak-Orlicz function, *Siberian Adv. Math.* 24: 193–203.

[24] Sönmez, A. 2013. Almost convergence and triple band matrix, *Math. Comput. Modelling* 57: 2393–2402.

[25] Subramanian, N. and Esi, A. 2010. The Nörlund space of double entire sequences, *Fasc. Math.* 43: 147–153.

[26] Wang, C.S. 1978. On Nörlund sequence spaces, *Tamkang J. Math.* 9: 269–274.

[27] Wilansky, A. 1984. *Summability through Functional Analysis*, North-Holland Math. Stud., Vol. 85.

Chapter 5

Factorization of the infinite Hilbert and Cesàro operators

Hadi Roopaei

5.1	Introductions and preliminaries	71
5.2	Hilbert matrix	72
5.3	Hausdorff matrix	73
5.4	Cesàro matrix of order n	73
5.5	Copson matrix	75
5.6	Gamma matrix of order n	75
5.7	Factorization of the infinite Hilbert operator	76
	5.7.1 Factorization of the Hilbert operator based on Cesàro operator	77
	5.7.2 Factorization of the Hilbert operator based on gamma operator	81
	5.7.3 Factorization of the Hilbert operator based on the generalized Cesàro operator	86
5.8	Factorization of the Cesàro operator	90
	Bibliography	93

5.1 Introductions and preliminaries

Let $p > 1$ and ω denote the set of all real-valued sequences. The Banach space ℓ_p is the set of all real sequences $x = (x_k)_{k=0}^{\infty} \in \omega$ such that

$$\|x\|_{\ell_p \to \ell_p} = \left(\sum_{k=0}^{\infty} |x_k|^p \right)^{1/p} < \infty.$$

Norm of operators. The operator T is called bounded, if the inequality $\|Tx\|_{\ell_p} \leq K\|x\|_{\ell_p}$ holds for all sequences $x \in \ell_p$, while the constant K is not depending on x. The constant K is called an upper bound for operator T and the smallest possible value of K is called the norm of T. One of the

DOI: 10.1201/9781003330868-5

advantages of finding the norm of an operators is obtaining useful inequalities. For instance, the Cesàro matrix C, which has the norm $\|C\|_{\ell_p \to \ell_p} = \frac{p}{p-1}$, results the well-known Hardy's inequality

$$\sum_{n=0}^{\infty} \left(\sum_{k=0}^{n} \frac{|x_k|}{n+1} \right)^p \leq \left(\frac{p}{p-1} \right)^p \sum_{n=0}^{\infty} |x_k|^p.$$

Also by knowing the Hilbert's norm, we have the following inequality

$$\sum_{n=0}^{\infty} \left(\sum_{k=0}^{\infty} \frac{|x_k|}{n+k+1} \right)^p \leq (\pi \csc(\pi/p))^p \sum_{n=0}^{\infty} |x_k|^p,$$

also known as Hilbert's inequality.

5.2 Hilbert matrix

The Hilbert matrix H, was introduced (1894) by David Hilbert to study a question in approximation theory and is defined by $[H]_{j,k} = \frac{1}{j+k+1}$ for all non-negative integers j, k. That is

$$H = \begin{pmatrix} 1 & 1/2 & 1/3 & \cdots \\ 1/2 & 1/3 & 1/4 & \cdots \\ 1/3 & 1/4 & 1/5 & \cdots \\ \vdots & \vdots & \vdots & \ddots \end{pmatrix}.$$

From [7] Theorem 323, we know that the Hilbert matrix is a bounded operator on ℓ_p and

$$\|H\|_{\ell_p \to \ell_p} = \pi \csc(\pi/p).$$

For a non-negative integer n, we define the Hilbert matrix of order n, H_n, by

$$[H_n]_{j,k} = \frac{1}{j+k+n+1} \quad (j, k = 0, 1, \cdots).$$

Note that for $n = 0$, $H_0 = H$ is the Hilbert matrix. For more examples:

$$H_1 = \begin{pmatrix} 1/2 & 1/3 & 1/4 & \cdots \\ 1/3 & 1/4 & 1/5 & \cdots \\ 1/4 & 1/5 & 1/6 & \cdots \\ \vdots & \vdots & \vdots & \ddots \end{pmatrix}, \quad H_2 = \begin{pmatrix} 1/3 & 1/4 & 1/5 & \cdots \\ 1/4 & 1/5 & 1/6 & \cdots \\ 1/5 & 1/6 & 1/7 & \cdots \\ \vdots & \vdots & \vdots & \ddots \end{pmatrix}.$$

5.3 Hausdorff matrix

Consider the Hausdorff matrix H^μ, with entries of the form:

$$[H^\mu]_{j,k} = \begin{cases} \binom{j}{k} \int_0^1 \theta^k (1-\theta)^{j-k} d\mu(\theta), & 0 \leq k \leq j, \\ 0, & k > j \end{cases}$$

for all $j, k \in \mathbb{N}_0$, where μ is a probability measure on $[0, 1]$. The Hausdorff matrix contains the famous classes of matrices. For $n > 0$, some of these classes are as follows:

- The choice $d\mu(\theta) = n(1-\theta)^{n-1} d\theta$ gives the Cesàro matrix of order n.
- The choice $d\mu(\theta) = n\theta^{n-1} d\theta$ gives the Gamma matrix of order n.
- The choice $d\mu(\theta) = \frac{|\log \theta|^{n-1}}{\Gamma(n)} d\theta$ gives the Hölder matrix of order n.
- The choice $d\mu(\theta) = $ point evaluation at $\theta = r$, $0 < r < 1$, gives the Euler matrix of order r.

Hardy's formula ([7], Theorem 216) states that the Hausdorff matrix is a bounded operator on ℓ_p if and only if $\int_0^1 \theta^{\frac{-1}{p}} d\mu(\theta) < \infty$ and

$$\|H^\mu\|_{\ell_p \to \ell_p} = \int_0^1 \theta^{\frac{-1}{p}} d\mu(\theta) \quad (1 < p < \infty). \tag{5.1}$$

Hausdorff operator has the following norm property.

Theorem 5.3.1 ([3], Theorem 9). *Let $p \geq 1$ and H^μ, H^φ, and H^ν be Hausdorff matrices such that $H^\mu = H^\varphi H^\nu$. Then H^μ is bounded on ℓ_p if and only if both H^φ and H^ν are bounded on ℓ_p. Moreover, we have*

$$\|H^\mu\|_{\ell_p \to \ell_p} = \|H^\varphi\|_{\ell_p \to \ell_p} \|H^\nu\|_{\ell_p \to \ell_p}.$$

5.4 Cesàro matrix of order n

By letting $d\mu(\theta) = n(1-\theta)^{n-1} d\theta$ in the definition of the Hausdorff matrix, the Cesàro matrix of order n, C_n, is defined by

$$[C_n]_{j,k} = \begin{cases} \dfrac{\binom{n+j-k-1}{j-k}}{\binom{n+j}{j}}, & 0 \leq k \leq j, \\ 0, & \text{otherwise} \end{cases} \tag{5.2}$$

which according to the relation (5.1) has the ℓ_p-norm

$$\|C_n\|_{\ell_p \to \ell_p} = \frac{\Gamma(n+1)\Gamma(1/p^*)}{\Gamma(n+1/p^*)}. \tag{5.3}$$

Note that, $C_0 = I$, where I is the identity matrix and C_1 is the well-known Cesàro matrix C who has ℓ_p-norm $\|C\|_{\ell_p \to \ell_p} = p^*$. For more examples

$$C_2 = \begin{pmatrix} 1 & 0 & 0 & \cdots \\ 2/3 & 1/3 & 0 & \cdots \\ 3/6 & 2/6 & 1/6 & \cdots \\ \vdots & \vdots & \vdots & \ddots \end{pmatrix} \quad \text{and} \quad C_3 = \begin{pmatrix} 1 & 0 & 0 & \cdots \\ 3/4 & 1/4 & 0 & \cdots \\ 6/10 & 3/10 & 1/10 & \cdots \\ \vdots & \vdots & \vdots & \ddots \end{pmatrix}.$$

For more information about the Cesàro operators and their associated sequence spaces the readers can refer to [9–11].

Lemma 5.4.1. *The Cesàro matrix of order n is invertible and it's inverse, C_n^{-1}, is*

$$[C_n^{-1}]_{j,k} = \begin{cases} (-1)^{(j-k)}\binom{n}{j-k}\binom{n+k}{k}, & k \leq j \leq n+k, \\ 0, & \text{otherwise.} \end{cases}$$

Proof. By applying the identity

$$\sum_{k=0}^{j}(-1)^k \binom{n}{k}\binom{n+j-k-1}{j-k} = 0, \quad j = 1, 2, \ldots$$

we have

$$\begin{aligned}
[C_n C_n^{-1}]_{i,j} &= \frac{\binom{n+j}{j}}{\binom{n+i}{i}} \sum_{k=j}^{i} (-1)^{k-j} \binom{k}{k-j}\binom{n+i-k-1}{i-k} \\
&= \frac{\binom{n+j}{j}}{\binom{n+i}{i}} \sum_{k=0}^{i-j} (-1)^{k} \binom{k}{j}\binom{n+i-j-k-1}{i-j-k} = \frac{\binom{n+j}{j}}{\binom{n+i}{i}}[I]_{i,j}.
\end{aligned}$$

$\square$

In the following, we present another proof for obtaining the inverse of the Cesàro matrix of order n.

Remark 5.4.2. *Let us recall the backward difference matrix of order n, Δ_n, which is a lower triangular matrix with entries*

$$[\Delta_n]_{j,k} = \begin{cases} (-1)^{(j-k)}\binom{n}{j-k}, & k \leq j \leq n+k, \\ 0, & \text{otherwise.} \end{cases}$$

This matrix is invertible which we denote it by Δ_n^{-1} and has the following entries

$$[\Delta_n^{-1}]_{j,k} = \begin{cases} \binom{n+j-k-1}{j-k}, & j \geq k, \\ 0, & \text{otherwise.} \end{cases}$$

From the relation (5.2), one can see that the Cesàro matrix of order n and its inverse can be rewritten based on the backward difference operator as below

$$[C_n]_{j,k} = \frac{\binom{n+j-k-1}{j-k}}{\binom{n+j}{j}} = \frac{[\Delta_n^{-1}]_{j,k}}{\binom{n+j}{j}}, \quad \text{and} \quad [C_n^{-1}]_{j,k} = [\Delta_n]_{j,k}\binom{n+k}{k}.$$

Now, by a simple calculation we deduce that

$$[C_n C_n^{-1}]_{i,j} = \frac{\binom{n+j}{j}}{\binom{n+i}{i}} \sum_{k=j}^{i} [\Delta_n^{-1}]_{i,k}[\Delta_n]_{k,j} = \frac{\binom{n+j}{j}}{\binom{n+i}{i}} [\Delta_n^{-1}\Delta_n]_{i,j} = \frac{\binom{n+j}{j}}{\binom{n+i}{i}} [I]_{i,j},$$

which completes the proof.

5.5 Copson matrix

Transposing the Cesàro matrix results the Copson matrix, which has the ℓ_p-norm $\|C^t\|_{\ell_p \to \ell_p} = p$ by Hellinger-Toeplitz theorem which is

Theorem 5.5.1 ([1], Proposition 7.2). *Suppose that $1 < p, q < \infty$. A matrix A maps ℓ_p into ℓ_q if and only if the transposed matrix, A^t, maps ℓ_{q^*} into ℓ_{p^*}. We then have*

$$\|A\|_{\ell_p \to \ell_q} = \|A^t\|_{\ell_{q^*} \to \ell_{p^*}}.$$

More information about the Copson operators are found in [12, 13].

5.6 Gamma matrix of order n

By letting $d\mu(\theta) = n\theta^{n-1}d\theta$ in the definition of the Hausdorff matrix, the Gamma matrix of order n, G_n, is

$$[G_n]_{j,k} = \begin{cases} \dfrac{\binom{n+k-1}{k}}{\binom{n+j}{j}}, & 0 \leq k \leq j \\ 0, & \text{otherwise,} \end{cases}$$

which according to the Hardy's formula has the ℓ_p-norm

$$\|G_n\|_{\ell_p \to \ell_p} = \frac{np}{np-1}. \tag{5.4}$$

Note that G_1 is the well-known Cesàro matrix.

Although, the Cesàro and the Gamma matrices of order n have the same entries in a row, they have the inverse order in columns. For example,

$$C_2 = \begin{pmatrix} 1 & 0 & 0 & \cdots \\ 2/3 & 1/3 & 0 & \cdots \\ 3/6 & 2/6 & 1/6 & \cdots \\ \vdots & \vdots & \vdots & \ddots \end{pmatrix}, \quad G_2 = \begin{pmatrix} 1 & 0 & 0 & \cdots \\ 1/3 & 2/3 & 0 & \cdots \\ 1/6 & 2/6 & 3/6 & \cdots \\ \vdots & \vdots & \vdots & \ddots \end{pmatrix}.$$

The following identity reveals the relation between these two matrices.

Lemma 5.6.1 ([14], Lemma 2.7). *The Cesàro matrix of order n, has a factorization of the form*

$$C_n = G_n C_{n-1} = C_{n-1} G_n.$$

Proof. Since every two Hausdorff matrices commute ([7], Theorem 197), we only prove the first equality. By using the identity $\sum_{j=0}^{k} \binom{n+j-1}{j} = \binom{n+k}{k}$, we have

$$\begin{aligned} [G_n C_{n-1}]_{i,k} &= \sum_{j=k}^{i} [G_n]_{i,j} [C_{n-1}]_{j,k} = \sum_{j=k}^{i} \frac{\binom{n+j-1}{j} \binom{n+j-k-2}{j-k}}{\binom{n+i}{i} \binom{n+j-1}{j}} \\ &= \frac{1}{\binom{n+i}{i}} \sum_{j=0}^{i-k} \binom{n-1+j-1}{j} = \frac{\binom{n+i-k-1}{i-k}}{\binom{n+i}{i}} = [C_n]_{i,k}. \end{aligned}$$

$\square$

For more information about the Gamma operators and their associated sequence spaces the readers can refer to [14, 15].

5.7 Factorization of the infinite Hilbert operator

In sequel, we state several factorizations for the Hilbert matrix based on the Cesàro and Gamma matrices of order n. There are two important techniques in the theory of ℓ_p spaces, the Schur and the Hellinger-Toeplitz theorems. In sequel, we need the Schur's theorem which is

Theorem 5.7.1 ([8], Theorem 275). *Let $p > 1$ and T be a matrix operator with $[T]_{j,k} \geq 0$ for all j, k. Suppose that C, R are two strictly positive numbers such that*

$$\sum_{j=0}^{\infty} [T]_{j,k} \leq C \quad \text{for all } k, \qquad \sum_{k=0}^{\infty} [T]_{j,k} \leq R \quad \text{for all } j,$$

(bounds for column and row sums respectively). Then

$$\|T\|_{\ell_p \to \ell_p} \leq R^{1/p^*} C^{1/p}.$$

5.7.1 Factorization of the Hilbert operator based on Cesàro operator

According to Bennett [2], H admits a factorization of the form $H = BC$, where C is the Cesàro matrix and the matrix B is defined by

$$[B]_{j,k} = \frac{k+1}{(j+k+1)(j+k+2)} \qquad (j, k = 0, 1, \ldots). \tag{5.5}$$

The matrix B is a bounded operator on ℓ_p and $\|B\|_{\ell_p \to \ell_p} = \frac{\pi}{p^*} \csc(\pi/p)$, ([2], Proposition 2).

For non-negative integers n, j, and k, let us define the matrix B_n by

$$\begin{aligned}[][B_n]_{j,k} &= \frac{(k+1)\cdots(k+n)}{(j+k+1)\cdots(j+k+n+1)} \\ &= \binom{n+k}{k} \beta(j+k+1, n+1) \qquad (j, k = 0, 1, \ldots),\end{aligned}$$

where the β function is

$$\beta(m, n) = \int_0^1 z^{m-1}(1-z)^{n-1} dz \qquad (m, n = 1, 2, \ldots).$$

Consider that for $n = 0$, $B_0 = H$, where H is the Hilbert matrix and for $n = 1$, $B_1 = B$ which was defined by relation (5.5).

We are ready to generalize Bennett's factorization for the Hilbert matrix based on the Cesàro matrix of order n, but we need first the following lemma.

Lemma 5.7.2. *For $|z| < 1$, we have*

$$(1-z)^{-n} = \sum_{j=0}^{\infty} \binom{n+j-1}{j} z^j.$$

Proof. By differentiating $n-1$ times the identity $(1-z)^{-1} = \sum_{j=0}^{\infty} z^j$, we obtain the result. $\square$

78   Advances in Mathematical Analysis and its Applications

Theorem 5.7.3. *The Hilbert matrix H and the Hilbert matrix of order n, H_n, have the following factorizations based on the Cesàro matrix of order n:*

(a) $H = B_n C_n$

(b) $H_n = C_n B_n$

(c) $C_n H = H_n C_n$

(d) B_n is a bounded operator on ℓ_p and
$$\|B_n\|_{\ell_p \to \ell_p} = \frac{\Gamma(n+1/p^*)\Gamma(1/p)}{\Gamma(n+1)}.$$

Proof. (a) By applying Lemma 5.7.2, we deduce that

$$\begin{aligned}
[B_n C_n]_{j,k} &= \sum_{i=k}^{\infty} \binom{n+i}{i} \beta(j+i+1, n+1) \frac{\binom{n+i-k-1}{i-k}}{\binom{n+i}{i}} \\
&= \sum_{i=0}^{\infty} \binom{n+i-1}{i} \beta(j+i+k+1, n+1) \\
&= \int_0^1 \sum_{i=0}^{\infty} \binom{n+i-1}{i} z^i z^{j+k} (1-z)^n dz \\
&= \int_0^1 z^{j+k} dz = \frac{1}{j+k+1} = [H]_{j,k}.
\end{aligned}$$

(b) For convenience, let $\lambda = \frac{\binom{n+k}{k}}{\binom{n+i}{i}}$. The factorization by the following calculations will be obtained.

$$\begin{aligned}
[C_n B_n]_{i,k} &= \sum_{j=0}^{i} \frac{\binom{n+i-j-1}{i-j}}{\binom{n+i}{i}} \binom{n+k}{k} \beta(j+k+1, n+1) \\
&= \lambda \Bigg\{ \binom{n+i-1}{i} \beta(k+1, n+1) + \cdots \\
&\quad + \binom{n}{1} \beta(i+k, n+1) + \binom{n-1}{0} \beta(i+k+1, n+1) \Bigg\} \\
&= \lambda \Bigg\{ \binom{n+i-1}{i} \beta(k+1, n+1) + \cdots + \binom{n+1}{2} \beta(i+k-1, n+1) \\
&\quad + \frac{(n+1)!(i+k-1)!}{(i+k+n-1)!(i+k+n+1)} \Bigg\} \\
&= \lambda \Bigg\{ \binom{n+i-1}{i} \beta(k+1, n+1) + \cdots + \binom{n+2}{3} \beta(i+k-2, n+1) \\
&\quad + \frac{(n+2)!(i+k-2)!}{2!(i+k+n-2)!(i+k+n+1)} \Bigg\}
\end{aligned}$$

⋮

$$= \lambda \left\{ \binom{n+i-1}{i} \beta(k+1, n+1) + \frac{(n+i-1)!(k+1)!}{(i-1)!(k+n+1)!(i+k+n+1)} \right\}$$

$$= \lambda \left\{ \frac{(n+i-1)!k!}{i!(n+k+1)!} \frac{(n+k+1)(n+i)}{i+k+n+1} \right\} = \frac{1}{i+k+n+1} = [H_n]_{i,k}.$$

(c) This is obvious by parts (a) and (b). (d) For computing the ℓ_p-norm of B_n, we introduce a family of matrices, $B(w)$, $0 < w \le 1$, given by

$$b(w)_{j,k} = \binom{j+k}{k} w^j (1-w)^{n+k}.$$

Since

$$\sum_{k=0}^{\infty} b(w)_{j,k} = w^j (1-w)^n \sum_{k=0}^{\infty} \binom{j+1+k-1}{k} (1-w)^k$$

$$= w^j (1-w)^n (1-(1-w))^{-(j+1)} = \frac{(1-w)^n}{w}$$

and

$$\sum_{j=0}^{\infty} b(w)_{j,k} = (1-w)^{n+k} \sum_{j=0}^{\infty} \binom{k+1+j-1}{j} w^j = (1-w)^{n-1},$$

the row sums are all $\frac{(1-w)^n}{w}$ and the column sums are all $(1-w)^{n-1}$. Thus Schur's theorem results

$$\|B(w)\|_{\ell_p \to \ell_p} \le (1-w)^{n-\frac{1}{p}} w^{\frac{-1}{p^*}}.$$

On the other hand,

$$\int_0^1 b(w)_{j,k} dw = \binom{j+k}{k} \int_0^1 w^j (1-w)^{n+k} dw$$

$$= \binom{j+k}{k} \beta(j+1, n+k+1)$$

$$= \binom{n+k}{k} \beta(j+k+1, n+1) = [B_n]_{j,k}.$$

Now,

$$\|B_n\|_{\ell_p \to \ell_p} = \left\| \int_0^1 B(w) dw \right\|_{\ell_p \to \ell_p} \le \int_0^1 \|B(w)\|_{\ell_p \to \ell_p} dw$$

$$\le \int_0^1 (1-w)^{n-1/p} w^{-1/p^*} dw = \beta(n-1/p+1, 1-1/p^*)$$

$$= \frac{\Gamma(n+1/p^*)\Gamma(1/p)}{\Gamma(n+1)}.$$

Also the factorization $H = B_n C_n$ results $\|H\|_{\ell_p \to \ell_p} \leq \|B_n\|_{\ell_p \to \ell_p} \|C_n\|_{\ell_p \to \ell_p}$. Therefore,

$$\|B_n\|_{\ell_p \to \ell_p} \geq \frac{\|H\|_{\ell_p \to \ell_p}}{\|C_n\|_{\ell_p \to \ell_p}} = \frac{\Gamma(n+1/p^*)\Gamma(1/p)}{\Gamma(n+1)},$$

which completes the proof. $\square$

Lemma 5.7.4. *We have*

$$\sum_{j=0}^{n}(-1)^j \binom{n}{j}\frac{1}{j+m} = \int_0^1 x^{m-1}(1-x)^n dx = \beta(m, n+1).$$

Proof. By multiplying both sides of the identity $(1-x)^n = \sum_{j=0}^{n}(-1)^j\binom{n}{j}x^j$, in x^{m-1} and also integrating from 0 to 1, we have the desired result. $\square$

Remark 5.7.5. *By using the inverse of Cesàro matrix and applying Lemma 5.7.4 one can directly obtain the factor B_n in the factorization $H = B_n C_n$ in Theorem 5.7.3 of the form*

$$\begin{aligned}[][B_n]_{j,k} &= \sum_{i=k}^{n+k}[H]_{j,i}[C_n^{-1}]_{i,k} = \sum_{i=k}^{n+k}\frac{1}{j+i+1}(-1)^{i-k}\binom{n}{i-k}\binom{n+k}{k} \\ &= \binom{n+k}{k}\sum_{i=0}^{n}(-1)^i\binom{n}{i}\frac{1}{j+i+k+1} = \binom{n+k}{k}\beta(j+k+1, n+1) \\ &= \frac{(k+1)\cdots(k+n)}{(j+k+1)\cdots(j+k+n+1)}.\end{aligned}$$

We are ready to generalize the inequality

$$\|Hx\|_{\ell_p} \leq \pi \csc(\pi/p)\|x\|_{\ell_p},$$

also known as Hilbert's inequality and the inequality

$$\|Hx\|_{\ell_p} \leq \frac{\pi}{p^*}\csc(\pi/p)\|Cx\|_{\ell_p}, \tag{5.6}$$

which is the Hardy's inequality versus Hilbert's, introduced in [2].

Corollary 5.7.6. *Let $p > 1$ and $x \in \ell_p$. Then*
(a)

$$\|Hx\|_{\ell_p} \leq \frac{\Gamma(n+1/p^*)\Gamma(1/p)}{\Gamma(n+1)}\|C_n x\|_{\ell_p}.$$

In particular, for $n = 0$ Hilbert's inequality occurs and for $n = 1$ we have the inequality (5.6).

Factorization of the infinite Hilbert 81

(b)
$$\|H_n x\|_{\ell_p} \leq \frac{\Gamma(n+1/p)\Gamma(1/p^*)}{\Gamma(n+1)}\|C_n^t x\|_{\ell_p},$$

In particular, for $n = 0$, the Hilbert's inequality occurs and for $n = 1$, we have the inequality

$$\|H_1 x\|_{\ell_p} \leq \pi/p\,\csc(\pi/p)\|C^t x\|_{\ell_p},$$

or

$$\sum_{j=0}^{\infty}\left|\sum_{k=0}^{\infty}\frac{x_k}{j+k+2}\right|^p \leq (\pi/p\,\csc(\pi/p))^p \sum_{j=0}^{\infty}\left|\sum_{k=j}^{\infty}\frac{x_k}{1+k}\right|^p.$$

Proof. (a) According to Theorem 5.7.3, $H = B_n C_n$ which results in

$$\|Hx\|_{\ell_p} = \|B_n C_n x\|_{\ell_p} \leq \frac{\Gamma(n+1/p^*)\Gamma(1/p)}{\Gamma(n+1)}\|C_n x\|_{\ell_p}.$$

Consider that for $n = 0$, $C_0 = I$, hence we have the Hilbert's inequality.
(b) By transposing part (b) of Theorem 5.7.3, we have the result. □

5.7.2 Factorization of the Hilbert operator based on gamma operator

Let us recall the definition of weighted mean matrix. Suppose that $a = (a_j)_{j=0}^{\infty}$ is a non-negative sequence with $a_0 > 0$ and $A_j = a_0 + a_1 + \cdots + a_j$. The weighted mean matrix M_a is a lower triangular matrix which is defined as

$$[M_a]_{j,k} = \begin{cases} \frac{a_k}{A_j} & 0 \leq k \leq j, \\ 0 & otherwise. \end{cases}$$

The sequence $(a_j)_{j=0}^{\infty}$ is called the "symbol" of weighted mean matrix. It must be mentioned that although the Gamma matrix of order n, G_n, is a Hausdorff matrix (let $d\mu(\theta) = n\theta^{n-1}d\theta$), it is also a weighted mean matrix with symbol $a_j^n = \binom{n+j-1}{j}$. For example, for $n = 1$ and $n = 2$, the sequences $a_j^1 = 1$ and $a_j^2 = 1 + j$ are the symbols of the Gamma matrix of order 1 (well-known Cesàro matrix) and the Gamma matrix of order 2, that are

$$G_1 = C = \begin{pmatrix} 1 & 0 & 0 & \cdots \\ 1/2 & 1/2 & 0 & \cdots \\ 1/3 & 1/3 & 1/3 & \cdots \\ \vdots & \vdots & \vdots & \ddots \end{pmatrix} \quad \text{and} \quad G_2 = \begin{pmatrix} 1 & 0 & 0 & \cdots \\ 1/3 & 2/3 & 0 & \cdots \\ 1/6 & 2/6 & 3/6 & \cdots \\ \vdots & \vdots & \vdots & \ddots \end{pmatrix}.$$

In sequel, we introduce the necessary and sufficient conditions for the factorizing an operator based on a weighted mean matrix then, by choosing $a_j^n = \binom{n+j-1}{j}$ as the symbol, we obtain two factorization for the Cesàro and Hilbert matrices based on the Gamma matrix.

Theorem 5.7.7. *Let T be a matrix and M_a is a weighted mean matrix with symbol $a = (a_j)_{j=0}^{\infty}$. Suppose that $a_j \neq 0$ for all $j \geq 0$. Then there is a matrix S such that $T = SM_a$ if and only if*

(i) $\frac{[T]_{j,k}}{a_k} \to 0$ as $k \to \infty \quad \forall j \geq 0$,

(ii) $[S]_{j,k} = A_k \left(\frac{[T]_{j,k}}{a_k} - \frac{[T]_{j,k+1}}{a_{k+1}} \right) \quad \forall j, k \geq 0.$

Proof. Let $j, k \geq 0$. Since $T = SM_a$ we have that

$$[T]_{j,k} = \sum_{n=0}^{\infty} [S]_{j,n} [M_a]_{n,k} = \sum_{n=k}^{\infty} [S]_{j,n} \frac{a_k}{A_n},$$

hence

$$\frac{[T]_{j,k}}{a_k} = \sum_{n=k}^{\infty} [S]_{j,n} \frac{1}{A_n} \to 0,$$

as $k \to \infty$, which gives (i). Also $\left(\frac{[T]_{j,k}}{a_k} - \frac{[T]_{j,k+1}}{a_{k+1}} \right) = [S]_{j,k} \frac{1}{A_k}$, which gives (ii).

Now, let $j, k \geq 0$, $N \geq k$. Then we have that

$$\sum_{n=0}^{N} [S]_{j,n} [M_a]_{n,k} = \sum_{n=k}^{N} [S]_{j,n} \frac{a_k}{A_n}$$

$$= a_k \sum_{n=k}^{N} \left(\frac{[T]_{j,n}}{a_n} - \frac{[T]_{j,n+1}}{a_{n+1}} \right) \quad (by\,(ii))$$

$$= a_k \left(\frac{[T]_{j,k}}{a_k} - \frac{[T]_{j,N+1}}{a_{N+1}} \right)$$

$$\to a_k \frac{[T]_{j,k}}{a_k} = [T]_{j,k} \quad (as\ N \to \infty\ by\,(i)).$$

This implies that SM_a exists and $T = SM_a$. $\square$

Corollary 5.7.8. *The Hilbert matrix has a factorization of the form $H = S_n G_n$, where S_n has the entries*

$$[S_n]_{j,k} = \frac{(1 - 1/n)(j+1) + (k+1)}{(j+k+1)(j+k+2)} \quad (j, k = 0, 1, \ldots),$$

and is a bounded operator on ℓ_p with $\|S_n\|_{\ell_p \to \ell_p} = \pi \left(1 - \frac{1}{np}\right) \csc(\pi/p)$. In particular, for $n = 1$, $H = BC$, where C is the Cesàro matrix and B is a bounded operator with $\|B\|_{\ell_p \to \ell_p} = \pi/p^* \csc(\pi/p)$.

Proof. Let n be fixed. By choosing $a_k^n = \binom{n+k-1}{k}$, the weighted mean matrix corresponding to this weight is the Gamma matrix of order n, i.e., $M_{a^n} = G_n$. Since for every sequence $a_k^n = \binom{n+k-1}{k}$

$$\frac{[H]_{j,k}}{a_k^n} = \frac{1}{(j+k+1)\binom{n+k-1}{k}} \to 0 \quad as \quad k \to \infty,$$

hence M_{a^n} satisfies the condition of Theorem 5.7.7 and therefore, H has a factorization of the form $H = S_n G_n$. By applying the identity $\sum_{j=0}^{k} \binom{n+j-1}{j} = \binom{n+k}{k}$, $A_k^n = \binom{n+k}{k}$ and the claimed S_n in Theorem 5.7.7 is

$$[S_n]_{j,k} = A_k^n \left(\frac{[H]_{j,k}}{a_k^n} - \frac{[H]_{j,k+1}}{a_{k+1}^n} \right)$$

$$= \binom{n+k}{k} \left[\frac{1}{(j+k+1)\binom{n+k-1}{k}} - \frac{1}{(j+k+2)\binom{n+k}{k+1}} \right]$$

$$= \frac{(1-1/n)(j+1) + (k+1)}{(j+k+1)(j+k+2)}, \quad (j,k = 0, 1, \ldots).$$

The factorization $H = S_n G_n$ and relation (5.6.1) result that

$$\|S_n\|_{\ell_p \to \ell_p} \geq \pi \left(1 - \frac{1}{np}\right) \csc(\pi/p),$$

while for proving the other side of the above inequality, we can rewrite

$$S_n = \left(1 - \frac{1}{n}\right) B^t + B,$$

where B is the matrix who was defined in relation (5.5). Now, by applying the Hellinger-Toeplitz theorem we have

$$\|S_n\|_{\ell_p \to \ell_p} \leq \left(1 - \frac{1}{n}\right) \|B^t\|_{\ell_p \to \ell_p} + \|B\|_{\ell_p \to \ell_p} = \pi \left(1 - \frac{1}{np}\right) \csc(\pi/p),$$

which completes the proof. In special case $n = 1$, $S_1 = B$, $\|S_1\|_{\ell_p \to \ell_p} = \|B\|_{\ell_p \to \ell_p} = \frac{\pi}{p^*} \csc(\pi/p)$, $G_1 = C$ and the Hilbert matrix has the factorization $H = BC$. □

As an immediate consequence of the above theorem, we generalize the inequality (5.6) again.

Corollary 5.7.9. *Let $p > 1$ and $x \in \ell_p$. Then*

$$\|Hx\|_{\ell_p} \leq \pi(1 - 1/np) \csc(\pi/p) \|G_n x\|_{\ell_p}.$$

In particular, for $n = 1$ inequality (5.6) occurs.

Proof. According to the above theorem, we have

$$\|Hx\|_{\ell_p} = \|S_n G_n x\|_{\ell_p} \leq \pi(1 - 1/np)\csc(\pi/p)\|G_n x\|_{\ell_p}.$$

□

In the following we bring another proof for the relation between the Cesàro and Gamma matrices in the Lemma 5.6.1.

Remark 5.7.10. *The Cesàro matrix of order n has a factorization of the form*

$$C_n = C_{n-1} G_n = G_n C_{n-1},$$

where C_{n-1} is the Cesàro matrix of order $n-1$.

Proof. Again choosing $a_k^n = \binom{n+k-1}{k}$ in Theorem 5.7.7 results in $M_{a^n} = G_n$. In this case, for $[T]_{j,k} = [C_n]_{j,k} = \frac{\binom{n+j-k-1}{j-k}}{\binom{n+j}{j}}$ we have

$$\frac{[C_n]_{j,k}}{a_k^n} = \frac{\binom{n+j-k-1}{j-k}}{\binom{n+j}{j}\binom{n+k-1}{k}} \to 0 \quad as \quad k \to \infty.$$

Since M_{a^n} satisfies the condition of Theorem 5.7.7, hence C_n has a factorization of the form $C_n = S_n G_n$. Now, the factor S_n in Theorem 5.7.7 is

$$[S_n]_{j,k} = A_k^n \left(\frac{[C_n]_{j,k}}{a_k^n} - \frac{[C_n]_{j,k+1}}{a_{k+1}^n} \right)$$

$$= \binom{n+k}{k} \left[\frac{\binom{n+j-k-1}{j-k}}{\binom{n+j}{j}\binom{n+k-1}{k}} - \frac{\binom{n+j-k-2}{j-k-1}}{\binom{n+j}{j}\binom{n+k}{k+1}} \right]$$

$$= \frac{\binom{n+j-k-2}{j-k}}{\binom{n+j-1}{j}} = [C_{n-1}]_{j,k}.$$

Hence, we have proved $C_n = C_{n-1} G_n$. But, Since every two Hausdorff matrices commute ([7], Theorem 197), we have the desired result. □

Corollary 5.7.11. *Let $p > 1$ and $x \in \ell_p$. Then*

$$\|C_n x\|_{\ell_p} \leq \frac{\Gamma(n)\Gamma(1/p^*)}{\Gamma(n-1/p)} \|G_n x\|_{\ell_p}.$$

In particular, for $n = 1$ equality occurs.

Proof. According to the above theorem, we have

$$\|C_n x\|_{\ell_p} = \|C_{n-1} G_n x\|_{\ell_p} \leq \frac{\Gamma(n)\Gamma(1/p^*)}{\Gamma(n-1/p)} \|G_n x\|_{\ell_p}.$$

□

Remark 5.7.12. Applying Theorem 5.7.3 and Lemma 5.6.1 result that $H = B_n C_n = B_n C_{n-1} G_n = S_n G_n$, where $S_n = B_n C_{n-1}$. Now, using Lemma 17.9 and the identity $\binom{n}{k} = \frac{n}{n-k}\binom{n-1}{k}$, we have

$$[S_n]_{j,k} = [B_n C_{n-1}]_{j,k}$$

$$= \sum_{i=k}^{\infty} \binom{n+i}{i} \beta(j+i+1, n+1) \frac{\binom{n+i-k-2}{i-k}}{\binom{n+i-1}{i}}$$

$$= \sum_{i=k}^{\infty} \frac{n+i}{n} \binom{n+i-k-2}{i-k} \beta(j+i+1, n+1)$$

$$= \sum_{i=0}^{\infty} \frac{n+i-1+k+1}{n} \binom{n+i-2}{i} \beta(j+i+k+1, n+1)$$

$$= \frac{n-1}{n} \sum_{i=0}^{\infty} \frac{n+i-1}{n-1} \binom{n+i-2}{i} \beta(j+i+k+1, n+1)$$

$$+ \frac{k+1}{n} \sum_{i=0}^{\infty} \binom{n+i-2}{i} \beta(j+i+k+1, n+1)$$

$$= \frac{n-1}{n} \int_0^1 \sum_{i=0}^{\infty} \binom{n+i-1}{i} z^i z^{j+k} (1-z)^n dz$$

$$+ \frac{k+1}{n} \int_0^1 \sum_{i=0}^{\infty} \binom{n-1+i-1}{i} z^i z^{j+k} (1-z)^n dz$$

$$= \left(1 - \frac{1}{n}\right) \int_0^1 z^{j+k} dz + \frac{k+1}{n} \int_0^1 z^{j+k}(1-z) dz$$

$$= \left(1 - \frac{1}{n}\right) \frac{1}{j+k+1} + \frac{k+1}{n} \frac{1}{(j+k+1)(j+k+2)}$$

$$= \frac{(1-1/n)(j+1) + (k+1)}{(j+k+1)(j+k+2)}.$$

For computing the norm of matrix S_n, by the identity $S_n = B_n C_{n-1}$ we obtain

$$\|S_n\|_{\ell_p \to \ell_p} \leq \|B_n\|_{\ell_p \to \ell_p} \|C_{n-1}\|_{\ell_p \to \ell_p}$$

$$= \frac{\Gamma(n+1/p^*)\Gamma(1/p)}{\Gamma(n+1)} \frac{\Gamma(n)\Gamma(1/p^*)}{\Gamma(n-1+1/p^*)}$$

$$= \pi(1 - 1/np) \csc(\pi/p).$$

The other side of the above inequality is because of the factorization $H = S_n G_n$.

Remark 5.7.13. Let G_n^{-1} be the inverse of the Gamma matrix which is a bidiagonal matrix with the entries

$$[G_n^{-1}]_{j,k} = \begin{cases} 1 + \frac{j}{n}, & j = k, \\ -\frac{j}{n}, & j = k+1, \\ 0, & otherwise. \end{cases}$$

Now, the factor S_n in the factorization $H = S_n G_n$ has the entries

$$\begin{aligned}[][S_n]_{j,k} &= \sum_{i=k,k+1} [H]_{j,i}[G_n^{-1}]_{i,k} \\ &= \frac{1}{j+k+1}\left(1 + \frac{k}{n}\right) + \frac{1}{j+k+2}\left(-\frac{k+1}{n}\right) \\ &= \frac{(1-1/n)(j+1)+(k+1)}{(j+k+1)(j+k+2)}. \end{aligned}$$

5.7.3 Factorization of the Hilbert operator based on the generalized Cesàro operator

Suppose that $N \geq 1$ is a real number. The generalized Cesàro matrix, C_N, is defined by

$$[C_N]_{j,k} = \begin{cases} \frac{1}{j+N} & 0 \leq k \leq j \\ 0 & otherwise, \end{cases}$$

That is,

$$C_N = \begin{pmatrix} \frac{1}{N} & 0 & 0 & \cdots \\ \frac{1}{1+N} & \frac{1}{1+N} & 0 & \cdots \\ \frac{1}{2+N} & \frac{1}{2+N} & \frac{1}{2+N} & \cdots \\ \vdots & \vdots & \vdots & \ddots \end{pmatrix},$$

who has the ℓ_p-norm $\|C_N\|_{\ell_p \to \ell_p} = p^*$ ([5], Lemma 2.3).
Note that C_1 is the well-known Cesàro matrix C. For more examples,

$$C_2 = \begin{pmatrix} 1/2 & 0 & 0 & \cdots \\ 1/3 & 1/3 & 0 & \cdots \\ 1/4 & 1/4 & 1/4 & \cdots \\ \vdots & \vdots & \vdots & \ddots \end{pmatrix} \quad and \quad C_3 = \begin{pmatrix} 1/3 & 0 & 0 & \cdots \\ 1/4 & 1/4 & 0 & \cdots \\ 1/5 & 1/5 & 1/5 & \cdots \\ \vdots & \vdots & \vdots & \ddots \end{pmatrix}.$$

Let us define the matrix B_N by

$$[B_N]_{j,k} = \frac{k+N}{(j+k+1)(j+k+2)} \qquad (j,k = 0,1,\ldots).$$

Note that for $N = 1$, $B_1 = B$, where B was defined in relation (5.5). It is obvious that $\|B_N\|_{\ell_p \to \ell_p} \geq \|B\|_{\ell_p \to \ell_p}$, but we will prove that the matrix B_N is a bounded operator on ℓ_p.

Theorem 5.7.14. *The Hilbert matrix admits a factorization of the form $H = B_N C_N$, where B_N is a bounded operator on ℓ_p and*

$$\frac{\pi}{p^*}\csc(\pi/p) \le \|B_N\|_{\ell_p \to \ell_p} \le \frac{N\pi}{p^*}\csc(\pi/p).$$

In particular, for $N = 1$, $H = BC$ and $\|B\|_{\ell_p \to \ell_p} = \frac{\pi}{p^*}\csc(\pi/p)$.

Proof. Consider that

$$[B_N C_N]_{i,k} = \sum_{j=k}^{\infty} \frac{j+N}{(i+j+1)(i+j+2)}\frac{1}{j+N} = \frac{1}{i+k+1} = [H]_{i,k},$$

which proves the factorization $H = B_N C_N$. Let us define the diagonal matrix D_N by

$$[D_N]_{j,k} = \begin{cases} \frac{j+N}{j+1} & j = k = 0, 1, \ldots \\ 0 & otherwise, \end{cases}$$

who has the following matrix representation:

$$D_N = \begin{pmatrix} N & 0 & 0 & \cdots \\ 0 & \frac{N+1}{2} & 0 & \cdots \\ 0 & 0 & \frac{N+2}{3} & \cdots \\ \vdots & \vdots & \vdots & \ddots \end{pmatrix}.$$

Note that for $N = 1$, $D_1 = I$, where I is the identity matrix. Since D_N is diagonal, hence $\|D_N\|_{\ell_p \to \ell_p} = \sup_j |[D_N]_{j,j}| = N$. On the other hand, it is easy to see $B_N = B D_N$, where the matrix B was defined in relation (5.5). Thus we have

$$\|B\|_{\ell_p \to \ell_p} \le \|B_N\|_{\ell_p \to \ell_p} \le \|B\|_{\ell_p \to \ell_p}\|D_N\|_{\ell_p \to \ell_p} = N\|B\|_{\ell_p \to \ell_p}.$$

Now, for $N = 1$, $B_1 = B$ and $C_1 = C$ which result $H = BC$ and $\|B_1\|_{\ell_p \to \ell_p} = \|B\|_{\ell_p \to \ell_p} = \frac{\pi}{p^*}\csc(\pi/p)$. Therefore, the proof is complete. $\square$

As an immediate result of the above theorem we generalize Hilbert's inequality versus Hardy's as follows:

Corollary 5.7.15. *Let $p > 1$. For every $x \in \ell_p$ we have*

$$\|Hx\|_{\ell_p} \le \frac{N\pi}{p^*}\csc(\pi/p)\|C_N x\|_{\ell_p}.$$

In particular, for $N = 1$, inequality (5.6) occurs.

Proof. Since $H = B_N C_N$, hence

$$\|Hx\|_{\ell_p} = \|B_N C_N x\|_{\ell_p} \le \frac{\pi N}{p^*}\csc(\pi/p)\|C_N x\|_{\ell_p}.$$

Now, for $N = 1$, $B_1 = B$ and $C_1 = C$ which results in the claimed inequality. $\square$

We define the non-negative symmetric matrix U_N by

$$[U_N]_{j,k} = \frac{2(j+N)(k+N)}{(j+k+1)(j+k+2)(j+k+3)} \qquad (j,k=0,1,\ldots).$$

In special case $N=1$, $U_1 = U$, where U has the following entries

$$[U]_{j,k} = \frac{2(j+1)(k+1)}{(j+k+1)(j+k+2)(j+k+3)} \qquad (j,k=0,1,\ldots). \qquad (5.7)$$

Theorem 5.7.16. *The Hilbert matrix has a factorization of the form* $H = C_N^t U_N C_N$, *where* U_N *is a bounded operator on* ℓ_p *with*

$$\frac{\pi}{pp^*} \csc(\pi/p) \leq \|U_N\|_{\ell_p \to \ell_p} \leq \frac{\pi}{pp^*} \csc(\pi/p) + (N-1) + \frac{(N-1)^2}{2}.$$

In particular for $N=1$, $U_1 = U$, $\|U\|_{\ell_p \to \ell_p} = \frac{\pi}{pp^*} \csc(\pi/p)$, *and* $H = C^t U C$.

Proof. To prove the factorization, we first prove that $U_N C_N = B_N^t$, where B_N^t is the transpose of matrix B_N. We have

$$\begin{aligned}(U_N C_N)_{j,k} &= \sum_{i=k}^{\infty} \frac{2(j+N)(i+N)}{(j+i+1)(j+i+2)(j+i+3)} \frac{1}{i+N} \\ &= (j+N) \sum_{i=k}^{\infty} \left\{ \frac{1}{(j+i+1)(j+i+2)} - \frac{1}{(j+i+2)(j+i+3)} \right\} \\ &= \frac{j+N}{(j+k+1)(j+k+2)} = [B_N^t]_{j,k}.\end{aligned}$$

Now,

$$C_N^t U_N C_N = C_N^t B_N^t = (B_N C_N)^t = H^t = H.$$

It is obvious that $U_N = U + 2(N-1)P + 2(N-1)^2 Q$, where U is defined as in relation (5.7) and P and Q are as follows

$$[P]_{j,k} = \frac{1}{(j+k+1)(j+k+3)} \qquad (j,k=0,1,\ldots),$$

and

$$[Q]_{j,k} = \frac{1}{(j+k+1)(j+k+2)(j+k+3)} \qquad (j,k=0,1,\ldots).$$

By a simple calculation

$$p_k = \sum_{j=0}^{\infty} p_{j,k} = \frac{1}{2(k+1)},$$

where p_k is the kth column sum of P. Since $\frac{1}{2} = p_0 > p_1 > \cdots$ and P is symmetric, hence R and K are both $\frac{1}{2}$ in Schur's theorem which results in $\|P\|_{\ell_p \to \ell_p} \leq \frac{1}{2}$. Similar reasoning for the matrix Q shows that $\|Q\|_{\ell_p \to \ell_p} \leq \frac{1}{4}$. For computing the norm of U we introduce a family of matrices, $U(w)$, $0 < w \leq 1$, given by

$$u(w)_{j,k} = 2\binom{j+k}{k} w^{j+1}(1-w)^{k+1}.$$

By applying Lemma 17.9 we have

$$\sum_{k=0}^{\infty} u(w)_{j,k} = 2w^{j+1}(1-w) \sum_{k=0}^{\infty} \binom{j+1+k-1}{k}(1-w)^k = 2(1-w),$$

and

$$\sum_{j=0}^{\infty} u(w)_{j,k} = 2w(1-w)^{k+1} \sum_{j=0}^{\infty} \binom{k+1+j-1}{j} w^j = 2w.$$

Hence, the row sums and the column sums are $2(1-w)$ and $2w$, respectively. As the result of Schur's theorem we have

$$\|U(w)\|_{\ell_p \to \ell_p} \leq 2w^{1/p}(1-w)^{1/p^*}.$$

On the other hand, it is not difficult to show

$$\int_0^1 u(w)_{j,k} dw = u_{j,k}.$$

Hence,

$$\begin{aligned}
\|U\|_{\ell_p \to \ell_p} &= \left\| \int_0^1 U(w) dw \right\| \leq \int_0^1 \|U(w)\| dw \\
&\leq \int_0^1 2w^{1/p}(1-w)^{1/p^*} dw = 2\beta(1+1/p, 1+1/p^*) \\
&= \frac{\pi}{pp^*} \csc(\pi/p).
\end{aligned}$$

But since $UC = B^t$ we deduce $\|U\|_{\ell_p \to \ell_p} \geq \frac{\|B^t\|_{\ell_p \to \ell_p}}{\|C\|_{\ell_p \to \ell_p}} = \frac{\pi}{pp^*} \csc(\pi/p)$, which proves $\|U\|_{\ell_p \to \ell_p} = \frac{\pi}{pp^*} \csc(\pi/p)$. Thus

$$\begin{aligned}
\|U_N\|_{\ell_p \to \ell_p} &\leq \|U\|_{\ell_p \to \ell_p} + 2(N-1)\|P\|_{\ell_p \to \ell_p} + 2(N-1)^2 \|Q\|_{\ell_p \to \ell_p} \\
&\leq \frac{\pi}{p^*} \csc(\pi/p) + (N-1) + \frac{(N-1)^2}{2}.
\end{aligned}$$

Also since $[U]_{j,k} \leq [U_N]_{j,k}$, hence $\|U\|_{\ell_p \to \ell_p} \leq \|U_N\|_{\ell_p \to \ell_p}$.
Now for $N = 1$, Hilbert matrix has a factorization of the form $C^t U C$, where C^t is the Copson matrix and U, defined by relation (5.7), has the norm $\|U\|_{\ell_p \to \ell_p} = \frac{\pi}{pp^*} \csc(\pi/p)$. Hence we have the desired result. $\square$

Remark 5.7.17. *One can find more factorizations for the infinite Hilbert operator in [16, 17].*

5.8 Factorization of the Cesàro operator

In this section, we introduce some factorization for the Cesàro matrix based on Cesàro and Gamma matrices which results in some interesting inequalities and inclusions.

Theorem 5.8.1. *Let $n \geq m \geq 1$, and let C_n and G_m be respectively the Cesàro and Gamma matrices of order n and m. Then the following assertions hold.*

(a) $C_n = G_1 G_2 \ldots G_n$.

(b) $C_n = R_{n,m} G_m$, *where*

$$R_{n,m} = \prod_{i=1, i \neq m}^{n} G_i.$$

Moreover, $R_{n,m}$ is a bounded operator on ℓ_p with the norm

$$\|R_{n,m}\|_{\ell_p \to \ell_p} = \frac{(1 - 1/mp)\Gamma(n+1)\Gamma(1/p^*)}{\Gamma(n + 1/p^*)}. \tag{5.8}$$

In particular, $C_n = U_n C$, where $\|U_n\|_{\ell_p \to \ell_p} = \frac{\Gamma(n+1)\Gamma(1+1/p^*)}{\Gamma(n+1/p^*)}$.

(c) $C_n = S_{n,m} C_m$, *where*

$$S_{n,m} = \prod_{i=m+1}^{n} G_i. \tag{5.9}$$

Moreover, $S_{n,m}$ is a bounded operator on ℓ_p with the norm

$$\|S_{n,m}\|_{\ell_p \to \ell_p} = \frac{\Gamma(n+1)\Gamma(m+1/p^*)}{\Gamma(m+1)\Gamma(n+1/p^*)}.$$

In particular, $C_n = U_n C$, where $\|U_n\|_{\ell_p \to \ell_p} = \frac{\Gamma(n+1)\Gamma(1+1/p^*)}{\Gamma(n+1/p^*)}$.

Proof. (a) This is a direct consequence of the identity $C_n = C_{n-1} G_n$ and induction. (b) According to part (a)

$$C_n = G_1 \ldots G_{m-1} G_{m+1} \cdots G_n \times G_m = R_{n,m} G_m.$$

Now, according the norm separating property of Hausdorff matrices (5.3.1), we have

$$\begin{aligned}
\|R_{n,m}\|_{\ell_p \to \ell_p} &= \|G_1\|_{\ell_p \to \ell_p} \cdots \|G_{m-1}\|_{\ell_p \to \ell_p} \|G_{m+1}\|_{\ell_p \to \ell_p} \cdots \|G_n\|_{\ell_p \to \ell_p} \\
&= \frac{\prod_{i=1}^n \|G_i\|_{\ell_p \to \ell_p}}{\|G_m\|_{\ell_p \to \ell_p}} \\
&= \frac{\|C_n\|_{\ell_p \to \ell_p}}{\|G_m\|_{\ell_p \to \ell_p}} \\
&= \frac{(1 - 1/mp)\Gamma(n+1)\Gamma(1/p^*)}{\Gamma(n+1/p^*)}.
\end{aligned}$$

In special case $m = 1$, by letting $R_{n,1} = U_n$, we have $G_1 = C$, where C is the well-known Cesàro matrix and $\|U_n\|_{\ell_p \to \ell_p} = \frac{\Gamma(n+1)\Gamma(1+1/p^*)}{\Gamma(n+1/p^*)}$.

(c) According to part (b), we have $C_n = C_m G_{m+1} \ldots G_n$. Hence, by the definition $S_{n,m} = G_{m+1} \ldots G_n$, we have the claimed factorization. For computing the norm of $S_{n,m}$, note thhat

$$\|S_{n,m}\|_{\ell_p \to \ell_p} = \frac{\|C_n\|_{\ell_p \to \ell_p}}{\|C_m\|_{\ell_p \to \ell_p}} = \frac{\Gamma(n+1)\Gamma(m+1/p^*)}{\Gamma(m+1)\Gamma(n+1/p^*)}.$$

For the case $m = 1$, by letting $S_{n,1} = U_n$, we have the desired result. □

Bennett found the corresponding measure $d\mu(\theta)$ for the matrix $S_{n,m} = G_{m+1} \cdots G_n$ which we obtained through the Lemma 5.8.1. He showed

$$d\mu(\theta) = \frac{\Gamma(n+1)}{\Gamma(m+1)\Gamma(n-m)} \theta^m (1-\theta)^{n-m-1} d\theta. \tag{5.10}$$

The matrix $S_{n,m}$ has the entries

$$[S_{n,m}]_{j,k} = \begin{cases} \dfrac{\binom{m+k}{k}\binom{n-m+j-k-1}{j-k}}{\binom{n+j}{j}} & , \; j \geq k \geq 0, \\ 0 & , \; \text{otherwise.} \end{cases}$$

See [2, 3].

We can rewrite the factorization discussed in Theorem 5.8.1 as

$$\begin{aligned}
\|C_n x\|_{\ell_p} &\leq \|C_{n-1}\|_{\ell_p \to \ell_p} \|G_n x\|_{\ell_p}, \\
\|C_n x\|_{\ell_p} &\leq \|G_n\|_{\ell_p \to \ell_p} \|C_{n-1} x\|_{\ell_p}, \\
\|C_n x\|_{\ell_p} &\leq \|R_{n,m}\|_{\ell_p \to \ell_p} \|G_m x\|_{\ell_p}, \\
\|C_n x\|_{\ell_p} &\leq \|S_{n,m}\|_{\ell_p \to \ell_p} \|C_m x\|_{\ell_p}.
\end{aligned}$$

We express a more explicit account of the above inequalities.

Corollary 5.8.2. *Let (x_n) be a sequence of real numbers. Then the following statements hold.*

- For $n \geq 1$

$$\sum_{j=0}^{\infty} \left| \sum_{k=0}^{j} \frac{\binom{n+j-k-1}{j-k}}{\binom{n+j}{j}} x_k \right|^p \leq \left(\frac{np}{np-1} \right)^p \sum_{j=0}^{\infty} \left| \sum_{k=0}^{j} \frac{\binom{n+j-k-2}{j-k}}{\binom{n+j-1}{j}} x_k \right|^p.$$

In particular, for $n = 1$, we have Hardy's inequality.

$$\sum_{j=0}^{\infty} \left| \sum_{k=0}^{j} \frac{x_k}{1+j} \right|^p \leq \left(\frac{p}{p-1} \right)^p \sum_{j=0}^{\infty} |x_k|^p.$$

- For $n \geq m \geq 1$

$$\sum_{j=0}^{\infty} \left| \sum_{k=0}^{j} \frac{\binom{n+j-k-1}{j-k}}{\binom{n+j}{j}} x_k \right|^p$$

$$\leq \left[\frac{(1-1/mp)\Gamma(n+1)\Gamma(1/p^*)}{\Gamma(n+1/p^*)} \right]^p \sum_{j=0}^{\infty} \left| \sum_{k=0}^{j} \frac{\binom{m+k-1}{k}}{\binom{m+j}{j}} x_k \right|^p.$$

In particular, for $m = n$,

$$\sum_{j=0}^{\infty} \left| \sum_{k=0}^{j} \frac{\binom{n+j-k-1}{j-k}}{\binom{n+j}{j}} x_k \right|^p \leq \left[\frac{\Gamma(n)\Gamma(1/p^*)}{\Gamma(n-1/p)} \right]^p \sum_{j=0}^{\infty} \left| \sum_{k=0}^{j} \frac{\binom{n+k-1}{k}}{\binom{n+j}{j}} x_k \right|^p.$$

- For $n \geq m \geq 0$

$$\sum_{j=0}^{\infty} \left| \sum_{k=0}^{j} \frac{\binom{n+j-k-1}{j-k}}{\binom{n+j}{j}} x_k \right|^p$$

$$\leq \left[\frac{\Gamma(n+1)\Gamma(m+1/p^*)}{\Gamma(m+1)\Gamma(n+1/p^*)} \right]^p \sum_{j=0}^{\infty} \left| \sum_{k=0}^{j} \frac{\binom{m+j-k-1}{j-k}}{\binom{m+j}{j}} x_k \right|^p.$$

In particular, for $n = 1, m = 0$, we have Hardy's inequality.

See [4, 6] for similar inequalities of the above type.

In the following we intend to introduce a factorization for the Cesàro matrix of order n based on the generalized Cesàro matrix.

Theorem 5.8.3. *For $n \geq 1$, Cesàro matrix of order n, C_n, has a factorization of the form $C_n = R_{n,N} C_N$, where C_N is the generalized Cesàro matrix of order N and $R^{n,N}$ is a bounded operator on ℓ_p with*

$$\|R_{n,N}\|_{\ell_p \to \ell_p} \leq \frac{N\Gamma(n+1)\Gamma(1+1/p^*)}{\Gamma(n+1/p^*)}.$$

In particular, $C_n = U_n C$, where $\|U_n\|_{\ell_p \to \ell_p} = \frac{\Gamma(n+1)\Gamma(1+1/p^*)}{\Gamma(n+1/p^*)}$.

Proof. According to Theorem 5.8.1, $C_n = U_n C$. Now, by the identity $C = D_N C_N$-the diagonal matrix D_N was defined in relation (5.10), we have

$$C_n = U_n C = U_n D_N C_N = R_{n,N} C_N,$$

where $R_{n,N} = U_n D_N$.

Since $\|D_N\|_{\ell_p \to \ell_p} = N$, as we stated in Theorem 5.7.16, hence

$$\|R_{n,N}\|_{\ell_p \to \ell_p} \le \|U_n\|_{\ell_p \to \ell_p} \|D_N\|_{\ell_p \to \ell_p} = \frac{N\Gamma(n+1)\Gamma(1+1/p^*)}{\Gamma(n+1/p^*)}.$$

In special case $N = 1$, by letting $R_{n,1} = U_n$, we have $C_n = C$, where C is the well-known Cesàro matrix and $\|U_n\|_{\ell_p \to \ell_p} = \frac{\Gamma(n+1)\Gamma(1+1/p^*)}{\Gamma(n+1/p^*)}$. □

Bibliography

[1] Bennett, G. 1996. *Factorizing the classical inequalities*, Mem. Amer. Math. Soc. 576: 130 pp.

[2] Bennett, G. 1986. Lower bounds for matrices, *Linear Algebra Appl.* 82: 81–98.

[3] Bennett, G. 1992. Lower bounds for matrices II, *Canad. J. Math.* 44: 54–74.

[4] Bennett, G. 1999. An inequality for Hausdorff means, *Houston J. Math.* 25:4: 709–744.

[5] Chen, C.P., Luor, D.C. and Ou, Z.Y. 2022. Extensions of Hardy inequality, *J. Math. Anal. Appl.* 273: 160–171.

[6] Hardy, G.H. 1943. An inequality for Hausdorff means, *J. London Math. Soc.* 18: 46–50.

[7] Hardy, G.H. 1973. *Divergent series*, Oxford University Press.

[8] Hardy, G.H., Littlewood, J.E. and Polya, G. 2001. *Inequalities*, 2nd edition, Cambridge University Press, Cambridge.

[9] Roopaei, H. 2020. Norm of Hilbert operator on sequence spaces, *J. Inequal. Appl.* 117:2020.

[10] Roopaei, H., Foroutannia, D., İlkhan, Kara, M. and E.E. 2020. Cesàro spaces and norm of operators on these matrix domains, *Mediterr. J. Math.* 17:121.

[11] Roopaei, H. and Başar, F. 2020. On the spaces of Cesàro absolutely p-summable, null and convergent sequences, *Math. Methods. Appl. Sci.* 44:5: 3670–3685.

[12] Roopaei, H. 2020. A study on Copson operator and its associated matrix domains, *J. Inequal. Appl.* 120:2020.

[13] Roopaei, H. 2020. A study on Copson operator and its associated matrix domains II, *J. Inequal. Appl.* 239:2020.

[14] Roopaei, H. and Başar, F. 2022. On the gamma spaces including the spaces of absolutely p-summable, null, convergent and bounded sequences, *Numer. Funct. Anal. Optim.* 43:6: 723–754.

[15] Kara, M.I. and Roopaei, H. 2022. A weighted mean Hausdorff type operator and its summability matrix domain, *J. Inequal. Appl.* 27, https://doi.org/10.1186/s13660-022-02760-w

[16] Roopaei, H. 2020. Factorization of the Hilbert matrix based on Cesàro and Gamma matrices, *Results Math.* 2020; 75:3, https://doi.org/10.1007/s00025-019-1129-1.

[17] Roopaei, H. 2020. Factorization of Cesàro and Hilbert matrices based on generalized Cesàro matrix, *Linear Multilinear Algebra*, 68:1: 193–204.

Chapter 6

On theorems of Galambos-Bojanić-Seneta type

Dragan Djurčić

Ljubiša D.R. Kočinac

6.1	Introduction	95
6.2	Known results	97
	6.2.1 Classes ORV$_s$ and ORV$_f$ and their subclasses	97
	6.2.2 Rapid and related variations	100
6.3	New result	106
	Bibliography	109

6.1 Introduction

Working on Tauberian theory in the 1930's, J. Karamata initiated investigation in asymptotic analysis of divergent processes, nowadays known as *Karamata's theory of regular variation* (see [26–30], and also [4, 34]).

In 1970, de Haan [25] defined and investigated rapid variation and so initiated further development in asymptotic analysis.

The theory of regular and rapid variability has many applications in different mathematical disciplines: differential and difference equations, in particular in description of asymptotic properties of solutions of these equations, dynamic equations, number theory, probability theory, time scales theory, selection principles theory, game theory and so on (see, for example, [31] and references therein and also [19, 21, 24, 35]).

In this chapter we consider only measurable functions $\varphi : [a, \infty) \to (0, \infty)$, $a > 0$ (the set of such functions we denote by $\mathbb{F}$), and sequences of positive real numbers (the set of such sequences is denoted by $\mathbb{S}$). We use the notation $\mathbf{x} = (x_n)_{n \in \mathbb{N}}$, $\mathbf{y} = (y_n)_{n \in \mathbb{N}}$, and so on, for sequences from $\mathbb{S}$. We will also need the following class $\mathbb{A}$ of functions:

$$\mathbb{A} = \{\varphi \in \mathbb{F} : \varphi \text{ is nondecreasing and unbounded}\}.$$

If $\varphi \in \mathbb{A}$, then the function $\varphi^{\leftarrow}$ defined by

$$\varphi^{\leftarrow}(t) := \inf\{u \geq a : \varphi(u) > t\}, \ t \geq \varphi(a)$$

is the *generalized inverse of φ* [4].

Notice the following two facts:

(1) If $\varphi \in \mathbb{A}$ is continuous and increasing, then $\varphi^{\leftarrow}$ is the inverse function φ^{-1} of φ, i.e., $\varphi(t) = \varphi^{-1}(t), \ t \geq \varphi(a)$;

(2) If $\varphi \in \mathbb{A}$, then $\varphi^{\leftarrow} \in \mathbb{A}$.

To each function $\varphi \in \mathbb{F}$, we assign the following three functions depending on $\lambda > 0$:

$$k_\varphi(\lambda) := \lim_{t \to \infty} \frac{\varphi(\lambda t)}{\varphi(t)|}, \ \lambda > 0;$$

$$\overline{k}_\varphi(\lambda) := \limsup_{t \to \infty} \frac{\varphi(\lambda t)}{\varphi(t)}, \ \lambda > 0;$$

$$\underline{k}_\varphi(\lambda) := \liminf_{t \to \infty} \frac{\varphi(\lambda t)}{\varphi(t)}, \ \lambda > 0.$$

Similarly, to each sequence $\mathbf{x} = (x_n)_{n \in \mathbb{N}} \in \mathbb{S}$ and each $\lambda > 0$ one assigns the following three functions:

$$k_\mathbf{x}(\lambda) := \lim_{n \to \infty} \frac{x_{[\lambda n]}}{x_n}; \ \overline{k}_\mathbf{x}(\lambda) := \limsup_{n \to \infty} \frac{x_{[\lambda n]}}{x_n}; \ \underline{k}_\mathbf{x}(\lambda) := \liminf_{n \to \infty} \frac{x_{[\lambda n]}}{x_n}.$$

The function $\overline{k}_\varphi(\lambda), \ \lambda > 0$, is called the *index function of φ*, and the function $\underline{k}_\varphi(\lambda)$ is called the *auxiliary index function of φ*. Karamata theory of regular variation has two basic lines of investigation: functional and sequential.

Definition 6.1.1. (1) A function $\varphi \in \mathbb{F}$, is said to be *regularly varying* if for each $\lambda > 0$ it satisfies the condition

$$k_\varphi(\lambda) < \infty.$$

The class of these functions is denoted by RV_f. When $k_\varphi(\lambda) = 1$, one obtains the class SV_f of *slowly varying functions* .

(2) A sequence $\mathbf{x} = (x_n)_{n \in \mathbb{N}} \in \mathbb{S}$ is said to be *regularly varying* if for each $\lambda > 0$ it satisfies the condition

$$k_\mathbf{x}(\lambda) < \infty.$$

The class of these sequences is denoted by RV_s.

In the case $k_\mathbf{x}(\lambda) = 1$ we have the class $\mathsf{SV_s}$ of *slowly varying sequences*. These two types of research developed independently of each other until the results by Galambos-Seneta [23] and Bojanić-Seneta [5].

In these two papers the authors proved the following result giving a natural connection between the two theories and nowadays these results are called Galambos-Bojanić-Seneta type theorems.

Theorem GBS. For a sequence $\mathbf{x} = (x_n)_{n \in \mathbb{N}} \in \mathbb{S}$ the following are equivalent:

(a) $\mathbf{x}$ is slowly varying (respectively, regularly varying);

(b) The function $\varphi_\mathbf{x}$, $\varphi_\mathbf{x}(t) = x_{[t]}$, $t \geq 1$, is slowly varying (respectively, regularly varying).

During the years Karamata's theory of regular variation was extended and modified in different directions (see [4, 31]) and a natural question arose: what about Galambos-Bojanić-Seneta result in the context of those modifications? Several such results, showing that similar assertions are true for the modifications of regular variation, have been proved. We are going to review these known results obtained by several authors and to prove a new result of this type for one of modifications of regular variation. We also give (without proof) a new result for a subclass of the class of rapidly varying sequences.

Some known theorems will be proven in order to demonstrate the methods of proving results of Galambos-Bojanić-Seneta type. Let us mention that in all these proofs one uses the classical (topological) Baire category theorem [22].

6.2 Known results

6.2.1 Classes ORV$_s$ and ORV$_f$ and their subclasses

In this subsection we present results on the classes of $\mathcal{O}$-regularly varying sequences and functions and their important subclasses.

Definition 6.2.1. ([3]) A function $\varphi \in \mathbb{F}$ is said to be $\mathcal{O}$-*regularly varying* if for each $\lambda > 0$ satisfies the condition

$$\overline{k}_\varphi(\lambda) < \infty.$$

or, equivalently,

$$\underline{k}_\varphi(\lambda) > 0.$$

The class of all these functions is denoted by ORV$_f$.

Definition 6.2.2. A sequence $\mathbf{x} = (x_n)_{n \in \mathbb{N}} \in \mathbb{S}$ is said to be $\mathcal{O}$-*regularly varying* for each $\lambda > 0$ satisfies the condition

$$\overline{k}_\mathbf{x}(\lambda) < \infty.$$

The class of these sequences is denoted by ORV$_s$.

The classes ORV$_f$ and ORV$_s$ have been studied in a number of papers (see, for instance, [1, 2, 8, 9]).

Observe that when $\overline{k}_\varphi(\lambda) = \underline{k}_\varphi(\lambda)$ (respectively, $\overline{k}_\mathbf{x}(\lambda) = \underline{k}_\mathbf{x}(\lambda)$), we have the class RV$_f$ of regularly varying functions (respectively, the class of regularly varying sequences) in the sense of Karamata.

When the index function $\overline{k}_\varphi(\lambda)$, $\lambda > 0$, of a function $\varphi \in$ ORV$_f$ is continuous, we say that φ belongs to the class CRV$_f$. Similarly, one defines the class CRV$_s$ of sequences. Notice that a function φ belongs to the class CRV$_f$ if and only if $\lim_{\lambda \to 1} \overline{k}_\varphi(\lambda) = 1$.

We begin with the Galambos-Bojanić-Seneta type theorem for the classes ORV$_s$ and ORV$_f$.

Theorem 6.2.3. ([9]) *Let* $\mathbf{x} = (x_n)_{n \in \mathbb{N}}$ *be a sequence in* $\mathbb{S}$. *Then the following assertions are equivalent:*

(a) $\mathbf{x} \in$ ORV$_s$;

(b) *The function* $\varphi_\mathbf{x}(t) = x_{[t]}$, $t \geq 1$, *belongs to* ORV$_f$.

We also have a similar result for the classes CRV$_s$ and CRV$_f$.

Theorem 6.2.4. ([17]) *For a sequence* $\mathbf{x} = (x_n)_{n \in \mathbb{N}} \in \mathbb{S}$ *the following assertions are equivalent:*

(a) $\mathbf{x} \in$ CRV$_s$;

(b) $\varphi_\mathbf{x}(t) = x_{[t]}$, $t \geq 1$, *belongs to* CRV$_f$.

An important subclass of of the class ORV$_f$ of $\mathcal{O}$-regularly varying functions is the class of Seneta functions defined as follows.

For a given $\beta \geq 1$, denote by SO$_f^\beta$ the set of all functions $\varphi \in$ ORV$_f$ such that $\overline{k}_\varphi(\lambda) \leq \beta$ for all $\lambda > 0$. Put SO$_f = \bigcup_{\beta \geq 1}$ SO$_f^\beta$. Functions in the class SO$_f$ are called the *Seneta functions*.

Similarly, a sequence $\mathbf{x} = (x_n)_{n \in \mathbb{N}}$ belongs to the class SO$_s^\beta$ for a given $\beta \geq 1$ if $\overline{k}_\mathbf{x}(\lambda) \leq \beta$ for each $\lambda > 0$. The set SO$_s = \bigcup_{\beta \geq 1}$ SO$_s^\beta$ is called the class of *Seneta sequences*.

Theorem 6.2.5. ([18]) *Let* $\mathbf{x} = (x_n)_{n \in \mathbb{N}}$ *be a sequence in* $\mathbb{S}$. *Then the following are equivalent:*

(a) $\mathbf{x} \in$ SO$_s$;

(b) $\varphi_\mathbf{x}(t) = x_{[t]} \in$ SO$_f$ *on the interval* $[1, \infty)$.

In fact, we have the following more precise result whose consequence is the above theorem.

Theorem 6.2.6. ([18, Peroposition 1]) *The following hold:*

(a) *If a sequence* $\mathbf{x} = (x_n)_{n \in \mathbb{N}}$ *belongs to the class* SO$_s^\beta$, *then the function* $\varphi_\mathbf{x}(t) = x_{[t]}$, $t \geq 1$, *belongs to the class* SO$_f^{\beta^2}$;

(b) *If* $\varphi_\mathbf{x}(t) = x_{[t]}$, $t \geq 1$, *belongs to the class* SO$_f^\beta$, *then* $\mathbf{x} \in$ SO$_s^\beta$.

Remark 6.2.7. (1) From the above theorem one concludes that a sequence **x** belongs to SO_s^1 if and only if $\varphi_{\mathbf{x}} \in SO_f^1$. But, it is not so for $\beta > 1$ (see [18, Proposition 2]). For each $\beta > 1$ consider the function

$$g(t) = \exp\{2\ln(\beta)\sqrt{|\sin(\ln t)|}\}, \, t > 0$$

which is O-regularly varying, hence, by Theorem 6.2.3, the sequence $\mathbf{x} = (x_n)_{n \in \mathbb{N}}$, $x_n = g(n)$, is in the class ORV_s. Moreover, $\mathbf{x} \in SO_s^{\beta^2} \setminus SO_s^{\beta}$. On the other hand, the function $\varphi_{\mathbf{x}}(t) = x_{[t]}$, $t \geq 1$, belongs to $SO_f^{\beta^2}$.

(2) Following [5], call a sequence **x** *embedable* in the function $g(t)$, $t \geq a, a > 0$, if $g(n) = x_n$ for each $n \geq [a] + 1$. A sequence **x** is embedable in a Seneta function $g(t)$, $t \geq a, a > 0$, if and only if it is a Seneta sequence.

We present now results on another subclasses of ORV_f and ORV_s introduced by Matuszewska.

Definition 6.2.8. ([32,33]) *A function $\varphi \in \mathbb{F}$ is said to be in the class* ERV_f *of Matuszewska if for each $\lambda \geq 1$ there are $c, d \in \mathbb{R}$, $c \leq d$, such that*

$$\lambda^c \leq \underline{k}_\varphi(\lambda) \leq \overline{k}_\varphi(\lambda) \leq \lambda^d.$$

Similarly one defines the class ERV_s of sequences.

Definition 6.2.9. *A sequence* $\mathbf{x} = (x_n)_{n \in \mathbb{N}} \in \mathbb{S}$ *belongs to the class* ERV_s *if*

$$\lambda^c \leq \underline{k}_{\mathbf{x}}(\lambda) \leq \overline{k}_{\mathbf{x}}(\lambda) \leq \lambda^d$$

for each $\lambda \geq 1$ and some $c, d \in \mathbb{R}$ with $c \leq d$.

Notice that the following inclusions hold:

$$RV_f \subsetneq ERV_f \subsetneq CRV_f \subsetneq ORV_f \supsetneq SO_f$$

and

$$RV_s \subsetneq ERV_s \subsetneq CRV_s \subsetneq ORV_s \supsetneq SO_s.$$

The following result holds for the ERV classes of sequences and functions.

Theorem 6.2.10. ([17]) *For a sequence* $\mathbf{x} = (x_n)_{n \in \mathbb{N}} \in \mathbb{S}$ *the following assertions are equivalent:*

(a) $\mathbf{x} \in ERV_s$;

(b) *The function* $\varphi_{\mathbf{x}}(t) = x_{[t]}$, $t \geq 1$, *belongs* ERV_f.

6.2.2 Rapid and related variations

In this section we quote Galambos-Bojanić-Seneta type results for rapidly varying sequences and functions and their variants.

Definition 6.2.11. ([25]) A function $\varphi \in \mathbb{F}$ is said to be *rapidly varying of index of variability* ∞ if it satisfies the asymptotic condition

$$k_\varphi(\lambda) = \infty, \ \lambda > 1.$$

The class of rapidly varying functions of index of variability ∞ we denote by $\mathsf{R}_{f,\infty}$.

Definition 6.2.12. A sequence $\mathbf{x} = (x_n)_{n\in\mathbb{N}} \in \mathbb{S}$ is *rapidly varying* (of index of variability ∞) if the following asymptotic condition is satisfied:

$$k_\mathbf{x}(\lambda) = \infty, \ \lambda > 1.$$

or, equivalently.

$$\lim_{n\to\infty} \frac{x_{[\lambda n]}}{x_n} = 0, \ 0 < \lambda < 1.$$

$\mathsf{R}_{s,\infty}$ denotes the class of rapidly varying sequences of index of variability ∞.

Properties of the important class of rapidly varying sequences have been studied in [11] where a result of Galambos-Bojanić-Seneta type theorem was proved.

Theorem 6.2.13. *For a sequence* $\mathbf{x} = (x_n)_{n\in\mathbb{N}} \in \mathbb{S}$ *the following are equivalent:*

(a) $\mathbf{x}$ *belongs to the class* $\mathsf{R}_{s,\infty}$;

(b) *The function* φ *defined by* $\varphi(t) = x_{[t]}$, $t \geq 1$, *is in the class* $\mathsf{R}_{f,\infty}$.

Proof. (a) $\Rightarrow$ (b): Let $\lambda \in (0,1)$. Then for every $\alpha \in (\lambda, 1)$ we have $\lim_{n\to\infty} \frac{x_{[\alpha n]}}{x_n} = 1$. We prove that for a given $\epsilon > 0$ there exist an interval $[A, B]$ which is a proper subset of $(\lambda, 1)$ and $n_0 \in \mathbb{N}$ such that $\frac{x_{[\alpha n]}}{x_n} < \epsilon$ for each $n \geq n_0$ and each $\alpha \in [A, B]$. For an arbitrary and fixed $\alpha \in (\lambda, 1)$ define $n_\alpha \in \mathbb{N}$ in the following way:

$$n_\alpha = \begin{cases} 1, & \text{if } \frac{x_{[\alpha n]}}{x_n} < \epsilon \text{ for each } n \in \mathbb{N}; \\ 1 + \max\{n \in \mathbb{N} : \frac{x_{[\alpha n]}}{x_n} \geq \epsilon\}, & \text{otherwise.} \end{cases}$$

It is easy to see that $1 \leq n_\alpha < \infty$ for each considered α. For each $k \in \mathbb{N}$ define

$$A_k = \{\alpha \in (\lambda, 1) : n_\alpha > k\}.$$

Then $(A_k)_{k \in \mathbb{N}}$ is a non-increasing sequence of sets such that $\bigcap_{k \in \mathbb{N}} A_k = \emptyset$. We prove that not all sets A_k are dense in $(\lambda, 1)$. If $k \in \mathbb{N}$ is fixed and $\alpha \in A_k$, then
$$\frac{x_{[\alpha(n_\alpha - 1)]}}{x_{n_\alpha - 1}} \geq \epsilon$$
and there is some $\delta_\alpha > 0$ such that for each $t \in [\alpha, \alpha + \delta_\alpha) \subset (\lambda, 1)$ we have
$$\frac{x_{[t(n_\alpha - 1)]}}{x_{n_\alpha - 1}} = \frac{x_{[\alpha(n_\alpha - 1)]}}{x_{n_\alpha - 1}} \geq \epsilon$$
This means that each $t \in (\alpha, \alpha + \delta_\alpha)$ belongs to the set A_k, since $n_t \geq (n_\alpha - 1) + 1 > k$. It follows that if $\alpha \in A_k$, then $(\alpha, \alpha + \delta_\alpha) \subset A_k$. If we assume that some of the sets A_k is dense in $(\lambda, 1)$, then the set $\text{Int}(A_k)$ is also dense in $(\lambda, 1)$. If, on the other side, we suppose that all the sets A_k are dense in $(\lambda, 1)$, then $(\text{Int}(A_k))_{k \in \mathbb{N}}$ is a sequence of open, dense subsets of the set $(\lambda, 1)$ of the second category. It follows that the set $\bigcap_{k \in \mathbb{N}} A_k$ is dense in $(\lambda, 1)$ and thus nonempty, and we have a contradiction. Therefore, there is $n_0 \in \mathbb{N}$ such that the set A_{n_0} is not dense in $(\lambda, 1)$. Consequently, there is an interval $[A, B]$, a proper subset of $(\lambda, 1)$, such that
$$[A, B] \subset (\lambda, 1) \setminus A_{n_0} = \{\alpha \in (\lambda, 1) : n_\alpha \leq n_0\}.$$
From here it follows that $n_\alpha \leq n_0$ for each $\alpha \in [A, B]$, and thus for each $n \geq n_0 \geq n_\alpha$ and each $\alpha \in [A, B]$ it holds $\frac{x_{[\alpha n]}}{x_n} < \epsilon$.
We conclude that for $\lambda \in (0, 1)$ and each $t \in [1, \infty)$ large enough, we have
$$\frac{x_{[\lambda t]}}{x_{[t]}} = \frac{x_{[u[\eta[t]]]}}{x_{[\eta[t]]}} \cdot \frac{x_{[\eta[t]]}}{x_{[t]}},$$
where $u = u(x) \in [A, B]$ and $\eta = \frac{2\lambda}{A+B}$.
Since $\eta \in (0, 1)$ we have
$$\limsup_{t \to \infty} \frac{x_{[\lambda t]}}{x_{[t]}} \leq \epsilon \cdot \limsup_{t \to \infty} \frac{x_{[\eta[t]]}}{x_{[t]}} = 0.$$
This means that the function φ defined by $\varphi(t) = x_{[t]}$, $t \geq 1$, belongs to the class $\mathsf{R}_{f,\infty}$.

(b) $\Rightarrow$ (a): It is trivial, because for an arbitrary and fixed $\lambda \in (0, 1)$ we have
$$\lim_{n \to \infty} \frac{x_{[\lambda n]}}{x_n} = \lim_{x \to \infty} \frac{x_{[\lambda t]}}{c_{[t]}} = 0.$$
$\square$

There is another class of rapidly varying functions (see [25] and also [4]).

Definition 6.2.14. A function $\varphi \in \mathbb{F}$ is said to be *rapidly varying of index of variability* $-\infty$ if for each $\lambda > 1$ it satisfies
$$\lim_{t \to +\infty} \frac{\varphi(\lambda t)}{\varphi(t)} = 0.$$
$\mathsf{R}_{f,-\infty}$ denotes the class of rapidly varying functions of index $-\infty$.

Definition 6.2.15. A sequence $\mathbf{x} = (x_n)_{n\in\mathbb{N}} \in \mathbb{S}$ is said to belong to the class $\mathsf{R}_{s,-\infty}$ of *rapidly varying sequences of index of variability* $-\infty$ if for each $\lambda > 1$ the following condition is satisfied:

$$\lim_{n\to\infty} \frac{x_{[\lambda n]}}{x_n} = 0.$$

The class of rapidly varying sequences of index of variability $-\infty$ is denoted b $\mathsf{R}_{s,-\infty}$.

We have the following result which is parallel to Theorem 6.2.13.

Theorem 6.2.16. ([12]) *For a sequence* $\mathbf{x} = (x_n)_{n\in\mathbb{N}}$ *in* $\mathbb{S}$ *the following are equivalent:*

(a) $\mathbf{x}$ *belongs to the class* $\mathsf{R}_{s,-\infty}$;

(b) *The function* $\varphi_{\mathbf{x}}$ *defined by* $\varphi_{\mathbf{x}}(t) = x_{[t]}$, $t \geq 1$, *is in the class* $\mathsf{R}_{f,-\infty}$;

(c) $\lim_{n\to\infty} \frac{x_{[\lambda n]}}{x_n} = \infty$, $0 < \lambda < 1$.

The following is one more kind of rapid variation.

Definition 6.2.17. A function $\varphi \in \mathbb{F}$ belongs to the class $\mathsf{Tr}(\mathsf{R}_{f,\infty})$ of *translationally rapidly varying functions* if for each $\lambda \geq 1$, the following condition holds:

$$\lim_{n\to\infty} \frac{\varphi(t+\lambda)}{\varphi(t)} = \infty.$$

Definition 6.2.18. A sequence $\mathbf{x} = (x_n)_{n\in\mathbb{N}} \in \mathbb{S}$ is in the class $\mathsf{Tr}(\mathsf{R}_{s,\infty})$ of *translationally rapidly varying sequences* if for each $\lambda \geq 1$, the following condition holds:

$$\lim_{n\to\infty} \frac{x_{[n+\lambda]}}{x_n} = \infty.$$

Note that
$$\mathsf{Tr}(\mathsf{R}_{f,\infty}) \subsetneq \mathsf{R}_{f,\infty} \quad \text{and} \quad \mathsf{Tr}(\mathsf{R}_{s,\infty}) \subsetneq \mathsf{R}_{s,\infty}.$$

The class $\mathsf{Tr}(\mathsf{R}_{s,\infty})$ (and its subclasses) was studied in [13–15], in particular in connection with selection principles and game theory. The following new result of Galambos-Bojanić-Seneta type we give without proof because it is similar to proof of Theorem 6.2.22.

Theorem 6.2.19. *For a sequence* $\mathbf{x} = (x_n)_{n\in\mathbb{N}} \in \mathbb{S}$ *the following are equivalent:*

(1) $\mathbf{x} \in \mathsf{Tr}(\mathsf{R}_{s,\infty})$;

(2) *The function* $\varphi(t) = x_{[t]}$, $t \geq 1$, *belongs to the class* $\mathsf{Tr}(\mathsf{R}_{f,\infty})$.

We consider now an important subclass of the class $\mathsf{R}_{s,\infty}$, that we denote by $\mathsf{KR}_{s,\infty}$.

Definition 6.2.20. For a sequence $\mathbf{x} = (x_n)_{n \in \mathbb{N}} \in \mathbb{S}$ the *lower Matuszewska index* $d(\mathbf{x})$ is defined as the supremum of all $d \in \mathbb{R}$ such that for each $\Lambda > 1$

$$\frac{x_{[\lambda n]}}{x_n} \geq \lambda^d (1 + o(1)) \quad (n \to \infty)$$

holds uniformly (with respect to λ) on the segment $[1, \Lambda]$. The sequence $\mathbf{x}$ belongs to the *class* $\mathsf{KR}_{\mathsf{s},\infty}$ if $d(\mathbf{x}) = \infty$.

The definition of lower Matuszewska index for functions can be found in [4, p. 68]. By a result from [4] we have $\mathsf{KR}_{\mathsf{f},\infty} \subsetneq \mathsf{R}_{\mathsf{f},\infty}$.

Lemma 6.2.21. *For a sequence* $\mathbf{x} = (x_n)_{n \in \mathbb{N}} \in \mathbb{S}$ *the following are equivalent:*

(1) $\mathbf{x} \in \mathsf{KR}_{\mathsf{s},\infty}$;

(2) *For each* $d \in \mathbb{R}$ *it holds* $\liminf_{n \to \infty} \inf_{\lambda \geq 1} \frac{x_{[\lambda n]}}{\lambda^d x_n} \geq 1$.

Proof. (1) $\Rightarrow$ (2) From $d(\mathbf{x}) = \infty$, it follows that for every $d \in \mathbb{R}$, every $\Lambda > 1$, and sufficiently large n, we have $\frac{x_{[\lambda n]}}{x_n} \geq \lambda^d(1 + o(1))$, where $\lambda \in [1, \Lambda]$ is an arbitrary fixed element. For the same d, λ, Λ, for sufficiently large n we have $\inf_{\lambda \in [1, \Lambda]} \frac{x_{[\lambda n]}}{\lambda^d x_n} \geq 1 + o(1)$. In other words, for each $\varepsilon > 0$ there is $n_0 = n_0(\varepsilon) \in \mathbb{N}$ such that $\inf_{\lambda \in [1, \Lambda]} \frac{x_{[\lambda n]}}{\lambda^d x_n} \geq 1 - \varepsilon$ for each $n \geq n_0$. Because the last inequality is true for each $\Lambda > 1$, it follows that (for the same d) for each $\lambda \geq 1$ we have $\inf_{\lambda \geq 1} \frac{x_{[\lambda n]}}{\lambda^d x_n} \geq 1 - \varepsilon$. As ε was arbitrary (2) follows. (2) $\Rightarrow$ (1) Suppose that for an arbitrarily fixed $d \in \mathbb{R}$, $\liminf_{n \to \infty} \inf_{\lambda \geq 1} \frac{x_{[\lambda n]}}{\lambda^d x_n} \geq 1$ is satisfied. Then for the same d and each $\varepsilon > 0$ there exists $n_0 = n_0(\varepsilon) \in \mathbb{N}$ such that $\inf_{\lambda \geq 1} \frac{x_{[\lambda n]}}{\lambda^d x_n} \geq 1 - \varepsilon$ for each $n \geq n_0$. In other words, for the same d, ε, n_0, and for each $\lambda \geq 1$, especially for $\lambda \in [1, \Lambda]$, $\Lambda > 1$ an arbitrary real number, it holds $\frac{x_{[\lambda n]}}{x_n} \geq \lambda^d (1 - \varepsilon)$ for each $n \geq n_0$. This means that for each $\Lambda > 1$ we have $\frac{x_{[\lambda n]}}{x_n} \geq \lambda^d (1 + o(1))$ uniformly with respect to $\lambda \in [1, \Lambda]$ for $n \to \infty$. Since d was arbitrary, (1) follows. $\square$

The next statement is a result of the Galambos-Bojanić-Seneta type.

Theorem 6.2.22. ([10]) *For a sequence* $\mathbf{x} = (x_n)_{n \in \mathbb{N}} \in \mathbb{S}$ *the following are equivalent:*

(1) $\mathbf{x} \in \mathsf{KR}_{\mathsf{s},\infty}$;

(2) *The function* $\varphi(t) = x_{[t]}$, $t \geq 1$, *belongs to the class* $\mathsf{KR}_{\mathsf{f},\infty}$

Proof. (1) $\Rightarrow$ (2) Let $\mathbf{x} = (x_n)_{n \in \mathbb{N}} \in \mathsf{KR}_{\mathsf{s},\infty}$. Then by Lemma 6.2.21 we have $\liminf_{n \to \infty} \inf_{\lambda \geq 1} \frac{x_{[\lambda n]}}{\lambda^d x_n} \geq 1$ for each $d \in \mathbb{R}$. This means that for the same d and each $\varepsilon > 0$ there is $n_0 = n_0(d, \varepsilon) \in \mathbb{N}$ such that $\inf_{\lambda \geq 1} \frac{x_{[\lambda [t]]}}{\lambda^d x_{[t]}} \geq 1 - \varepsilon$ for each $t \geq n_0$ (≥ 1). Therefore, for the same d, ε, n_0 it is true

$$\inf_{\lambda \geq 1} \frac{x_{[\lambda t]}}{\lambda^d x_{[t]}} = \inf_{\lambda \geq 1} \frac{x_{[\frac{t}{[t]} \cdot [t] \cdot \lambda]}}{\lambda^d x_{[t]}} \geq \inf_{\lambda \geq 1} \frac{x_{[[t] \cdot \lambda]}}{\lambda^d x_{[t]}} \geq 1 - \varepsilon,$$

i.e., (for this d)
$$\liminf_{t\to\infty} \inf_{\lambda\geq 1} \frac{x_{[\lambda t]}}{\lambda^d x_{[t]}} \geq 1.$$

By [4, Proposition 2.4.3(ii)] it follows that the function $x_{[t]}$ belongs to the class $\mathsf{KR}_{\mathsf{f},\infty}$. (2) $\Rightarrow$ (1) From $\liminf_{t\to\infty} \inf_{\lambda\geq 1} \frac{x_{[\lambda t]}}{\lambda^d x_{[t]}} \geq 1$ it follows that for this d and each $\varepsilon > 0$ there is $t_0 = t_0(d, \varepsilon) \geq 1$ such that $\inf_{\lambda\geq 1} \frac{x_{[\lambda t]}}{\lambda^d x_{[t]}} \geq 1 - \varepsilon$ for all $t \geq t_0$. Since
$$\inf_{\lambda\geq 1} \frac{x_{[\lambda n]}}{\lambda^d x_{[n]}} \geq \inf_{\lambda\geq 1} \frac{x_{[\lambda t]}}{\lambda^d x_{[t]}} \quad \text{for } n \geq [t_0] + 1,$$
one obtains
$$\liminf_{n\to\infty} \inf_{\lambda\geq 1} \frac{x_{[\lambda n]}}{\lambda^d x_{[n]}} \geq 1,$$
i.e., (1) is true. □

Then following are the definitions of classes of functions and sequences containing the classes of rapidly varying functions and sequences of index of variability ∞ (see ([7] and also [12, 13]).

Definition 6.2.23. A function $\varphi \in \mathbb{F}$ is said to be in the *class* $\mathsf{ARV_f}$ if for each $\lambda > 1$ it satisfies
$$\underline{k}_\varphi(\lambda) > 1.$$

Definition 6.2.24. A sequence $\mathbf{x} = (x_n)_{n\in\mathbb{N}} \in \mathbb{F}$ belongs to the *class* $\mathsf{ARV_s}$ if for each $\lambda > 1$ it satisfies
$$\underline{k}_\mathbf{x}(\lambda) > 1.$$

The next theorem show the relationship between the classes $\mathsf{CRV_f}$ and $\mathsf{ARV_f}$.

Theorem 6.2.25. ([20, Theorems 3 and 4]). *Let $\varphi \in \mathbb{A}$. Then:*

(a) $\varphi \in \mathsf{ARV_f}$ *if and only if* $\varphi^\leftarrow \in \mathsf{CRV_f}$;

(b) $\varphi \in \mathsf{CRV_f}$ *if and only if* $\varphi^\leftarrow \in \mathsf{ARV_f}$.

Proof. (a) ($\Rightarrow$) Let $\varphi \in \mathbb{A} \cap \mathsf{ARV_f}$. Then for all $\lambda > 1$ and all $t \geq t_0 = t_0(\lambda)$ we have $\varphi(t) \geq c(\lambda)\varphi(t)$, where $c(\lambda)$ is a function depending on φ, such that $c(\lambda) > 1$, $\lambda > 1$. It follows $\frac{\varphi(\lambda t)}{c(\lambda)} \geq \varphi(t)$ for $\lambda > 1$, hence $\frac{\varphi^\leftarrow(c(\lambda)t)}{\lambda} \leq \varphi^\leftarrow(t)$. Therefore, for all $\lambda > 1$, we have
$$\overline{k}_{\varphi^\leftarrow}(c(\lambda)) = \limsup_{t\to\infty} \frac{\varphi^\leftarrow(c(\lambda)t)}{\varphi^\leftarrow(t)} \leq \lambda.$$

As $\varphi^\leftarrow$ is nondecreasing, its index function $\overline{k}_{\varphi^\leftarrow}(c(\overline{\lambda}))$, $\overline{\lambda} > 0$, is also nondecreasing in $\overline{\mathbb{R}}$. The facts that $\overline{k}_{\varphi^\leftarrow}(c(\overline{\lambda}))$ is defined for $\overline{\lambda} \in (0, c(\lambda))$, and $c(\lambda) > 1$ imply $\varphi^\leftarrow \in \mathsf{ORV_f}$. It follows
$$1 \leq \liminf_{\lambda\to 1+} \underline{k}_{\varphi^\leftarrow}(c(\lambda)) \leq \limsup_{\lambda\to 1+} \overline{k}_{\varphi^\leftarrow}(c(\lambda)) \leq 1,$$

which gives $\lim_{\lambda \to 1+} \underline{k}_{\varphi^{\leftarrow}}(c(\lambda)) = 1$. Let $A = \liminf_{\lambda \to 1+} c(\lambda)$ we have $A \geq 1$. There is a sequence $(\lambda_n)_{n \in \mathbb{N}}$ with $\lambda_n > 1$ for all n, $\lim_{n \to \infty} \lambda_n = 1+$ and $\lim_{n \to \infty} c(\lambda_n) = A$. Define $c_n = c(\lambda_n)$, $n \in \mathbb{N}$. Then $\lim_{n \to \infty} \overline{k}_{\varphi^{\leftarrow}}(c_n) = 1$. If $A = 1$, then $\lim_{\lambda \to 1+} \overline{k}_{\varphi^{\leftarrow}}(\lambda) = 1$ since $\overline{k}_{\varphi^{\leftarrow}}$ is nondecreasing. So, in this case $\varphi^{\leftarrow} \in \mathsf{CRV_f}$. If $A > 1$, then similarly $\lim_{\lambda \to A-} \overline{k}_{\varphi^{\leftarrow}}(\lambda) = 1$. So, if $\lambda \in [1, (A+1)/2]$, then $\overline{k}_{\varphi^{\leftarrow}}(\lambda) = 1$, and by [4], $\varphi^{\leftarrow} \in \mathsf{SV_f} \subset \mathsf{CRV_f}$. ($\Leftarrow$) Suppose now $\varphi \in \mathbb{A}$ and $\varphi^{\leftarrow} \in \mathsf{CRV}$. Then by [8] it holds

$$\lim_{t \to \infty, \lambda \to 1} \frac{\varphi^{\leftarrow}(\lambda t)}{\varphi^{\leftarrow}(t)} = 1.$$

Therefore, for each $\varepsilon > 1$ there are $t_0 = t_0(\varepsilon) > 0$ and $\delta_0 = \delta_0(\varepsilon) > 0$ such that

$$\frac{1}{\varepsilon} \leq \frac{\varphi^{\leftarrow}(\lambda t)}{\varphi^{\leftarrow}(t)} \leq \varepsilon$$

for each $t \geq t_0$ and each $\lambda \in [1 - \delta_0, 1 + \delta_0]$. Hence, for these λ and t we have

$$\frac{\varphi^{\leftarrow}(\lambda t)}{\varepsilon} \leq \varphi^{\leftarrow}(t) \text{ and } (\varphi(\varepsilon t)/\lambda)^{\leftarrow} \leq \varphi^{\leftarrow}(t)$$

so that

$$((\varphi(\varepsilon t)/\lambda)^{\leftarrow})^{\leftarrow} \geq (\varphi^{\leftarrow}(t))^{\leftarrow}.$$

Since for every function h in $\mathbb{A}$ and each $\beta > 1$ it holds $h(t) \leq ((h(t)^{\leftarrow}))^{\leftarrow} \leq h(\beta t)$, $t \geq a$, we obtain $\varphi(t) \leq \varphi(\varepsilon^2 t)/\lambda$, i.e., $\varphi(\varepsilon^2 t) \geq \lambda \varphi(t)$. So $\varphi(\varepsilon^2 t) \geq (1 + \delta_0(\varepsilon))\varphi(t)$ for $t \geq t_0$.

If $\alpha > 1$, take $\varepsilon = \sqrt{\alpha}$. So, we have

$$\varphi(\alpha t) \geq (1 + \delta_0(\sqrt{\alpha}))\varphi(t)$$

for $t \geq t_0(\sqrt{\alpha}) > 0$, which means $\varphi \in \mathsf{ARV_f}$. (b) ($\Rightarrow$) Since $\varphi \in \mathbb{A} \cap \mathsf{CRV_f}$ we have

$$\lim_{t \to \infty, \lambda \to 1} \frac{\varphi(\lambda t)}{\varphi(t)} = 1.$$

It follows that for each $\varepsilon > 1$ there exist $t_0 = t_0(\varepsilon)$ and $\delta_0 = \delta_0(\varepsilon)$ such that $\frac{1}{\varepsilon} \leq \frac{\varphi(\lambda t)}{\varphi(t)} \leq \varepsilon$ for each $t \geq t_0$ and each $\lambda \in [1 - \delta_0, 1 + \delta_0]$, so that for these λ and t, $\varphi(\lambda t)/\varepsilon \leq \varphi(t)$, and consequently $\varphi^{\leftarrow}(\varepsilon t) \geq \lambda \varphi^{\leftarrow}(t)$. Therefore,

$$\varphi^{\leftarrow}(\varepsilon t) \geq (1 + \delta_0)\varphi^{\leftarrow}(t) \text{ for } t \geq t_0$$

which means that $\varphi^{\leftarrow} \in \mathsf{ARV_f}$.

($\Leftarrow$) Let now $\varphi \in \mathbb{A}$ and $\varphi^{\leftarrow} \in \mathsf{ARV_f}$. Then for each $\lambda > 1$ and each $t \geq t_0 = t_0(\lambda)$, we have $\varphi^{\leftarrow}(\lambda t) \geq c(\lambda)\varphi^{\leftarrow}(t)$, where $c(\lambda) > 1$ depends on φ. So, for these λ and t, $\varphi^{\leftarrow}(\lambda t)/c(\lambda) \geq \varphi(t)$, and thus $(\varphi(c(\lambda)t)/\lambda)^{\leftarrow} \geq \varphi^{\leftarrow}(t)$. Similarly to the proof of the second part of the previous theorem we get

$$\frac{\varphi(c(\lambda)t)}{\lambda} \leq \varphi(t\sqrt{c(\lambda)}), \text{ i.e. } \frac{\varphi(c(\lambda)t)}{\varphi(t\sqrt{c(\lambda)})} \leq \lambda.$$

Therefore, for each $\lambda > 1$ we have $\varphi(u\sqrt{c(\lambda)})/\varphi(u) \leq \lambda$ for each $u \geq t_0\sqrt{c(\lambda)} = u_0(\lambda)$ which means that $\overline{k}_\varphi(\sqrt{c(\lambda)}) \leq \lambda$ (for each $\lambda > 1$). This implies
$$1 \leq \liminf_{\lambda \to 1+} \overline{k}_\varphi(\sqrt{c(\lambda)}) \leq \limsup_{\lambda \to 1+} \overline{k}_\varphi(\sqrt{c(\lambda)}) \leq 1,$$
which gives
$$\lim_{\lambda \to 1+} \overline{k}_\varphi(\sqrt{c(\lambda)}) = 1.$$
Set $A = \liminf_{\lambda \to 1+} \sqrt{c(\lambda)} \geq 1$. There exists a sequence $(\lambda_n)_{n \in \mathbb{N}}$ with $\lambda_n > 1$ for all n, $\lim_{n \to \infty} \lambda_n = 1+$ and $\lim_{n \to \infty} \sqrt{c(\lambda_n)} = A$.

Define $c_n = \sqrt{c(\lambda_n)}$, $n \in \mathbb{N}$. Then $\lim_{n \to \infty} \overline{k}_\varphi(c_n) = 1$. If $A = 1$, then $\lim_{\lambda \to 1+} \overline{k}_\varphi(\lambda) = 1$ since $\overline{k}_\varphi$ is nondecreasing. So, $\varphi \in \mathsf{CRV}_\mathsf{f}$. If $A > 1$, then $\lim_{\lambda \to A-} \overline{k}_\varphi(\lambda) = 1$. So, if $\lambda \in [1, (A+1)/2]$, then $\overline{k}_\varphi(\lambda) = 1$, and thus $\varphi \in \mathsf{SV}_\mathsf{f} \subsetneq \mathsf{CRV}_\mathsf{f}$. □

The following Galambos-Bojanić-Seneta type result is true.

Theorem 6.2.26. ([7]) *Let* $\mathbf{x} = (x_n)_{n \in \mathbb{N}}$ *be a sequence in* $\mathbb{S}$. *Then the following assertions are equivalent:*

(a) $\mathbf{x} \in \mathsf{ARV}_\mathsf{s}$;

(b) $\varphi_\mathbf{x}(t) = x_{[t]}$, $t \geq 1$, *belongs to* ARV_f.

6.3 New result

In this section we prove a new result of Galambos-Bojanić-Seneta type.

Definition 6.3.1. ([6,7]) *A function* $\varphi \in \mathbb{F}$ *is said to be in the class* Pl_f^* *if there exists* $\lambda_0 \geq 1$ *such that for each* $\lambda > \lambda_0$
$$\underline{k}_\varphi(\lambda) := \liminf_{x \to \infty} \frac{\varphi(\lambda x)}{\varphi(x)} > 1.$$

Definition 6.3.2. *A sequence* $\mathbf{x} = (x_n)_{n \in \mathbb{N}} \in \mathbb{S}$ *is said to belong to the class* Pl_s^* *if there is* $\lambda_0 \geq 1$ *such that for each* $\lambda > \lambda_0$ *it holds*
$$\underline{k}_x(\lambda) > 1.$$

Clearly, if in the previous two definitions $\lambda_0 = 1$ we have the classes ARV_f and ARV_s, respectively, from the previous section.
Observe that the following holds:
$$\mathsf{R}_{\mathsf{s},\infty} \subsetneq \mathsf{ARV}_\mathsf{s} \subsetneq \mathsf{Pl}_\mathsf{s}^*.$$

The next theorem shows the importance of the class Pl_f^* because it is conjugate (by the generalized inverse) with the very important class ORV_f.

On Theorems of Galambos-Bojanić-Seneta Type

Theorem 6.3.3. ([16, Propositions 3 and 4]) *Let $\varphi \in \mathbb{A}$. Then:*

(a) $\varphi \in \mathsf{PI}_f^*$ *if and only if* $\varphi^{\leftarrow} \in \mathsf{ORV}_f$;

(b) $\varphi \in \mathsf{ORV}_f$ *if and only if* $\varphi^{\leftarrow} \in \mathsf{PI}_f^*$.

Proof. (a) First assume $\varphi \in \mathbb{A} \cap \mathsf{PI}_f^*$. Then for some $\lambda_0 \geq 1$ and for some $\lambda > \lambda_0$ it holds
$$\varphi(\lambda t) \geq c(\lambda)(t), \ t \geq t_0 = t_0(\lambda),$$
where $c(\lambda) = c_\varphi(\lambda) > 1$ for $\lambda > \lambda_0$. Therefore, for these λ and t we have $\varphi(\lambda t)/c(\lambda) \geq \varphi(t)$. It follows $\varphi^{\leftarrow}(c(\lambda)t)/\lambda \leq \varphi^{\leftarrow}(t)$. Hence, for this λ we have
$$\overline{k}_{\varphi^{\leftarrow}}(c(\lambda)) \leq \lambda < \infty$$
which means $\varphi^{\leftarrow} \in \mathsf{ORV}_f$.

Conversely, assume $\varphi^{\leftarrow} \in \mathsf{ORV}_f \cap \mathbb{A}$. Then by [1] we have
$$\limsup_{t \to \infty} \sup_{\lambda \in [1,2]} \frac{\varphi^{\leftarrow}(\lambda t)}{\varphi^{\leftarrow}(t)} = \limsup_{t \to \infty} \frac{\varphi^{\leftarrow}(2x)}{\varphi^{\leftarrow}(t)} = \overline{k}_{\varphi^{\leftarrow}}(2) \geq 1.$$

For each $\varepsilon > 0$ there is a $t_0 = t_0(\varepsilon) > 0$ such that
$$\sup_{\lambda \in [1,2]} \frac{\varphi^{\leftarrow}(\lambda t)}{\varphi^{\leftarrow}(t)} \leq \overline{k}_{\varphi^{\leftarrow}}(2) + \varepsilon = M(\varepsilon), \ t \geq t_0,$$
so that for each $t \geq t_0$ and each $\lambda \in [1,2]$ we have
$$\frac{\varphi^{\leftarrow}(\lambda t)}{\varphi^{\leftarrow}(t)} \leq M(\varepsilon).$$

It follows
$$\frac{\varphi^{\leftarrow}(\lambda t)}{M(\varepsilon)} \leq \varphi^{\leftarrow}(t) \ \Rightarrow \ \left(\left(\frac{f(M(\varepsilon)t)}{\lambda}\right)^{\leftarrow}\right)^{\leftarrow} \geq (\varphi^{\leftarrow}(t))^{\leftarrow}$$
$$\Rightarrow \ \varphi(t) \leq \frac{\varphi(M^2(\varepsilon)t)}{\lambda}$$
$$\Rightarrow \ \frac{\varphi(M^2(\varepsilon)t)}{\varphi(t)} \geq \lambda$$
$$\Rightarrow \ \frac{\varphi(M^2(\varepsilon)t)}{\varphi(t)} \geq 2 > 1$$
$$\Rightarrow \ \liminf_{t \to \infty} \frac{\varphi(M^2(\varepsilon)t)}{\varphi(t)} = \underline{k}_\varphi(M^2(\varepsilon)) \geq 2 > 1.$$

Since $\underline{k}_\varphi(u)$ is nondecreasing for $u > 0$, we find that $\underline{k}_\varphi(\lambda) > 1$, for $\lambda > M^2(\varepsilon) > 1$. Hence, $\varphi \in \mathsf{Pl}_f^{n*} \cap \mathsf{A}$.

(b) First assume $\varphi \in \mathsf{A} \cap \mathsf{ORV}_f$. By [1]

$$\limsup_{t \to \infty} \sup_{\lambda \in [1,2]} \frac{\varphi(\lambda t)}{\varphi(t)} = \limsup_{t \to \infty} \frac{\varphi(2t)}{\varphi(t)} = \overline{k}_\varphi(2) \geq 1.$$

For each $\varepsilon > 0$, there is $t_0 = t_0(\varepsilon) > 0$ such that

$$\sup_{\lambda \in [1,2]} \frac{\varphi(\lambda t)}{\varphi(t)} \leq \overline{k}_\varphi(2) + \varepsilon = m(\varepsilon), \quad \text{for all } t \geq t_0.$$

So, for the same t and for each $\lambda \in [1,2]$ we have $\varphi(\lambda t)/\varphi(t) \leq m(\varepsilon)$. Therefore, $\varphi(\lambda t)/m(\varepsilon) \leq \varphi(t)$. It follows

$$\frac{\varphi^{\leftarrow}(m(\varepsilon)t)}{\lambda} \geq \varphi^{\leftarrow}(t) \Rightarrow \quad \varphi^{\leftarrow}(m(\varepsilon)t) \geq \lambda \varphi^{\leftarrow}(t)$$

$$\Rightarrow \quad \varphi^{\leftarrow}(m(\varepsilon)t) \geq 2\varphi^{\leftarrow}(t)$$

$$\Rightarrow \quad \frac{\varphi^{\leftarrow}(m(\varepsilon)t)}{\varphi^{\leftarrow}(t)} \geq 2 > 1$$

$$\Rightarrow \quad \liminf_{t \to \infty} \frac{\varphi^{\leftarrow}(m(\varepsilon)t)}{\varphi^{\leftarrow}(t)} \geq 2 > 1$$

$$\Rightarrow \quad \underline{k}_{\varphi^{\leftarrow}}(m(\varepsilon)) < 1.$$

Hence, $\underline{k}_{\varphi^{\leftarrow}}(\lambda) > 1$ for $\lambda > m(\varepsilon) = \lambda_0 \geq 1$, which means $\varphi^{\leftarrow} \in \mathsf{Pl}_f^*$.

Conversely, assume $\varphi^{\leftarrow} \in \mathsf{Pl}_f^* \cap \mathsf{A}$. Then for some $\lambda_0 \geq 1$ and all $\lambda > \lambda_0$ we have $\varphi^{\leftarrow}(\lambda t) \geq c(\lambda)\varphi^{\leftarrow}(t)$, for all $t \geq t_0 = t_0(\lambda)$, where $c(\lambda) = c_\varphi(\lambda) > 1$, $\lambda > \lambda_0$. Hence, for those λ and t we have $\varphi^{\leftarrow}(\lambda t)/c(\lambda) \geq \varphi^{\leftarrow}(t)$, so that $\left(\varphi(c(\lambda)t)/\lambda\right)^{\leftarrow} \geq \varphi^{\leftarrow}(t)$. As in the previous proof, we have $\varphi(c(\lambda)t)/\lambda \leq \varphi(\sqrt{c(\lambda)}t)$. Therefore, $\varphi(c(\lambda)t)/\varphi(\sqrt{c(\lambda)}t) \leq \lambda$, and consequently, for a fixed $\lambda > \lambda_0$, we obtain $\overline{k}_\varphi(\sqrt{c(\lambda)}) \leq \lambda < \infty$. In other words, $\varphi \in \mathsf{ORV}_f$. □

We prove now a new result of the Galambos-Bojanić-Seneta type for the classes Pl^*.

Theorem 6.3.4. *Let* $\mathbf{x} = (x_n)_{n \in \mathbb{N}}$ *be a sequence in* $\mathbb{S}$. *Then the following are equivalent:*

(a) $\mathbf{x}$ *belongs to the class* Pl_s^*;

(b) *The function* $\varphi(t) = x_{[t]}$, $t \geq 1$, *belongs to the class* Pl_f^*.

Proof. (a) ⇒ (b): Let the sequence $\mathbf{x} = (x_n)_{n \in \mathbb{N}}$ belong to the class Pl_s^*. Then

$$\liminf_{n \to \infty} \frac{x_{[\lambda n]}}{x_n} > 1 \quad \text{for some } \lambda_0 \geq 1 \text{ and each } \lambda > \lambda_0.$$

Consider the interval (λ_0, λ_0^2). Then $k_\mathbf{x}(\lambda) > 1$ on the interval (λ_0, λ_0^2). For each $\lambda \in (\lambda_0, \lambda_0^2)$ define $n_\lambda \in \mathbb{N}$ in the following way:

$$n_\lambda = \begin{cases} 1, & \text{if } \frac{x_{[\lambda n]}}{x_n} > 1 \text{ for each } n \in \mathbb{N}; \\ 1 + \max\{n \in \mathbb{N}: \frac{x_{[\lambda n]}}{x_n} \leq 1\}, & \text{otherwise.} \end{cases}$$

Evidently, $1 \leq n_\lambda < \infty$ for each λ. Then one defines the sequence $(A_k)_{k \in \mathbb{N}}$ by

$$A_k = \{\lambda \in (\lambda_0, \lambda_0^2) : n_\lambda > k\}$$

which is non-increasing and $\bigcap_{k \in \mathbb{N}} A_k = \emptyset$.
We prove that not all sets A_k are dense in (λ_0, λ_0^2). If $\lambda \in A_k$ for some k, then $\frac{x_{[\lambda(n_\lambda - 1)]}}{x_{n_\lambda - 1}} \leq 1$ and there is $\delta_\lambda > 0$ such that

$$\frac{x_{[t(n_\lambda - 1)]}}{x_{n_\lambda - 1}} \leq 1 \text{ for each } t \in [\lambda, \lambda + \delta_\lambda) \subsetneq (\lambda_0, \lambda_0^2).$$

So, each $t \in (\lambda, \lambda + \delta_\lambda)$ belongs to A_k, hence $n_t \geq (n_\lambda - 1) + 1 > k$. We conclude that $(\lambda, \lambda + \delta_\lambda) \subset A_k$ whenever $\lambda \in A_k$. Suppose now that each set A_k is dense in (λ_0, λ_0^2). Then each set $\text{Int}(A_k)$ is also dense, hence we have the sequence $(\text{Int}(A_k))_{k \in \mathbb{N}}$ of open dense subsets of (λ_0, λ_0^2) which is a Baire second category set. By the Baire category theorem we have that the set $\bigcap_{k \in \mathbb{N}} \text{Int}(A_k)$ dense in (λ_0, λ_0^2) and so the set $\bigcap_{k \in \mathbb{N}} A_k$ is nonempty. It is a contradiction.
Therefore, there $k_0 \in \mathbb{N}$ so that the set A_{k_0} is not dense in (λ_0, λ_0^2). There is the closed interval $[A, B] \subsetneq (\lambda_0, \lambda_0^2)$ such that $[A, B] \subset (\lambda_0, \lambda_0^2) \setminus A_{k_0} == \{\lambda \in (\lambda_0, \lambda_0^2) : n_\lambda \leq k_0\}$. Therefore, for each $\lambda \in [A, B]$, $n_\lambda \leq k_0$. It follows that for each $\lambda \in [A, B]$ and each $k \geq k_0 \geq k_\lambda$, $\frac{x_{[\lambda k]}}{x_k} > 1$. Thus for each $\lambda > \lambda_0^3$ and each sufficiently large $t \geq t_0 \geq 1$, it holds

$$\frac{x_{[\lambda t]}}{x_t} = \frac{x_{z[\eta t]}}{x_{[\eta t]}} \cdot \frac{x_{[\eta t]}}{x_t} \geq k_\mathbf{x}(\eta) > 1.$$

This means that the function φ, $\varphi(t) = x_{[t]}$, $t \geq 1$, belongs to the class PI_f^*.
(b) $\Rightarrow$ (a): It is evident. □

Bibliography

[1] Aljančić, S. and Arandjelović, D. 1977. $\mathcal{O}$-regularly varying functions, *Publ. Inst. Math. (Beograd)* 22:36: 5–22.

[2] Arandjelović, D. 1990. $\mathcal{O}$-regularly variation and uniform convergence, *Publ. Inst. Math (Beograd)* 48:62: 25–40.

[3] Avakumović, V.G. 1936. Über einen $\mathcal{O}$-inversionssatz, *Bull. Int. Acad. Youg. Sci.* 2930: 107–117.

[4] Bingham, N.H., Goldie, C.M. and Teugels, J.L. 1987. *Regular Variation, Encyclopedia of Mathematics and its Applications*, Vol. 17, Cambridge University Press, Cambridge, UK.

[5] Bojanić, R. and Seneta, E. 1973. A unified theory of regularly varying sequences, *Math. Z.* 134: 91–106.

[6] Buldygin, V.V., Klesov, O.I. and Steinebach, J.G. 2005. On some properties of asymptotically quasi-inverse functions and their application-I, *Theory Probability Math. Stat.* 70: 11–28.

[7] Buldygin, V.V., Klesov, O.I. and Steinebach, J.G. 2008. On some properties of asymptotically quasi-inverse functions, *Theory Probability Math. Stat.* 77: 15–30.

[8] Djurčić, D. 1998. O-regularly varying functions and strong asymptotic equivalence, *J. Math. Anal. Appl.* 220: 451–461.

[9] Djurčić, D. and Božin, V. 1997. A proof of S. Aljančić hypothesis on O-regularly varying sequences, *Publ. Inst. Math.* 61:76: 46–52.

[10] Djurčić, D., Elez, N. and Kočinac, Lj.D.R. 2015. On a subclass of the class of rapidly varying sequences, *Appl. Math. Comput.* 251: 626–632.

[11] Djurčić, D., Kočinac, Lj.D.R. and Žižović, M.R. 2007. Some properties of rapidly varying sequences, *J. Math. Anal. Appl.* 327: 1297–1306.

[12] Djurčić, D., Kočinac, Lj.D.R. and Žižović, M.R. 2008. Rapidly varying sequences and rapid convergence, *Topol. Appl.* 155: 2143–2149.

[13] Djurčić, D., Kočinac, Lj.D.R. and Žižović, M.R. 2008. Classes of sequences of real numbers, games and selection properties, *Topol. Appl.* 156: 46–55.

[14] Djurčić, D., Kočinac, Lj.D.R. and Žižović, M.R. 2009. A few remarks on divergent sequences: rates of divergence, *J. Math. Anal. Appl.* 360: 588–598.

[15] Djurčić, D., Kočinac, Lj.D.R. and Žižović, M.R. 2010. A few remarks on divergent sequences: Rates of divergence II, *J. Math. Anal. Appl.* 367: 705–709.

[16] Djurčić, D., Nikolić, R. and Torgašev, A. 2010. The weak asymptotic equivalence and the generalized inverse, *Lithuanian Math. J.* 50: 34–42.

[17] Djurčić, D. and Torgašev, A. 2004. Representation theorems for sequences of the classes CRc and ERc, *Siberian Math. J.* 45: 834–838.

[18] Djurčić, D. and Torgašev, A. 2006. On the Seneta sequences, *Acta Math. Sinica* 22: 689–692.

[19] Djurčić, D. and Torgašev, A. 2009. A theorem of Galambos-Bojanić-Seneta type, *Abstr. Appl. Anal.* 20090: Art. ID 360794, 6 pages.

[20] Djurčić, D., Torgašev, A. and Ješić, S. 2008. The strong asymptotic equivalence and the generalized inverse, *Siberian Math. J.* 49: 628–636.

[21] Drasin, D.and Seneta, E. 1986. A generalization of slowly varying functions,*Proc. Amer. Math. Soc.* 96: 470–472.

[22] Engelking, R. 1989. *General Topology*, 2nd edition, Sigma Series in Pure Mathematics, Vol. 6, Heldermann, Berlin.

[23] Galambos, J. and Seneta, E. 1973. Regularly varying sequences, *Proc. Amer. Math. Soc.* 41: 110–116.

[24] Grow, D.E. and Stanojević, Č.V. 1995. Convergence and the Fourier character of trigonometric transforms with slowly varying convergence moduli, *Math. Annalen* 302: 433–472.

[25] Haan, L. de 1970. On Regular Variation and Its Application to the Weak Convergence of Sample Extremes, Mathematical Centre Tracts, Vol. 32, Mathematisch Centrum, Amsterdam, The Netherlands.

[26] Karamata, J. 1930. Sur certains "Tauberian theorems" de G.H. Hardy et Littlewood, *Mathematica (Cluj)* 3: 33–48.

[27] Karamata, J. 1930. Sur un mode de croissance régulière des fonctions, *Mathematica (Cluj)* 4: 38–53.

[28] Karamata, J. 1930. Über die Hardy-Littlewoodschen Umkehrungen des Abelschen Stetigkeitsätzes, *Math. Z.* 32: 319–320.

[29] Karamata, J. 1931. Neuer Beweis und Verallgemeinerung der Tauberschen Sätze, welche die Laplacesche un Stieltjessche Transformation betreffen, *J. fur Reine Angew. Math.* 164: 27–39.

[30] Karamata, J. 1933. Sur un mode de croissance régulière. Théorèmes fondamenteaux, *Bull. Soc. Math. France* 61: 55–62.

[31] Kočinac, Lj.D.R., Djurčić, D. and Manojlović, J.V. 2018. Regular and Rapid Variations and Some Applications, In: M. Ruzhansky, H. Dutta, R.P. Agarwal (eds.), Mathematical Analysis and Applications: Selected Topics, Chapter 12, John Wiley & Sons, Inc. pp 414–474.

[32] Matuszewska, W. 1964. On a generalization of regularly increasing functions, *Studia Math.* 24: 271–279.

[33] Matuszewska, W. and Orlicz, W. 1965. On some classes of functions with regard to their orders of growth, *Studia Math.* 26: 11–24.

[34] Seneta, E. 1976. *Functions of Regular Variation*, LNM, Springer, Vol. 506, New York.

[35] Stanojević, Č. 1988. Structure of Fourier and Fourier-Stieltjes coefficients of series with slowly varying convergence moduli, *Bull. Amer. Math. Soc.* 19: 283–286.

Chapter 7

On the spaces of absolutely p-summable and bounded q-Euler difference sequences

Taja Yaying

7.1	Introduction	113
	7.1.1 Euler matrix of order 1 and sequence spaces	114
	7.1.2 Quantum calculus	115
7.2	q-Euler difference sequence spaces $e_p^q(\nabla)$ and $e_\infty^q(\nabla)$	117
7.3	Alpha-, beta-, and gamma-duals	119
7.4	Matrix transformations	122
	Bibliography	125

7.1 Introduction

Let w denote the set of all real-valued sequences. Any linear subspace of w is called sequence space. Let $1 \leq p < \infty$, then by ℓ_p and ℓ_∞, we denote the spaces of all absolutely p-summable and bounded sequences, respectively. A Banach sequence space $\mathfrak{X}$ is called a BK-space if it has continuous coordinates. The spaces ℓ_p and ℓ_∞ are BK-spaces equipped with the norms

$$\|z\|_{\ell_p} = \left(\sum_{k=0}^{\infty} |z_k|^p\right)^{1/p} \quad \text{and} \quad \|z\|_{\ell_\infty} = \sup_{k \in \mathbb{N}_0} |z_k|,$$

respectively. Here, and in the rest of the paper, $\mathbb{N}_0 = \{0, 1, 2, \ldots\}$. By c and c_0, we denote the spaces of all convergent and null sequences, respectively. Moreover bs, cs, and cs_0 denote the spaces of bounded, convergent and null series, respectively.

DOI: 10.1201/9781003330868-7

Let $\mathfrak{X}$ and $\mathfrak{Y}$ be two sequence spaces and $\Omega = (\omega_{nk})$ be an infinite matrix of real entries. By Ω_n, we denote the nth row of the matrix Ω. We say that Ω defines a matrix mapping from $\mathfrak{X}$ to $\mathfrak{Y}$ if Ω-transform of sequence z i.e., $\Omega z = \{(\Omega z)_n\} = \left\{\sum_{k=0}^{\infty} \omega_{nk} z_k\right\} \in \mathfrak{Y}$ for every $z = (z_k) \in \mathfrak{X}$, provided that the series $\sum_{k=0}^{\infty} \omega_{nk} z_k$ exists for each n. In what follows, $(\mathfrak{X}, \mathfrak{Y})$ denotes the family of all matrices that map from $\mathfrak{X}$ to $\mathfrak{Y}$.

Define the sequence space $\mathfrak{X}_\Omega$ by

$$\mathfrak{X}_\Omega = \{z \in w : \Omega z \in \mathfrak{X}\}. \tag{7.1}$$

The set $\mathfrak{X}_\Omega$ is called the domain of matrix Ω in space $\mathfrak{X}$. The domain of a matrix plays an important role in defining sequence spaces. It is known that if $\mathfrak{X}$ is a Banach sequence space then the matrix domain $\mathfrak{X}_\Omega$ is also a Banach sequence space with the norm $\|z\|_{\mathfrak{X}_\Omega} = \|\Omega z\|_{\mathfrak{X}}$. With this concept, several authors have introduced new Banach sequence and series spaces using the domain of interesting triangular matrices. For relevant literature, we refer the papers [1, 2, 16, 17, 24].

7.1.1 Euler matrix of order 1 and sequence spaces

Let r, s be two non-zero real numbers, then the binomial matrix $B^{r,s} = (b_{nk}^{r,s})$ is defined by

$$b_{nk}^{r,s} = \begin{cases} \frac{1}{(r+s)^n} \binom{n}{k} r^k s^{n-k} & 0 \leq k \leq n, \\ 0 & otherwise. \end{cases}$$

Bişgin [6, 7] studied the domains $b_\infty^{r,s}$, $b_0^{r,s}$, $b_c^{r,s}$ and $b_p^{r,s}$ of the matrix $B^{r,s}$ in the spaces ℓ_∞, c, c_0, and ℓ_p, respectively.

We observe that when $s = 1 - r$, then the binomial matrix reduces to the Euler matrix E^r [1]. By using the matrix E^r, Altay et al. [1, 2] defined and studied the well known Euler sequence spaces $e_\infty^r = (\ell_\infty)_{E^r}$, $e_0^r = (c_0)_{E^r}$, $e_c^r = c_{E^r}$ and $e_p^r = (\ell_p)_{E^r}$.

Further, taking $r = s = 1$, the matrix $B^{r,s}$ contracts to the Euler matrix $E = (e_{nk})$ of order 1 defined by

$$e_{nk} = \begin{cases} \frac{\binom{n}{k}}{2^n} & 0 \leq k \leq n, \\ 0 & otherwise, \end{cases}$$

for all $n, k \in \mathbb{N}_0$. Not much studies related to sequence spaces obtained using the domain of the matrix E can be found in the literature. Recently, Başar and Braha [5] studied the spaces of Euler-Cesàro bounded, convergent, and null difference sequences. It is shown that these spaces are separable BK-spaces. Further the authors obtained certain inclusion relations, Schauder basis, Köthe duals, and characterized the certain classes of matrix transformatìons on these spaces. More recently, Ellidokuzoğlu and Demiriz [8] give a further generalization of the spaces defined in [5] by introducing Euler-Riesz bounded, convergent, and null difference spaces.

7.1.2 Quantum calculus

The q-calculus is a branch of mathematics that deals with the generalization of some well known mathematical expressions by using the parameter q. The generalized expression so obtained is called q-analog (or quantum analog) of the original expression. Further, q-analog returns the original expression when q approaches 1. Several researchers are engaged in the field of q-calculus due to its broad applications in mathematics, physics, and engineering sciences. It is widely used by researchers in operator theory, approximation theory, hypergeometric series, special functions, quantum algebras, combinatorics, etc. We refer the book [10] for details in q-calculus.

The following notations and definitions are well-known in the field of q-calculus:

Definition 7.1.1. *[10] Let $0 < q < 1$. Then the q-number is defined by*

$$[n]_q = \begin{cases} \frac{1-q^n}{1-q} & (n = 1, 2, 3, \cdots), \\ 0 & (n = 0). \end{cases}$$

Clearly, $[n]_q = n$ when $q \to 1^-$.

Definition 7.1.2. *[10] The q-binomial coefficient is defined by*

$$\begin{bmatrix} n \\ k \end{bmatrix}_q = \begin{cases} \frac{[n]_q!}{[n-k]_q! [k]_q!} & (n \geq k), \\ 0 & (n < k), \end{cases}$$

where $[n]_q!$ is called the q-factorial of n, and is defined by

$$[n]_q! = [n]_q [n-1]_q \ldots [2]_q [1]_q.$$

The application of q-theory in the field of sequence spaces has been realized very recently. By using quantum theory, many authors in recent times constructed q-analogue of well known sequence spaces like q-Cesàro sequence spaces [9, 24], q-Catalan sequence spaces [27], q-Pascal sequence spaces [22], Padovan q-difference sequence spaces [23], and (p,q)-Euler sequence spaces [25].

Define the operators Δ and ∇ by

$$(\Delta z)_k = z_k - z_{k+1} \text{ and } (\nabla z)_k = z_k - z_{k-1} \text{ for all } k \in \mathbb{N}_0,$$

which are well known as forward and backward difference operators, respectively, where we assumed that $z_k = 0$ for $k < 0$. The domains $\ell_\infty(\Delta)$, $c(\Delta)$

and $c_0(\Delta)$ are studied by Kızmaz [13]. Difference operators play a very crucial role in the theory of sequence spaces and summability. For example, the sequence $z = (z_k)$ such that $z_k = k$ for all k, is not convergent rather it diverges to ∞. However, the sequence $\nabla z = ((\nabla z)_k)$ where $(\nabla z)_k = -1$ for all k, is convergent to -1. The difference operator ∇ may also be represented in the matrix form $\nabla = (\delta_{nk})$ as follows:

$$\delta_{nk} = \begin{cases} (-1)^{n-k}, & n \leq k \leq n+1, \\ 0, & k > n. \end{cases}$$

Motivation: Several authors studied Euler difference sequence spaces in the literature. For instances, Altay and Polat [3] studied the Euler difference sequence spaces $e_\infty^r(\nabla) = (\ell_\infty)_{E^r\nabla}$, $e_0^r(\nabla) = (c_0)_{E^r\nabla}$ and $e_c^r(\nabla) = c_{E^r\nabla}$. More recently, Meng and Song [14] studied binomial difference sequence spaces $b_\infty^{r,s}(\nabla) = (\ell_\infty)_{B^{r,s}\nabla}$, $b_0^{r,s}(\nabla) = (c_0)_{B^{r,s}\nabla}$ and $b_c^{r,s}(\nabla) = c_{B^{r,s}\nabla}$. Besides, Song and Meng [19] studied binomial difference sequence space $b_p^{r,s}(\nabla) = (\ell_p)_{B^{r,s}\nabla}$. We refer to [11, 14, 15] for studies related to difference sequence spaces involving Euler (or binomial) matrix.

Yaying et al. studied the (p', q)-analog of the Euler sequence spaces as follows (see [25]):

$$e_p^{p',q} := \{z \in \omega : E^{r,s}(p',q)z \in \ell_p\},$$
$$e_\infty^{p',q} := \{z \in \omega : E^{r,s}(p',q)z \in \ell_\infty\},$$

where $E^{r,s}(p',q) = (e_{nk}^{r,s})$ is (p,q)-analogue of the Binomial matrix $B^{r,s}$. Considering $p' = r = s = 1$, the above sequence spaces reduce to q-Euler sequence spaces $e_p^q = (\ell_p)_{E^q}$ and $e_\infty^q = (\ell_\infty)_{E^q}$, respectively, of the first order. Here, $E^q = (e_{nk}^q)$ is the q-analog of the Euler matrix of order 1 defined by

$$e_{nk}^q = \begin{cases} \frac{[{}^n_k]_q q^{\binom{k}{2}}}{\pi^{(n)}(q)} & (0 \leq k \leq n), \\ 0 & (k > n), \end{cases}$$

where $\pi^{(n)}(q) = \Pi_{k=0}^{n-1}(1+q^k)$ with $\pi^{(0)}(q) = 1$. Besides, Yaying [20] studied q-analogue of Euler sequence spaces as follows:

$$e_c^q := \{z \in \omega : E^q z \in c\},$$
$$e_0^q := \{z \in \omega : E^q z \in c_0\}.$$

Equivalently $e_c^q = c_{E^q}$ and $e_0^q = (c_0)_{E^q}$.

As a natural continuation of above mentioned studies, we construct Euler difference sequence spaces $e_p^q(\nabla)$ and $e_\infty^q(\nabla)$ derived by the domain of the product matrix $E^q \nabla$ in the spaces ℓ_p and ℓ_∞, respectively. We determine Schauder bases for the spaces $e_p^q(\nabla)$. In Section 3, we determine α-, β-, and γ-duals of the spaces $e_p^q(\nabla)$ and $e_\infty^q(\nabla)$. In Section 4, we characterize certain classes of matrix mappings from the spaces $e_p^q(\nabla)$ and $e_\infty^q(\nabla)$ to any one the space ℓ_∞, c, c_0, or ℓ_1.

7.2 q-Euler difference sequence spaces $e_p^q(\nabla)$ and $e_\infty^q(\nabla)$

Define the sequence spaces $e_p^q(\nabla)$ and $e_\infty^q(\nabla)$ as follows:

$$e_p^q(\nabla) := \{z \in \omega : \nabla z \in e_p^q\},$$
$$e_\infty^q(\nabla) := \{z \in \omega : \nabla z \in e_\infty^q\}.$$

We define the product matrix $H := E^q \Delta = (e_{nk}^{q,\delta})$ by

$$e_{nk}^{q,\delta} = \begin{cases} \dfrac{1}{\pi^{(n)}(q)}\left\{\begin{bmatrix} n \\ k \end{bmatrix}_q q^{\binom{k}{2}} - \begin{bmatrix} n \\ k+1 \end{bmatrix}_q q^{\binom{k+1}{2}}\right\}, & 0 \le k \le n, \\ 0, & k > n. \end{cases}$$

We observe that the H-transform of the sequence z,

$$\begin{aligned} t = Hz &= \sum_{k=0}^n \frac{1}{\pi^{(n)}(q)}\left\{\begin{bmatrix} n \\ k \end{bmatrix}_q q^{\binom{k}{2}} - \begin{bmatrix} n \\ k+1 \end{bmatrix}_q q^{\binom{k+1}{2}}\right\} z_k \\ &= \sum_{k=0}^n \frac{1}{\pi^{(n)}(q)} \begin{bmatrix} n \\ k \end{bmatrix}_q q^{\binom{k}{2}} (\nabla z)_k = E^q(\nabla z). \end{aligned} \quad (7.2)$$

Thus, $e_p^q(\nabla)$ and $e_\infty^q(\nabla)$ may also be defined by

$$e_p^q(\nabla) := \{z \in \omega : Hz \in \ell_p\},$$
$$e_\infty^q(\nabla) := \{z \in \omega : Hz \in \ell_\infty\}.$$

Equivalently $e_p^q(\nabla) = (e_p^q)_\nabla = (\ell_p)_H$ and $e_\infty^q(\nabla) = (e_\infty^q)_\nabla = (\ell_\infty)_H$. We emphasize that $e_p^q(\nabla)$ and $e_\infty^q(\nabla)$ reduce to $b_p^{r,s}(\nabla)$ and $b_\infty^{r,s}(\nabla)$, respectively, as $q \to 1^-$, studied by Song and Meng [19, with $r = s = 1$].

Lemma 7.2.1. *The inverse of the matrix H is given by the matrix $H^{-1} = (h_{nk}^{-1})$ defined for all $n, k \in \mathbb{N}_0$ by*

$$h_{nk}^{-1} = \begin{cases} \sum_{j=k}^n (-1)^{j-k} \dfrac{\pi^{(k)}(q) \begin{bmatrix} j \\ k \end{bmatrix}_q q^{\binom{j-k}{2}}}{q^{\binom{j}{2}}}, & 0 \le k \le n, \\ 0, & k > n. \end{cases}$$

In view of Lemma 7.2.1, the sequence z can also be expressed in terms of the sequence t as follows:

$$z_n = (H^{-1}t)_n = \sum_{k=0}^n \sum_{j=k}^n (-1)^{j-k} \frac{\pi^{(k)}(q) \begin{bmatrix} j \\ k \end{bmatrix}_q q^{\binom{j-k}{2}}}{q^{\binom{j}{2}}} t_k \quad (7.3)$$

for all $n \in \mathbb{N}_0$. In the rest of the paper, the sequences z and t are connected by (7.2) or equivalently by (7.3).

Theorem 7.2.2. *The sequence spaces $e_p^q(\nabla)$ and $e_\infty^q(\nabla)$ are BK spaces under the norms defined by*

$$\|z\|_{e_p^q(\nabla)} = \|Hz\|_{\ell_p} = \left(\sum_{n=0}^{\infty} \left|\sum_{k=0}^{n} \frac{1}{\pi^{(n)}(q)} \left\{\begin{bmatrix}n\\k\end{bmatrix}_q q^{\binom{k}{2}} - \begin{bmatrix}n\\k+1\end{bmatrix}_q q^{\binom{k+1}{2}}\right\} z_k\right|^p\right)^{1/p}$$

and

$$\|z\|_{e_\infty^q(\nabla)} = \|Hz\|_{\ell_\infty} = \sup_{n \in \mathbb{N}_0} \left|\sum_{k=0}^{n} \frac{1}{\pi^{(n)}(q)} \left\{\begin{bmatrix}n\\k\end{bmatrix}_q q^{\binom{k}{2}} - \begin{bmatrix}n\\k+1\end{bmatrix}_q q^{\binom{k+1}{2}}\right\} z_k\right|,$$

respectively.

Proof. The proof is a routine exercise and hence omitted. □

Theorem 7.2.3. *The spaces $e_p^q(\nabla)$ and $e_\infty^q(\nabla)$ are linearly isomorphic to ℓ_p and ℓ_∞, respectively.*

Proof. We give the proof for the space $e_p^q(\nabla)$. Define the mapping $T : e_p^q(\nabla) \to \ell_p$ by $Tz = Hz$ for all $z \in e_p^q(\nabla)$. Clearly T is linear. Since H is invertible so T is invertible. Let $t = (t_k) \in \ell_p$ and $z = (z_k)$ is defined as in (7.3). Then, we have

$$\|z\|_{e_p^q(\nabla)} = \|Hz\|_{\ell_p} = \|t\|_{\ell_p} < \infty.$$

Thus $z \in e_p^q(\nabla)$ and the mapping $T : e_p^q(\nabla) \to \ell_p$ is onto and norm preserving. Hence, the space $e_p^q(\nabla)$ is linearly isomorphic to ℓ_p. This completes the proof. □

Now, we construct Schauder basis for the sequence spaces $e_p^q(\nabla)$ and $e_\infty^q(\nabla)$.

A sequence $z = (z_k)$ of a normed space $(\mathfrak{X}, \|\cdot\|)$ is called a Schauder basis if for every $u \in \mathfrak{X}$ there exists a unique sequence of scalars (α_k) such that $\|u - \sum_{k=0}^{n} \alpha_k z_k\| \to 0$, as $n \to \infty$.

It is known that the domain $\mathfrak{X}_\Omega$ of the triangle Ω in the space $\mathfrak{X}$ has a basis if and only if $\mathfrak{X}$ has a basis. In the light of this together with Theorem 7.2.3, we present the following result:

Theorem 7.2.4. *Consider the sequence $s^{(k)}(q) = (s_n^{(k)}(q))$ for every fixed $k \in \mathbb{N}_0$ defined by*

$$s_n^{(k)}(q) = \begin{cases} \sum_{j=k}^{n} (-1)^{j-k} \dfrac{\pi^{(k)}(q) \begin{bmatrix}j\\k\end{bmatrix}_q q^{\binom{j-k}{2}}}{q^{\binom{j}{2}}}, & k \leq n, \\ 0, & k > n. \end{cases}$$

Then the set $\{s^{(0)}(q), s^{(1)}(q), s^{(2)}(q), \ldots\}$ forms the basis for the space $e_p^q(\nabla)$ and every $z \in e_p^q(\nabla)$ has a unique representation of the form $z = \sum_{k=0}^{\infty} t_k s^{(k)}(q)$.

7.3 Alpha-, beta-, and gamma-duals

In this section, we compute the α-, β-, and γ-duals of the sequence spaces $e_p^q(\nabla)$ and $e_\infty^q(\nabla)$. We omit the proof for the case $p = 1$, since the proof is analogous with the case $p > 1$.

For the sequence spaces $\mathfrak{X}$ and $\mathfrak{Y}$, the set $M(\mathfrak{X}, \mathfrak{Y})$ defined by

$$M(\mathfrak{X}, \mathfrak{Y}) := \{d = (d_k) \in \omega : dz = (d_k z_k) \in \mathfrak{Y} \text{ for all } z = (z_k) \in \mathfrak{X}\}$$

is called the *multiplier space* of $\mathfrak{X}$ and $\mathfrak{Y}$. In particular, if $\mathfrak{Y}$ is ℓ_1, cs or bs, then the sets

$$\mathfrak{X}^\alpha = M(\mathfrak{X}, \ell_1), \quad \mathfrak{X}^\beta = M(\mathfrak{X}, cs) \text{ and } \mathfrak{X}^\gamma = M(\mathfrak{X}, bs)$$

are, respectively, termed as α-, β-, and γ-dual of the sequence space $\mathfrak{X}$.

The following lemmas are necessary for our examinations. Throughout $\mathcal{N}$ denotes the family of all finite subsets of $\mathbb{N}_0$.

Lemma 7.3.1. *[18] The following statements hold:*

(i) $\Omega = (\omega_{nk}) \in (\ell_p, \ell_1)$ *iff*

$$\sup_{N \in \mathcal{N}} \sum_{k=0}^{\infty} \left| \sum_{n \in N} \omega_{nk} \right|^{p^*} < \infty. \tag{7.4}$$

(ii) $\Omega = (\omega_{nk}) \in (\ell_p, c)$ *iff*

$$\exists \alpha_k \in \mathbb{C} \ni \lim_{n \to \infty} \omega_{nk} = \alpha_k \text{ for each } k \in \mathbb{N}_0, \tag{7.5}$$

$$\sup_{n \in \mathbb{N}_0} \sum_{k=0}^{\infty} |\omega_{nk}|^{p^*} < \infty. \tag{7.6}$$

(iii) $\Omega = (\omega_{nk}) \in (\ell_p, \ell_\infty)$ *iff (7.6) holds.*

(iv) $\Omega = (\omega_{nk}) \in (\ell_\infty, \ell_1)$ *iff (7.4) holds with $p^* = 1$.*

(v) $\Omega = (\omega_{nk}) \in (\ell_\infty, c)$ iff (7.5) holds and

$$\lim_{n\to\infty} \sum_{k=0}^{\infty} |\omega_{nk}| = \sum_{k=0}^{\infty} \left|\lim_{n\to\infty} \omega_{nk}\right|. \qquad (7.7)$$

(vi) $\Omega = (\omega_{nk}) \in (\ell_\infty, \ell_\infty)$ iff (7.6) holds with $p^* = 1$.

Theorem 7.3.2. Define the sets $\alpha_{p^*}(q)$ and $\alpha_\infty(q)$ by

$$\alpha_{p^*}(q) = \left\{ (a_k) \in w : \sup_{N \in \mathcal{N}} \sum_{k=0}^{\infty} \left| \sum_{n \in N} \sum_{j=k}^{n} (-1)^{j-k} \frac{\pi^{(k)}(q) \begin{bmatrix} j \\ k \end{bmatrix}_q q^{\binom{j-k}{2}}}{q^{\binom{j}{2}}} a_n \right|^{p^*} < \infty \right\}$$

and

$$\alpha_\infty(q) = \left\{ (a_k) \in w : \sup_{k \in \mathbb{N}_0} \sum_{n=0}^{\infty} \left| (-1)^{n-k} \frac{\pi^{(k)} \begin{bmatrix} n \\ k \end{bmatrix}_q q^{\binom{n-k}{2}}}{q^{\binom{n}{2}}} a_n \right| < \infty \right\}.$$

Then $[e_1^q(\nabla)]^\alpha = \alpha_\infty(q)$, $[e_p^q(\nabla)]^\alpha = \alpha_{p^*}(q)$ and $[e_\infty^q(\nabla)]^\alpha = \alpha_1(q)$.

Proof. Define the matrix $A(q) = (a_{nk}^q)$ by

$$a_{nk}^q = \begin{cases} \sum_{j=k}^{n} (-1)^{j-k} \dfrac{\pi^{(k)}(q) \begin{bmatrix} j \\ k \end{bmatrix}_q q^{\binom{j-k}{2}}}{q^{\binom{j}{2}}} a_n & (0 \leq k \leq n), \\ 0 & (k > n). \end{cases}$$

Then, we have

$$a_n z_n = \sum_{k=0}^{n} \sum_{j=k}^{n} (-1)^{j-k} \frac{\pi^{(k)}(q) \begin{bmatrix} j \\ k \end{bmatrix}_q q^{\binom{j-k}{2}}}{q^{\binom{j}{2}}} a_n t_k = (A(q)t)_n \qquad (7.8)$$

for all $n \in \mathbb{N}_0$. Thus, we deduce that $az = (a_n z_n) \in \ell_1$ whenever $z \in e_p^q(\nabla)$ (or $z \in e_\infty^q(\nabla)$) if and only if $A(q)t \in \ell_1$ whenever $t \in \ell_p$ (or $t \in \ell_\infty$). Thus, we obtain that $a = (a_n) \in [e_p^q(\nabla)]^\alpha$ (or $a = (a_n) \in [e_\infty^q(\nabla)]^\alpha$) if and only the matrix $A(q) \in (\ell_p, \ell_1)$ (or $A(q) \in (\ell_\infty, \ell_1)$). Thus, we conclude by applying Part (i) and Part (iv) of Lemma 7.3.1 that

$$[e_p^q(\nabla)]^\alpha = \alpha_{p^*}(q) \text{ and } [e_\infty^q(\nabla)]^\alpha = \alpha_\infty(q).$$

This completes the proof. □

Theorem 7.3.3. *Define the sets* $\beta_1(q)$, $\beta_2(q)$, $\beta_3(q)$, *and* $\beta_{p^*}(q)$ *by*

$$\beta_1(q) = \left\{ (a_k) \in w : \lim_{n \to \infty} \sum_{i=k}^{n} \sum_{j=k}^{n} (-1)^{j-k} \frac{\pi^{(k)}(q) \left[\begin{smallmatrix} j \\ k \end{smallmatrix} \right]_q q^{\binom{j-k}{2}}}{q^{\binom{j}{2}}} a_i \right.$$

$$\left. \text{exists for each } k \in \mathbb{N}_0 \right\},$$

$$\beta_2(q) = \left\{ (a_k) \in w : \sup_{n,k \in \mathbb{N}_0} \left| \sum_{i=k}^{n} \sum_{j=k}^{n} (-1)^{j-k} \frac{\pi^{(k)}(q) \left[\begin{smallmatrix} j \\ k \end{smallmatrix} \right]_q q^{\binom{j-k}{2}}}{q^{\binom{j}{2}}} a_i \right| < \infty \right\},$$

$$\beta_3(q) = \left\{ (a_k) \in w : \lim_{n \to \infty} \sum_{k=0}^{\infty} \left| \sum_{i=k}^{n} \sum_{j=k}^{n} (-1)^{j-k} \frac{\pi^{(k)}(q) \left[\begin{smallmatrix} j \\ k \end{smallmatrix} \right]_q q^{\binom{j-k}{2}}}{q^{\binom{j}{2}}} a_i \right| = \right.$$

$$\left. \sum_{k=0}^{\infty} \left| \lim_{n \to \infty} \sum_{i=k}^{n} \sum_{j=k}^{n} (-1)^{j-k} \frac{\pi^{(k)}(q) \left[\begin{smallmatrix} j \\ k \end{smallmatrix} \right]_q q^{\binom{j-k}{2}}}{q^{\binom{j}{2}}} a_i \right| \right\},$$

$$\beta_r(q) = \left\{ (a_k) \in w : \sup_{n \in \mathbb{N}_0} \sum_{k=0}^{n} \left| \sum_{i=k}^{n} \sum_{j=k}^{n} (-1)^{j-k} \frac{\pi^{(k)}(q) \left[\begin{smallmatrix} j \\ k \end{smallmatrix} \right]_q q^{\binom{j-k}{2}}}{q^{\binom{j}{2}}} a_i \right|^{p^*} < \infty \right\}.$$

Then $[e_1^q(\nabla)]^\beta = \beta_1(q) \cap \beta_2(q)$, $[e_p^q(\nabla)]^\beta = \beta_1(q) \cap \beta_{p^*}(q)$, *and* $[e_\infty^q(\nabla)]^\beta = \beta_1(q) \cap \beta_3(q)$.

Proof. Consider the following equality

$$\sum_{k=0}^{n} a_k z_k = \sum_{k=0}^{n} \left\{ \sum_{j=0}^{k} \sum_{i=j}^{k} (-1)^{i-j} \frac{\pi^{(j)}(q) \left[\begin{smallmatrix} i \\ j \end{smallmatrix} \right]_q q^{\binom{i-j}{2}}}{q^{\binom{i}{2}}} t_j \right\} a_k$$

$$= \sum_{k=0}^{n} \left\{ \sum_{i=k}^{n} \sum_{j=k}^{n} (-1)^{j-k} \frac{\pi^{(k)}(q) \left[\begin{smallmatrix} j \\ k \end{smallmatrix} \right]_q q^{\binom{j-k}{2}}}{q^{\binom{j}{2}}} a_i \right\} t_k = (B(q)t)_n \quad (7.9)$$

for each $n \in \mathbb{N}_0$, where the matrix $B(q) = (b_{nk}^q)$ is defined by

$$b_{nk}^q = \begin{cases} \sum_{i=k}^{n} \sum_{j=k}^{n} (-1)^{j-k} \frac{\pi^{(k)}(q) \left[\begin{smallmatrix} j \\ k \end{smallmatrix} \right]_q q^{\binom{j-k}{2}}}{q^{\binom{j}{2}}} a_i & (0 \le k \le n), \\ 0 & (k > n), \end{cases}$$

for all $n, k \in \mathbb{N}_0$. Thus, in view of (7.9), we observe that $az = (a_n z_n) \in cs$ whenever $z = (z_n) \in e_p^q(\nabla)$ (or $z = (z_n) \in e_\infty^q(\nabla)$) if and only if $B(q)t \in c$ whenever $t = (t_k) \in \ell_p$ (or $t = (t_k) \in \ell_\infty$). This yields that $a = (a_n) \in [e_p^q(\nabla)]^\beta$ (or $a = (a_n) \in [e_\infty^q(\nabla)]^\beta$) if and only $B(q) \in (\ell_p, c)$ (or $B(q) \in (\ell_\infty, c)$). By using Part (ii) and Part (v) of Lemma 7.3.1, we conclude that

$$[e_p^q(\nabla)]^\beta = \beta_1(q) \cap \beta_{p^*}(q) \text{ and } [e_\infty^q(\nabla)]^\beta = \beta_1(q) \cap \beta_3(q).$$

This completes the proof. □

Theorem 7.3.4. *Define the set* $\gamma_1(q)$ *by*

$$\gamma_1(q) = \left\{ a = (a_k) \in w : \sup_{n \in \mathbb{N}_0} \sum_{k=0}^{n} \left| \sum_{i=k}^{n} \sum_{j=k}^{n} (-1)^{j-k} \frac{\pi^{(k)}(q) \begin{bmatrix} j \\ k \end{bmatrix}_q q^{\binom{j-k}{2}}}{q^{\binom{j}{2}}} a_i \right| < \infty \right\}.$$

Then $[e_1^q]^\gamma = \beta_2(q)$, $[e_p^q]^\gamma = \beta_{p^*}(q)$, *and* $[e_\infty^q]^\gamma = \gamma_1(q)$.

Proof. This is similar to the proof of the previous theorem except that Parts (iii) and (vi) of Lemma 7.3.1 is utilized instead of Parts (ii) and (v) of Lemma 7.3.1. □

7.4 Matrix transformations

In the present section, we characterize matrix mappings from $e_p^q(\nabla)$ and $e_\infty^q(\nabla)$ to any one of the space ℓ_∞, c, c_0, or ℓ_1. The following theorem is fundamental in our investigation.

Theorem 7.4.1. *Let* $\mathfrak{Y} \subset w$ *and* $1 \leq p \leq \infty$. *Then* $\Omega = (\omega_{nk}) \in (e_p^q(\nabla), \mathfrak{Y})$ *iff* $\Lambda^{(n)} = (\lambda_{mk}^{(n)}) \in (\ell_p, c)$ *for each* $n \in \mathbb{N}_0$, *and* $\Lambda = (\lambda_{nk}) \in (\ell_p, \mathfrak{Y})$, *where*

$$\lambda_{mk}^{(n)} = \begin{cases} 0 & (k > m), \\ \sum_{i=k}^{m} \sum_{j=k}^{m} (-1)^{j-k} \frac{\pi^{(k)}(q) \begin{bmatrix} j \\ k \end{bmatrix}_q q^{\binom{j-k}{2}}}{q^{\binom{j}{2}}} \omega_{ni} & (0 \leq k \leq m), \end{cases} \quad (7.10)$$

$$\lambda_{nk} = \sum_{i=k}^{\infty} \sum_{j=k}^{\infty} (-1)^{j-k} \frac{\pi^{(k)}(q) \begin{bmatrix} j \\ k \end{bmatrix}_q q^{\binom{j-k}{2}}}{q^{\binom{j}{2}}} \omega_{ni} \quad (7.11)$$

for all $n, k \in \mathbb{N}_0$.

Proof. This is similar to the proof of Theorem 4.1 of [12], and hence details omitted. □

Now, using the results presented in the Stieglitz and Tietz [18] together with Theorem 7.4.1, we obtain the following results:

Corollary 7.4.2. *The following statements hold:*

1. $\Omega \in (e_1^q(\nabla), \ell_\infty)$ *iff*

$$\sup_{m,k \in \mathbb{N}_0} \left| \lambda_{mk}^{(n)} \right| < \infty, \quad (7.12)$$

$$\lim_{m \to \infty} \lambda_{mk}^{(n)} \text{ exists for all } k \in \mathbb{N}_0, \quad (7.13)$$

$$\sup_{n,k \in \mathbb{N}_0} |\lambda_{nk}| < \infty. \quad (7.14)$$

2. $\Omega \in (e_1^q(\nabla), c)$ iff (7.12) and (7.13) hold, and (7.14) and

$$\lim_{n \to \infty} \lambda_{nk} \text{ exists for all } k \in \mathbb{N}_0, \tag{7.15}$$

also hold.

3. $\Omega \in (e_1^q(\nabla), c_0)$ iff (7.12) and (7.13) hold, and (7.14) and

$$\lim_{n \to \infty} \lambda_{nk} = 0 \text{ for all } k \in \mathbb{N}_0, \tag{7.16}$$

also hold.

4. $\Omega \in (e_1^q(\nabla), \ell_1)$ iff (7.12) and (7.13) hold, and

$$\sup_{k \in \mathbb{N}_0} \sum_{n=0}^{\infty} |\lambda_{nk}| < \infty. \tag{7.17}$$

Corollary 7.4.3. *The following statements hold:*

1. $\Omega \in (e_p^q(\nabla), \ell_\infty)$ iff (7.13) holds, and

$$\sup_{m \in \mathbb{N}_0} \sum_{k=0}^{\infty} \left|\lambda_{mk}^{(n)}\right|^{p^*} < \infty, \tag{7.18}$$

$$\sup_{n \in \mathbb{N}_0} \sum_{k=0}^{\infty} |\lambda_{nk}|^{p^*} < \infty. \tag{7.19}$$

2. $\Omega \in (e_p^q(\nabla), c)$ iff (7.13) and (7.18) hold, and (7.15) and (7.19) also hold.

3. $\Omega \in (e_p^q(\nabla), c_0)$ iff (7.13) and (7.18) hold, and (7.16) and (7.19) also hold.

4. $\Omega \in (e_p^q(\nabla), \ell_1)$ iff (7.13) and (7.18) hold, and

$$\sup_{N \in \mathcal{N}} \sum_{k=0}^{\infty} \left|\sum_{n \in \mathbb{N}_0} \lambda_{nk}\right|^{p^*} < \infty. \tag{7.20}$$

Corollary 7.4.4. *The following statements hold:*

1. $\Omega \in (e_\infty^q(\nabla), \ell_\infty)$ iff (7.13) holds, and

$$\lim_{m \to \infty} \sum_{k=0}^{\infty} \left|\lambda_{mk}^{(n)}\right| = \sum_{k=0}^{\infty} \left|\lim_{m \to \infty} \lambda_{mk}^{(n)}\right|, \tag{7.21}$$

$$\sup_{n \in \mathbb{N}_0} \sum_{k=0}^{\infty} |\lambda_{nk}| < \infty. \tag{7.22}$$

2. $\Omega \in (e_\infty^q(\nabla), c)$ iff (7.13) and (7.21) hold, and (7.15) and

$$\lim_{n\to\infty} \sum_{k=0}^{\infty} |\lambda_{nk}| = \sum_{k=0}^{\infty} \left|\lim_{n\to\infty} \lambda_{nk}\right|, \qquad (7.23)$$

also hold.

3. $\Omega \in (e_\infty^q(\nabla), c_0)$ iff (7.13) and (7.21) hold, and

$$\lim_{n\to\infty} \sum_{k=0}^{\infty} |\lambda_{nk}| = 0. \qquad (7.24)$$

4. $\Omega \in (e_\infty^q(\nabla), \ell_1)$ iff (7.13) and (7.21) hold, and (7.20) holds with $p^* = 1$.

The following lemma is due to Başar and Altay [4]:

Lemma 7.4.5. *[4] Let $\mathfrak{X}$ and $\mathfrak{Y}$ be any two sequence spaces, Ω be an infinite matrix and Φ be a triangle. Then, $\Omega \in (\mathfrak{X}, \mathfrak{Y}_\Phi)$ if and only if $\Phi\Omega \in (\mathfrak{X}, \mathfrak{Y})$.*

We utilize this lemma together with Corollaries 7.4.2, 7.4.3, and 7.4.4 to give the characterizations of the following classes of matrix mappings:

Corollary 7.4.6. *Let $\Omega = (\omega_{nk})$ be an infinite matrix and define the matrix $C^q = (c_{nk}^q)$ by*

$$c_{nk}^q = \sum_{m=0}^{n} \frac{q^{m-1}}{[n+1]_q} \phi_{mk}, \ (q \in (0,1))$$

for all $n, k \in \mathbb{N}$, where $[n]_q$ is the q-analog of $n \in \mathbb{N}_0$. Then, the necessary and sufficient conditions that $\Omega \in (\mathfrak{X}, \mathfrak{Y})$, where $\mathfrak{X} \in \{e_1^q(\nabla), e_p^q(\nabla), e_\infty^q(\nabla)\}$ and $\mathfrak{Y} \in \{X_0^q, X_c^q, X_\infty^q\}$, are determined from the respective ones in Corollaries 7.4.2, 7.4.3, and 7.4.4 by replacing the elements of the matrix Ω by those of the matrix C^q, where X_0^q, X_c^q, and X_∞^q are q-Cesàro sequence spaces defined by Demiriz and Şahin [9] and Yaying et al. [24].

Corollary 7.4.7. *Let $\Omega = (\omega_{nk})$ be an infinite matrix and define the matrix $\tilde{C} = (\tilde{C}_{nk})$ by*

$$\tilde{C}_{nk} = \sum_{m=0}^{n} q^m \frac{C_m(q) C_{n-m}(q)}{C_{n+1}(q)} \phi_{mk}, \ (n,k \in \mathbb{N}_0),$$

where $(C_n(q))$ is a sequence of q-Catalan numbers. Then, the necessary and sufficient conditions that $\Omega \in (\mathfrak{X}, \mathfrak{Y})$, where $\mathfrak{X} \in \{e_1^q(\nabla), e_p^q(\nabla), e_\infty^q(\nabla)\}$ and $\mathfrak{Y} \in \{c_0(\tilde{C}), c(\tilde{C})\}$, are determined from the respective ones in Corollaries 7.4.2, 7.4.3, and 7.4.4, by replacing the elements of the matrix Ω by those of matrix $\tilde{C}$, where $c(\tilde{C})$ and $c_0(\tilde{C})$ are q-Catalan sequence spaces defined by Yaying et al. [27].

Corollary 7.4.8. *Let* $\Omega = (\omega_{nk})$ *be an infinite matrix and define the matrix* $T = (t_{nk})$ *by*

$$t_{nk} = \sum_{m=0}^{n} \frac{2t_m}{t_n t_{n+1}} \phi_{mk}, \ (n, k \in \mathbb{N}_0)$$

where (t_n) *is a sequence of tribonacci numbers. Then, the necessary and sufficient conditions that* $\Omega \in (\mathfrak{X}, \mathfrak{Y})$, *where* $\mathfrak{X} \in \{e_1^q(\nabla), e_p^q(\nabla), e_\infty^q(\nabla)\}$ *and* $\mathfrak{Y} \in \{\ell_\infty(T), c(T), c_0(T)\}$, *are determined from the respective ones in Corollaries 7.4.2, 7.4.3, and 7.4.4 by replacing the elements of the matrix* Ω *by those of matrix* T, *where* $\ell_\infty(T)$, $c(T)$ *and* $c_0(T)$ *are tribonacci sequence spaces defined by Yaying et al. [21, 26].*

Bibliography

[1] Altay, B. and Başar, F. 2005. On some Euler sequence spaces of non-absolute type, *Ukrainian Math. J.* 57: 1–17.

[2] Altay, B., Başar, F. and Mursaleen, M. 2006. On the Euler sequence spaces which include the spaces ℓ_p and ℓ_∞ I, *Inf. Sci.* 176: 1450–1462.

[3] Altay, B. and Polat, H. 2006. On some new Euler difference sequence spaces, *Southeast Asian Bull. Math.* 30: 209–220.

[4] Başar, F. and Altay, B. 2003. On the spaces of p-bounded variation and related matrix mappings, *Ukrainian Math. J.* 55:1: 136–147.

[5] Braha, N.L. and Başar, F. 2016. Euler-Cesàro difference spaces of bounded, convergent and null sequences, *Tamkang J. Math.* 47:4: 405–420.

[6] Bişgin, M.C. 2016. The binomial sequence spaces of nonabsolute type, *J. Inequal. Appl.* 2016: 309.

[7] Bişgin, M.C. 2016. The binomial sequence spaces which include the spaces ℓ_p and ℓ_∞ and geometric properties, *J. Inequal. Appl.* 2016: 304.

[8] Ellidokuzoğlu, H.B. and Demiriz, S. 2017. Euler-Riesz difference sequence spaces, *Türk. J. Math. Comput. Sci.* 7: 63–72.

[9] Demiriz, S. and Şahin, A. 2016. q-Cesàro sequence spaces derived by q-analogue. *Adv. Math.* 5:2: 97–110.

[10] Kac, V. and Cheung, P. 2002. *Quantum Calculus*, Springer, New York.

[11] Kara, E.E. and Başarır, M. 2011. On compact operators and some Euler $B^{(m)}$-difference sequence spaces, *J. Math. Anal. Appl.* 379: 499–511.

[12] Kirişçi, M. and Başar, F. 2010. Some new sequence spaces derived by the domain of generalized difference matrix, *Comput. Math. Appl.* 60: 1299–1309.

[13] Kızmaz, H. 1981. On certain sequence spaces, *Canad. Math. Bull.* 24:2: 169–176.

[14] Meng, J. and Song, M. 2017. On some binomial difference sequence spaces, *Kyungpook Math. J.* 57: 631–640.

[15] Meng, J. and Song, M. 2017. On some binomial $B^{(m)}$-differnce sequence spaces, *J. Inequal. Appl.* 2017: 194.

[16] Mursaleen, M., Başar, F. and Altay, B. 2006. On the Euler sequence spaces which include the spaces ℓ_p and ℓ_∞ II, *Nonlinear. Anal.* 65: 707–717.

[17] Polat, H. and Başar, F. 2007. Some Euler spaces of difference sequences of order m, *Acta Math. Sci. B Engl. Ed.* 27:2: 254–266.

[18] Stieglitz, M. and Tietz, H. 1977. Matrixtransformationen von Folgenräumen eine Ergebnisübersicht. *Math. Z.* 154: 1–16.

[19] Song, M. and Meng, J. 2017. Some normed binomial difference sequence spaces related to the ℓ_p space, *J. Inequal. Appl.* 2017: 128.

[20] Yaying, T. 2021. On the domain of q-Euler matrix in c_0 and c, In: O. Chadli, S. Das, R.N. Mohapatra, A. Swaminathan,*Mathematical Analysis and Applications*, Springer Proceedings in Mathematics and Statistics, Vol. 381, Springer Singapore.

[21] Yaying, T. and Hazarika, T. 2020. On sequence spaces defined by the domain of a regular Tribonacci matrix, *Math. Slovaca*, 70:3: 697–706.

[22] Yaying, T., Hazarika, B. and Başar, F. 2022. On some new sequence spaces defined by q-Pascal matrix, *Trans. A. Radmadze Math. Soc.* 176:1: 99–113.

[23] Yaying, T., Hazarika, B. and Mohiuddine, S.A. 2022. Domain of Padovan q-difference matrix in sequence spaces ℓ_p and ℓ_∞, *Filomat* 36:3: 905–919.

[24] Yaying, T., Hazarika, B. and Mursaleen, M. 2021. On sequence space derived by the domain of q-Cesàro matrix in ℓ_p space and the associated operator ideal, *J. Math. Anal. Appl.* 493:1: 124453.

[25] Yaying, T., Hazarika, B. and Mursaleen, M. 2021. On generalized (p,q)-Euler matrix and associated sequence spaces, *J. Function Spaces* 2021: 8899960.

[26] Yaying, T. and Kara, M.İ. 2021. On sequence spaces defined by the domain of tribonacci matrix in c_0 and c, *The Korean J. Math.* 29:1: 25–40.

[27] Yaying, T., Kara, M.İ., Hazarika, B. and Kara, E.E.: A study on q-analogue of Catalan sequence spaces, *Filomat*, accepted.

Chapter 8

Approximation by the double sequence of LPO based on multivariable q-Lagrange polynomials

Behar Baxhaku

P. N. Agrawal

8.1	Introduction	129
8.2	Double sequence of $\mathfrak{K}_{n,q}^{\beta^{(1)},\cdots,\beta^{(r)}}(.)(x)$	131
8.3	Approximation by using power series summability method (p.s.s.m)	132
	8.3.1 Illustrative example	136
8.4	$\mathcal{A}$-statistical convergence of operators $\mathfrak{K}_{n_1,q_{n_1}}^{n_2,q_{n_2}}(.)(\mathbf{x})$	139
	8.4.1 Application of Theorem 8.4.4	143
8.5	$\mathcal{A}$-statistical convergence by GBS operators	146
	Bibliography	151

8.1 Introduction

For $\mathcal{I} = [0, 1]$, let $\mathcal{C}(\mathcal{I})$ be the space of all continuous functions on $\mathcal{I}$, with the sup-norm $||.||$. With the help of the multivariate Lagrange polynomials proposed by Chan et al. [13] for $f \in \mathfrak{C}(\mathcal{I})$, Erkus et al. [20] defined the following sequence of LPOs:

$$\mathfrak{L}_n^{\Phi^{(1)},\cdots,\Phi^{(r)}}(f(s))(x)$$
$$= \left\{ \prod_{k=1}^{r}\left(1 - x\Phi_n^{(k)}\right)^n \right\} \sum_{p=0}^{\infty}\left\{ \sum_{l_1+l_2+\cdots+l_r=p} f\left(\frac{l_r}{n+l_r-1}\right) \cdot \prod_{s=1}^{r}\left(\Phi_n^{(s)}\right)^{l_s} \frac{(n)_{l_s}}{l_s!} \right\} x^p, \qquad (8.1)$$

DOI: 10.1201/9781003330868-8

where $x \in \mathcal{I}$, and $\Phi^{(j)} = \langle \Phi_n^{(j)} \rangle$ is a sequence in $(0,1)$ for each $j = 1, 2, \cdots r$. Altin et al. [3] proposed the q-multivariable Lagrange polynomials $h_{n,q}^{(\eta_1, \cdots \eta_r)}(z_1, z_2, \cdots, z_r)$ as follows:

$$h_{n,q}^{(\eta_1, \cdots \eta_r)}(z_1, z_2, \cdots, z_r) = \sum_{l_1+l_1+\cdots+l_r=n} \prod_{k=1}^{r} (q^{\eta_k}, q)_{l_k} \frac{(z_k)^{l_k}}{(q,q)_{l_k}}, \qquad (8.2)$$

where the q-Pochhammer symbol or the q-shifted factorial $(\kappa;q)_n$ is given by

$$(\kappa;q)_n = \begin{cases} 1, & \text{if } n = 0, \\ (1-\kappa)(1-\kappa q)\ldots(1-\kappa q^{n-1}), & \text{if } n \in \mathbb{N}. \end{cases}$$

The generating function (g.f.) for the above q-multivariate Lagrange polynomials has the following form

$$\frac{1}{(tz_k;q)_{\eta_k}} = \sum_{n=0}^{\infty} h_{n,q}^{(\eta_1, \cdots, \eta_r)}(z_1, z_2, \cdots, z_r) t^n, \qquad (8.3)$$

where $|t| < \min\{|z_1|^{-1}, \cdots, |z_r|^{-1}\}$.
With the help the g.f. given by (8.3), Erkuş-Duman [18] defined the following q-analogue of the operator $\mathfrak{L}_n^{\Phi^{(1)}, \cdots \Phi^{(r)}}$

$$\mathfrak{S}_{n,q}^{\Phi^{(1)}, \cdots, \Phi^{(r)}}(f)(x) = \left\{ \prod_{k=1}^{r} (x\phi_n^{(k)}; q)_n \right\} \sum_{p=0}^{\infty} \left\{ \sum_{l_1+l_2+\cdots+l_r=p} (q^n;q)_{l_1}(q^n;q)_{l_2} \right.$$

$$\left. \ldots (q^n;q)_{l_r} \frac{(\Phi_n^{(1)})^{l_1}(\Phi_n^{(2)})^{l_2}\ldots(\Phi_n^{(r)})^{l_r}}{(q;q)_{l_1}(q;q)_{l_2}\ldots(q;q)_{l_r}} f\left(\frac{[l_r]_q}{[n+l_r-1]_q}\right) \right\} x^p. \qquad (8.4)$$

We observe that the q-analogue of the operator (8.1) was also independently proposed by Mursaleen et al. [24]. Baxhaku et al. [9] considered the bivariate case of the operator (8.4) and studied the rate of convergence by means of the modulus of continuity and the Peetre's K-functional. The authors [9] also defined the associated GBS (Generalized Boolean Sum) operator and determined the degree of approximation with the aid of the mixed modulus of smoothness. For some recent works related to the GBS of LPO's (Linear Positive Operators), we refer to [2, 22, 23].

The aim of this chapter is to study the Korovkin type approximation theorems for the double sequence of the operators (8.4) by means of the power series summability method (pssm), which is a sequence to function transformation and includes the Abel and Borel methods. We also discuss the approximation properties of the operators (8.4) and their GBS counterparts, by using a sequence to sequence transformation, namely the $\mathcal{A}$-statistical convergence.

8.2 Double sequence of $\mathfrak{K}_{n,q}^{\beta^{(1)},\cdots,\beta^{(r)}}(.)(x)$

Let $\eta^{(i)} = \langle \eta_{n_1}^{(i)} \rangle_{n_1 \in \mathbb{N}}$, $(i = 1, \cdots r_1)$ and $\zeta^{(j)} = \langle \zeta_{n_2}^{(j)} \rangle_{n_2 \in \mathbb{N}}$, $(j = 1, \cdots r_2)$ be sequences of real numbers such that $0 < \eta_{n_1}^{(i)} < 1$, $0 < \zeta_{n_2}^{(j)} < 1$; ($i = 1, \cdots r_1$, $j = 1, \cdots r_2$, $r_1, r_2 \in \mathbb{N}$). Further, let us assume that $\langle q_{n_i} \rangle$ is a sequence such that $0 < q_{n_i} < 1$ satisfying $q_{n_i} \to 1$ and $q_{n_i}^{n_i} \to a_i$;, $0 \le a_i < 1$ as $n_i \to \infty$, for $i = 1, 2$.

For $f \in \mathcal{C}(\mathcal{I}^2)$, the space of continuous functions on $\mathcal{I}^2 = \mathcal{I} \times \mathcal{I}$ endowed with the norm $\|f\|_{\mathcal{C}((\mathcal{I}^2)} = \sup_{\mathbf{x} \in \mathcal{I}^2} |f(\mathbf{x})|$, the double sequence of (8.4) is given by:

$$\mathfrak{K}_{n_1,q_1}^{n_2,q_{n_2}}(f)(\mathbf{x}) = \left\{ \prod_{k_1=1}^{r_1} \prod_{k_2=1}^{r_2} (x_1 \eta_{n_1}^{(k_1)}; q_{n_1})_{n_1} (x_2 \zeta_{n_2}^{(k_2)}; q_{n_2})_{n_2} \right\}$$

$$\sum_{p_1,p_2=0}^{\infty} \left\{ \sum_{l_1+l_2+\cdots+l_{r_1}=p_1} \sum_{l_1^*+l_2^*+\cdots+l_{r_2}^*=p_2} \prod_{s_1=1}^{r_1} \prod_{s_2=1}^{r_2} (q_{n_1}^{n_1}; q_{n_1})_{l_{s_1}} (q_{n_2}^{n_2}; q_{n_2})_{l_{s_2}^*} \right.$$

$$\left. \frac{(\eta_{n_1}^{(s_1)})^{l_{s_1}}}{(q_{n_1}; q_{n_1})_{l_{s_1}}} \frac{(\zeta_{n_2}^{(s_2)})^{l_{s_2}^*}}{(q_{n_2}; q_{n_2})_{l_{s_2}^*}} f\left(\frac{[l_{r_1}]_{q_{n_1}}}{[n_1 + l_{r_1} - 1]_{q_{n_1}}}, \frac{[l_{r_2}^*]_{q_{n_2}}}{[n_2 + l_{r_2}^* - 1]_{q_{n_2}}} \right) x_1^{p_1} x_2^{p_2}, \quad (8.5)$$

where $\mathbf{x} = (x_1, x_2) \in \mathcal{I}^2$. For the simplicity, we adopt the following notations. Let

$$e_i^j(\mathbf{t}) = t_1^i t_2^j, \quad i, j \in \{0, 1, 2\}.$$

We denote:

$$\varkappa_{n_1,q_{n_1},r}^{\eta^{(1)},\eta^{(2)}\cdots,\eta^{(r_1)}}(x_1) = \mathfrak{K}_{n_1,q_{n_1}}^{n_2,q_{n_2}}((t_1 - x_1)^r)(\mathbf{x})$$

$$\varkappa_{n_2,q_{n_2},r}^{\zeta^{(1)},\zeta^{(2)}\cdots,\zeta^{(r_2)}}(x_2) = \mathfrak{K}_{n_1,q_{n_1}}^{n_2,q_{n_2}}((t_2 - x_2)^r)(\mathbf{x}),$$

and their supremum over $\mathcal{I}$ by $\Upsilon_{n_1,q_{n_1},r}^{\eta^{(1)},\eta^{(2)},\cdots,\eta^{(r_1)}}$ and $\Psi_{n_2,q_{n_2},r}^{\zeta^{(1)},\zeta^{(2)},\cdots,\zeta^{(r_2)}}$, respectively.

Lemma 8.2.1. *[9] The operators* $\mathfrak{K}_{n_1,q_{n_1},\eta^{(1)},\eta^{(2)},\cdots,\eta^{(r_1)}}^{n_2,q_{n_2},\zeta^{(1)},\zeta^{(2)},\cdots,\zeta^{(r_2)}}(.)(\mathbf{x})$ *satisfy the following*

(i) $\mathfrak{K}_{n_1,q_{n_1}}^{n_2,q_{n_2}}(1)(\mathbf{x}) = 1$;

(ii) $\mathfrak{K}_{n_1,q_{n_1}}^{n_2,q_{n_2}}(t_1)(\mathbf{x}) = x_1 \eta_{n_1}^{(r_1)}$;

(iii) $\mathfrak{K}_{n_1,q_{n_1}}^{n_2,q_{n_2}}(t_2)(\mathbf{x}) = x_2\zeta_{n_2}^{(r_2)};$

(iv) $\mathfrak{K}_{n_1,q_{n_1}}^{n_2,q_{n_2}}(t_1^2)(\mathbf{x}) \leq q_{n_1}(x_1\eta_{n_1}^{(r_1)})^2 + \dfrac{x_1\eta_{n_1}^{(r_1)}}{[n_1]_{q_{n_1}}}.$ Moreover,

$$|\mathfrak{K}_{n_1,q_{n_1}}^{n_2,q_{n_2}}(t_1^2)(\mathbf{x}) - x_1^2| \leq 2x_1^2(1 - \eta_{n_1}^{(r_1)}) + \dfrac{x_1\eta_{n_1}^{(r_1)}}{[n_1]_{q_{n_1}}}.$$

(v) $\mathfrak{K}_{n_1,q_{n_1}}^{n_2,q_{n_2}}(t_2^2)(\mathbf{x}) \leq q_{n_2}(x_2\zeta_{n_2}^{(r_2)})^2 + \dfrac{x_2\zeta_{n_2}^{(r_2)}}{[n_2]_{q_{n_2}}}.$ Moreover,

$$|\mathfrak{K}_{n_1,q_{n_1}}^{n_2,q_{n_2}}(t_2^2)(\mathbf{x}) - x_2^2| \leq 2x_2^2(1 - \zeta_{n_2}^{(r_2)}) + \dfrac{x_2\zeta_{n_2}^{(r_2)}}{[n_2]_{q_{n_2}}}.$$

8.3 Approximation by using power series summability method (p.s.s.m)

Assume that (ξ_i) be a sequence such that $\xi_1 > 0$ and $\xi_i \geq 0$, $\forall i = 2, 3, ...$ Let the radius of convergence r of the power series $\xi(\varpi) = \sum\limits_{i=1}^{\infty} \xi_i\varpi^{i-1}$ satisfy $0 < r \leq \infty$, we say that the sequence $\eta = (\eta_i)$ is convergent to l in the sense of p.s.s.m if $\lim\limits_{\varpi \to r-} \sum\limits_{i=1}^{\infty} \xi_i\eta_i\varpi^{i-1} = l$. The power series summability method (p.s.s.m) is said to be regular provided

$$\lim_{\varpi \to r-} \dfrac{\xi_i\varpi^{i-1}}{\xi(\varpi)} = 0, \ \forall i \in \mathbb{N}.$$

Several researchers have contributed to this area (cf. [7, 12, 25] and [28], etc.) Let $\langle \xi_{jk} \rangle$ be a double sequence of non-negative real numbers such that $\xi_{11} > 0$, and the power series $\xi(\varpi_1, \varpi_2) := \sum\limits_{j,k=1}^{\infty} \xi_{jk}\varpi_1^{j-1}\varpi_2^{k-1}$ has a radius of convergence $\mathcal{R}$ with $0 < \mathcal{R} \leq \infty$ and $\varpi_1, \varpi_2 \in (0, \mathcal{R})$. The number sequence $x = \langle x_{jk} \rangle$ is said to be convergent to L in the sense of p.s.s.m [7,14], if for all $\varpi_1, \varpi_2 \in (0, \mathcal{R})$,

$$\lim_{\varpi_1,\varpi_2 \to \mathcal{R}-} \dfrac{1}{\xi(\varpi_1,\varpi_2)} \sum_{j,k=1}^{\infty} \xi_{jk} x_{jk} \varpi_1^{j-1}\varpi_2^{k-1} = L, \quad (8.6)$$

holds. Further, the p.s.m is said to be regular if and only if,

$$\lim_{\varpi_1,\varpi_2 \to \mathcal{R}-} \dfrac{\sum\limits_{j=1}^{\infty} \xi_{j\lambda}\varpi_1^{j-1}}{\xi(\varpi_1,\varpi_2)} = 0, \quad \lim_{u,v \to \mathcal{R}-} \dfrac{\sum\limits_{k=1}^{\infty} \xi_{\mu k}\varpi_2^{k-1}}{\xi(\varpi_1,\varpi_2)} = 0 \quad (8.7)$$

hold for any $\lambda, \mu \in \mathbb{N}$.

Example 8.3.1. Let $x = \langle x_{jk} \rangle = 0$ when k is odd and 0 otherwise. Let $\xi_{jk} = 1$, $\forall j, k \in \mathbb{N}$, $|\varpi_1| < 1$ and $|\varpi_2| < 1$. Then it is easy to observe that

$$\lim_{\varpi_1, \varpi_2 \to 1^-} \frac{\sum_{j,k=1}^{\infty} \xi_{jk} x_{jk} u^{j-1} v^{k-1}}{\xi(\varpi_1, \varpi_2)}$$

$$= \lim_{\varpi_1, \varpi_2 \to 1^-} \prod_{i=1}^{2}(1-\varpi_i) \sum_{j=1}^{\infty} \varpi_1^{j-1} \sum_{k=1}^{\infty} \varpi_2^{2k-1} = \frac{1}{2}. \quad (8.8)$$

So, $x = \langle x_{jk} \rangle$ is not convergent in the Pringsheim sense but converges to $\frac{1}{2}$ in the sense of p.s.m.

Throughout our discussion, we use the test functions defined by $g_0(\mathbf{x}) = 1$, $g_1(\mathbf{x}) = x_1$, $g_2(\mathbf{x}) = x_2$ and $g_3(\mathbf{x}) = x_1^2 + x_2^2$, and note $\Delta_{t_1, x_1}^{t_2, x_2} = \sqrt{(t_1 - x_1)^2 + (t_2 - x_2)^2}$.

Let recall the definition of the function $w_f(\delta)$. For $\mathbf{x}, \mathbf{t} \in \mathcal{I}^2$, and a given $\delta > 0$ the full modulus of continuity of $f \in \mathcal{C}(\mathcal{I}^2)$ is defined by

$$w_f(\delta) := \sup_{\Delta_{t_1,x_1}^{t_2,x_2} \leq \delta} \{|f(\mathbf{t}) - f(\mathbf{x})|\}.$$

From the above definition, for any $\delta > 0$, and for all $\mathbf{x}, \mathbf{t} \in \mathcal{I}^2$, we have

$$|f(\mathbf{t}) - f(\mathbf{x})| \leq w_f(\delta)\left(1 + \left(\Delta_{t_1,x_1}^{t_2,x_2}\right)^2 \delta^{-2}\right). \quad (8.9)$$

First, we establish a Korovkin type approximation theorem for the operators $\mathfrak{K}_{n_1,q_{n_1}}^{n_2,q_{n_2}}$ by using the p.s.s.m.

Theorem 8.3.2. Let $f \in \mathcal{C}(\mathcal{I}^2)$, then

$$\lim_{\varpi_1,\varpi_2 \to R^-} \frac{1}{\xi(\varpi_1,\varpi_2)} \sum_{n_1,n_2=1}^{\infty} \xi_{n_1 n_2 k} \varpi_1^{n_1-1} \varpi_2^{n_2-1} \|\mathfrak{K}_{n_1,q_{n_1}}^{n_2,q_{n_2}}(f) - (f)\| = 0, \quad (8.10)$$

if and only if

$$\lim_{\varpi_1,\varpi_2 \to R^-} \frac{\sum_{n_1,n_2=1}^{\infty} \xi_{n_1 n_2 k} \varpi_1^{n_1-1} \varpi_2^{n_2-1} \|\mathfrak{K}_{n_1,q_{n_1}}^{n_2,q_{n_2}}(g_i) - (g_i)\|}{\xi(\varpi_1,\varpi_2)} \quad (8.11)$$
$= 0$, for $i = 1, 2, 3$.

Proof. First, let us assume that (8.10) is true. Then (8.11) is obvious. Conversely, suppose

$$\lim_{\varpi_1,\varpi_2 \to R^-} \frac{1}{\xi(\varpi_1,\varpi_2)} \sum_{n_1,n_2=1}^{\infty} \xi_{n_1 n_2 k} \varpi_1^{n_1-1} \varpi_2^{n_2-1} \|\mathfrak{K}_{n_1,q_{n_1}}^{n_2,q_{n_2}}(g_i) - (g_i)\| = 0",$$
for $i = 1, 2, 3$,

is true. Since $f \in \mathcal{C}(\mathcal{I}^2)$, it follows that for every $\epsilon > 0$, there exists a number $\delta > 0$ such that for all $\mathbf{t}, \mathbf{x} \in \mathcal{I}^2$, we have

$$|f(\mathbf{t}) - f(\mathbf{x})| < \epsilon + \frac{2M}{\delta^2}\left\{\left(\Delta_{t_1,x_1}^{t_2,x_2}\right)^2\right\}, \tag{8.12}$$

where $M := \sup_{\mathbf{x} \in \mathcal{I}^2} |f(\mathbf{x})|$.

Applying the LPOs $\mathfrak{K}_{n_1,q_{n_1}}^{n_2,q_{n_2}}(.)(\mathbf{x})$ to (8.12), one has

$$|\mathfrak{K}_{n_1,q_{n_1}}^{n_2,q_{n_2}}(f)(\mathbf{x}) - f(\mathbf{x})| \leq \epsilon + \frac{2M}{\delta^2}\Big\{|\mathfrak{K}_{n_1,q_{n_1}}^{n_2,q_{n_2}}(t_1^2 + t_2^2)(\mathbf{x})$$

$$- (x_1^2 + x_2^2)| + 2|x_1||\mathfrak{K}_{n_1,q_{n_1}}^{n_2,q_{n_2}}(t_1)(\mathbf{x}) - x_1| + 2|x_2||\mathfrak{K}_{n_1,q_{n_1}}^{n_2,q_{n_2}}(t_2)(\mathbf{x}) - x_2|\Big\}.$$

Hence, taking supremum over $(x_1, x_2) \in \mathcal{I}^2$ and letting $\varpi_1, \varpi_2 \to \mathcal{R}^-$, we conclude that

$$\lim_{\varpi_1,\varpi_2 \to \mathcal{R}^-} \frac{\sum_{n_1,n_2=1}^{\infty} \xi_{n_1 n_2 k} \varpi_1^{n_1-1} \varpi_2^{n_2-1} \|\mathfrak{K}_{n_1,q_{n_1}}^{n_2,q_{n_2}}(f)(\mathbf{x}) - f(\mathbf{x})\|}{\xi(\varpi_1, \varpi_2)} \leq \epsilon +$$

$$\lim_{\varpi_1,\varpi_2 \to \mathcal{R}^-} \frac{4M}{\delta^2} \frac{\sum_{n_1,n_2=1}^{\infty} \xi_{n_1 n_2 k} \varpi_1^{n_1-1} \varpi_2^{n_2-1}\Big\{\Big(\|\mathfrak{K}_{n_1,q_{n_1}}^{n_2,q_{n_2}}(t_1^2 + t_2^2)(\mathbf{x})}{\xi(\varpi_1, \varpi_2)}$$

$$-(x_1^2 + x_2^2)\| + \|\mathfrak{K}_{n_1,q_{n_1}}^{n_2,q_{n_2}}(t_1)(\mathbf{x}) - x_1\| + \|\mathfrak{K}_{n_1,q_{n_1}}^{n_2,q_{n_2}}(t_2)(\mathbf{x}) - x_2\|\Big)\Big\}.$$

Now, the assertion follows easily on using the hypothesis (8.11) and the arbitrariness of ϵ. □

Remark 8.3.3. Let us assume that $f \in \mathcal{C}(\mathcal{I}^2)$ and $\lim_{n_i \to \infty} \eta_{n_i}^{(r_i)} = 1$, for $i = 1, 2$. In order to show the convergence of $\langle \mathfrak{K}_{n_1,q_{n_1}}^{n_2,q_{n_2}}(f) \rangle$ to f on $\mathcal{I}^2$, by p.s.m, it is sufficient to establish the following;

$$\lim_{\varpi_1,\varpi_2 \to \mathcal{R}^-} \frac{\sum_{n_1,n_2=1}^{\infty} \|\mathfrak{K}_{n_1,q_{n_1}}^{n_2,q_{n_2}}(g_i) - g_i\| \xi_{n_1 n_2} \varpi_1^{n_1-1} \varpi_2^{n_2-1}}{\xi(\varpi_1, \varpi_2)}$$

$$= 0, \text{ for } i = 0, 1, 2, 3. \tag{8.13}$$

Using the Lemma 8.2.1, the condition (8.13) trivially holds. Again using Lemma 8.2.1, we have

$$\lim_{\varpi_1,\varpi_2 \to \mathcal{R}^-} \frac{1}{\xi(\varpi_1, \varpi_2)} \sum_{n_1,n_2=1}^{\infty} \|\mathfrak{K}_{n_1,q_{n_1}}^{n_2,q_{n_2}}(g_1) - g_1\| \xi_{n_1 n_2} \varpi_1^{n_1-1} \varpi_2^{n_2-1}$$

$$\leq \lim_{\varpi_1,\varpi_2 \to \mathcal{R}^-} \frac{1}{\xi(\varpi_1, \varpi_2)} \sum_{n_1,n_2=1}^{\infty} (1 - \eta_{n_1}^{(r_1)}) \xi_{n_1 n_2} \varpi_2^{n_1-1} \varpi_2^{n_2-1}$$

$$= J, \quad \text{say}.$$

Let $\epsilon > 0$ be an arbitrary. Then from our hypothesis $\lim_{n_1 \to \infty} \eta_{n_1}^{(r_1)} = 1$, $\exists n_0(\epsilon)$ such that $|((1 - \eta_{n_1}^{(r_1)}))| < \frac{\epsilon}{2}$, for all $n_1 > n_0(\epsilon)$. Hence,

$$J \leq \lim_{\varpi_1, \varpi_2 \to \mathcal{R}-} \frac{1}{\xi(\varpi_1, \varpi_2)} \sum_{n_1=1}^{n_0} \sum_{n_2=1}^{\infty} (1 - \eta_{n_1}^{(r_1)}) \xi_{n_1 n_2} \varpi_1^{n_1-1} \varpi_2^{n_2-1}$$

$$+ \lim_{\varpi_1, \varpi_2 \to \mathcal{R}-} \frac{\epsilon}{2\xi(\varpi_1, \varpi_2)} \sum_{n_1=n_0+1}^{\infty} \sum_{n_2=1}^{\infty} \xi_{n_1 n_2} \varpi_1^{n_1-1} \varpi_2^{n_2-1}$$

$$< \lim_{\varpi_1, \varpi_2 \to \mathcal{R}-} \frac{1}{\xi(\varpi_1, \varpi_2)} \sum_{n_1=1}^{n_0} \left(\sum_{n_2=1}^{\infty} \xi_{n_1 n_2} v^{n_2-1} \right) (1 - \eta_{n_1}^{(r_1)}) \varpi_1^{n_1-1}$$

$$+ \lim_{\varpi_1, \varpi_2 \to \mathcal{R}-} \frac{\epsilon}{2\xi(\varpi_1, \varpi_2)} \sum_{n_1, n_2=1}^{\infty} \xi_{n_1 n_2} \varpi_1^{n_1-1} \varpi_2^{n_2-1}.$$

Since $< (1 - \eta_{n_1}^{(r_1)}) >$ is a bounded sequence, $\exists M_1 > 0$ such that $M_1 = \max_{1 \leq n_1 \leq n_0} (1 - \eta_{n_1}^{(r_1)})$ and therefore,

$$J \leq \lim_{\varpi_1, \varpi_2 \to \mathcal{R}-} \frac{M_1}{\xi(\varpi_1, \varpi_2)} \sum_{n_1=1}^{n_0} \left(\sum_{n_2=1}^{\infty} \xi_{n_1 n_2} \varpi_2^{n_2-1} \right) \varpi_1^{n_1-1} + \frac{\epsilon}{2}.$$

In view of the regularity conditions given by (8.7) there exists $\delta_j(\epsilon) > 0$ such that $\frac{1}{\xi(\varpi_1, \varpi_2)} \sum_{n_2=1}^{\infty} \xi_{jn_2} \varpi_2^{n_2-1} < \frac{\epsilon}{2M_1 n_0}$, for all $\mathcal{R} - \delta_j(\epsilon) < \varpi_1, \varpi_2 < \mathcal{R}$, and $j = 1, 2, ... n_0(\epsilon)$. Let us consider $\delta(\epsilon) = \min(\delta_1(\epsilon), \delta_2(\epsilon), ..., \delta_{n_0}(\epsilon))$, then for every $\mathcal{R} - \delta(\epsilon) < \varpi_1, \varpi_2 < \mathcal{R}$ and for all $n = 1, 2, ... n_0$, we have

$$J < \frac{\epsilon}{2M_1 n_0} M_1 n_0 + \frac{\epsilon}{2} = \epsilon.$$

Hence due to the arbitrariness of $\epsilon > 0$, it follows that

$$\lim_{\varpi_1, \varpi_2 \to \mathcal{R}-} \frac{1}{\xi(\varpi_1, \varpi_2)} \sum_{n_1, n_2=1}^{\infty} \|\mathfrak{K}_{n_1, q_{n_1}}^{n_2, q_{n_2}}(g_1) - g_1\| \xi_{n_1 n_2} \varpi_1^{n_1-1} \varpi_2^{n_2-1} = 0.$$

Using a similar argument as given above, we can easily show that the condition (8.13) holds true for $i = 2, 3$.

Our next result determines the rate of convergence by the operators (8.5) considering the total modulus of continuity.

Theorem 8.3.4. Let $f \in C(\mathcal{I}^2)$, and $\Omega(\varpi_1, \varpi_2)$ be a positive function on $(0, \mathcal{R}) \times (0, \mathcal{R})$ such that

$$\frac{\sum_{n_1, n_2=1}^{\infty} \omega_f \left(\sqrt{\Upsilon_{n_1, q_{n_1}, 2}^{\eta^{(1)}, \eta^{(2)}, \cdots, \eta^{(r_1)}} + \Psi_{n_2, q_{n_2}, 2}^{\zeta^{(1)}, \zeta^{(2)}, \cdots, \zeta^{(r_2)}}} \right) \xi_{n_1 n_2} \varpi_1^{n_1-1} \varpi_2^{n_2-1}}{\xi(\varpi_1, \varpi_2)}$$

$$= O(\Omega(\varpi_1, \varpi_2)),$$

as $\varpi_1, \varpi_2 \to \mathcal{R}-$ then

$$\frac{\sum_{n_1,n_2=1}^{\infty} \|\mathfrak{K}_{n_1,q_{n_1}}^{n_2,q_{n_2}}(f) - f\| \xi_{n_1 n_2} \varpi_1^{n_1-1} \varpi_2^{n_2-1}}{\xi(\varpi_1, \varpi_2)}$$
$$= O(\Omega(\varpi_1, \varpi_1)), \text{ as } \varpi_1, \varpi_1 \to \mathcal{R}-.$$

Proof. For $f \in C(\mathcal{I}^2)$ and $\delta > 0$, exploiting $|f(\mathbf{t}) - f(\mathbf{x})| \leq \omega_f(f,\delta)\left(1 + \frac{(\Delta_{t_1,x_1}^{t_2,x_2})^2}{\delta^2}\right)$, and taking supremum over all $\mathbf{x} \in \mathcal{I}^2$, we have

$$\frac{1}{\xi(\varpi_1,\varpi_2)} \sum_{n_1,n_2=1}^{\infty} \|\mathfrak{K}_{n_1,q_{n_1}}^{n_2,q_{n_2}}(f) - f\| \xi_{n_1 n_2} \varpi_1^{n_1-1} \varpi_2^{n_2-1}$$
$$\leq \frac{1}{\xi(\varpi_1,\varpi_2)} \sum_{n_1,n_2=1}^{\infty} \left\{1 + \frac{1}{\delta^2}\|\mathfrak{K}_{n_1,q_{n_1}}^{n_2,q_{n_2}}(\Delta_{t_1,x_1}^{t_2,x_2})^2\|\right\} \omega_f(\delta) \xi_{n_1 n_2} \varpi_1^{n_1-1} \varpi_2^{n_2-1},$$

for every $\varpi_1, \varpi_2 \in (0, \mathcal{R})$. If we take $\delta = \sqrt{\Upsilon_{n_1,q_{n_1},2}^{\eta^{(1)},\eta^{(2)},\cdots,\eta^{(r_1)}} + \Psi_{n_2,q_{n_2},2}^{\zeta^{(1)},\zeta^{(2)},\cdots,\zeta^{(r_2)}}}$, we obtain

$$\frac{1}{\xi(\varpi_1,\varpi_2)} \sum_{n_1,n_2=1}^{\infty} \|\mathfrak{K}_{n_1,q_{n_1}}^{n_2,q_{n_2}}(f) - f\| \xi_{n_1 n_2} \varpi_1^{n_1-1} \varpi_2^{n_2-1}$$
$$\leq \frac{2}{\xi(u,v)} \sum_{n_1,n_2=1}^{\infty} \omega_f\left(\sqrt{\Upsilon_{n_1,q_{n_1},2}^{\eta^{(1)},\eta^{(2)},\cdots,\eta^{(r_1)}} + \Psi_{n_2,q_{n_2},2}^{\zeta^{(1)},\zeta^{(2)},\cdots,\zeta^{(r_2)}}}\right)$$
$$\times \xi_{n_1 n_2} \varpi_1^{n_1-1} \varpi_1^{n_2-1}.$$

This completes the proof. □

8.3.1 Illustrative example

The following example shows that our Theorem 8.3.2 is a non-trivial generalization of the classical Korovkin result given in [9].
Considering the operators $\mathfrak{K}_{n_1,q_{n_1}}^{n_2,q_{n_2}}$ defined by (8.5), for $f \in C(\mathcal{I}^2)$, we define the following LPOs:

$$\overline{\mathfrak{K}_{n_1,q_{n_1}}^{n_2,q_{n_2}}}(f)(\mathbf{x}) = (1 + x_{n_1 n_2}) \mathfrak{K}_{n_1,q_{n_1}}^{n_2,q_{n_2}}(f)(\mathbf{x}), \tag{8.14}$$

where the sequence $\langle x_{n_1 n_2} \rangle$ is defined by

$$\langle x_{n_1 n_2} \rangle = \begin{cases} 0, & \text{if } n_1 \text{ and } n_2 \text{ are squares of some positive integers,} \\ 1, & \text{otherwise.} \end{cases}$$

Now if we take $\xi_{n_1 n_2} = 1$ for all $n \in \mathbb{N}$, then

$$\xi(\varpi_1, \varpi_2) = \sum_{n_1, n_2=1}^{\infty} \xi_{n_1 n_2} \varpi_1^{n_1-1} \varpi_2^{n_2-1}$$

$$= \frac{1}{(1-\varpi_1)(1-\varpi_2)}, \quad |\varpi_1| < 1, |\varpi_2| < 1$$

which implies that $\mathcal{R} = 1$. Further, we note that

$$\frac{1}{\xi(\varpi_1, \varpi_2)} \sum_{n_1=1}^{\infty} \xi_{n_1 n_2} \varpi_1^{n_1-1} \varpi_2^{n_2-1} x_{n_1 n_2} = \frac{(1-\varpi_1)(1-\varpi_2)}{\varpi_1 \varpi_2} \sum_{p, q_1=1}^{\infty} \varpi_1^{p^2} \varpi_2^{q^2}.$$

From Exercise 35 on page 54 of [G. Pólya and G. Szegö, Problems and theorems in analysis I: Series. Integral Calculus. Theory of Functions, Springer-Verlag, 1972], we have

$$\lim_{\varpi_1, \varpi_2 \to 1^-} \frac{\sum_{n_1=1}^{\infty} \sum_{n_2=1}^{\infty} \xi_{n_1 n_2} \varpi_1^{n_1-1} \varpi_2^{n_2-1} x_{n_1 n_2}}{\xi(\varpi_1, \varpi_2)}$$

$$= \lim_{\varpi_1, \varpi_2 \to 1^-} \frac{(1-\varpi_1)(1-\varpi_2)}{\varpi_1 \varpi_2} \sum_{p, q=1}^{\infty} \varpi_1^{p^2} \varpi_2^{q^2} = 0. \tag{8.15}$$

Hence, the sequence $\langle x_{n_1 n_2} \rangle$ converges to zero in the sense of p.s.s.m. Since $\overline{\mathfrak{K}_{n_1, q_{n_1}}^{n_2, q_{n_2}}}(g_0)(\mathbf{x}) - g_0(\mathbf{x}) = x_{n_1 n_2} g_0(\mathbf{x})$, using relation (8.15) we conclude that

$$\lim_{\varpi_1, \varpi_2 \to 1^-} \frac{1}{\xi(\varpi_1, \varpi_2)} \sum_{n_1, n_2=1}^{\infty} \xi_{n_1 n_2} \varpi_1^{n_1-1} \varpi_2^{n_2-1} x_{n_1 n_2} ||\overline{\mathfrak{K}_{n_1, q_{n_1}}^{n_2, q_{n_2}}}(g_0) - g_0|| = 0.$$

Moreover, from Lemma 8.2.1 and (8.14) we have

$$||\overline{\mathfrak{K}_{n_1, q_{n_1}}^{n_2, q_{n_2}}}(g_1) - g_1|| \leq (1 - \eta_{n_1}^{(r)}) + x_{n_1 n_2} \eta_{n_1}^{(r_1)}.$$

Since $\langle 1 - \eta_{n_1}^{(r_1)} \rangle$, converges to 0, as $n_1 \to \infty$, it will also converge to 0, in the sense of p.s.s.m. On the other hand, since $0 < \eta_{n_1}^{(r)} \leq 1$ and from (8.15),

$$\lim_{\varpi_1, \varpi_2 \to 1^-} \frac{1}{\xi(\varpi_1, \varpi_2)} \sum_{n_1, n_2=1}^{\infty} \xi_{n_1 n_2} \varpi_1^{n_1-1} \varpi_2^{n_2-1} x_{n_1 n_2} = 0,$$

we get

$$\lim_{\varpi_1, \varpi_2 \to 1^-} \frac{1}{\xi(\varpi_1, \varpi_2)} \sum_{n_1, n_2=1}^{\infty} \xi_{n_1 n_2} \varpi_1^{n_1-1} \varpi_2^{n_2-1} ||\overline{\mathfrak{K}_{n_1, q_{n_1}}^{n_2, q_{n_2}}}(g_1) - g_1|| = 0.$$

Similarly,

$$\lim_{\varpi_1,\varpi_2\to 1^-} \frac{1}{\xi(\varpi_1,\varpi_2)} \sum_{n_1=1}^{\infty}\sum_{n_2=1}^{\infty} \xi_{n_1 n_2} \varpi_1^{n_1-1} \varpi_2^{n_2-1} ||\overline{\mathfrak{K}_{n_1,q_{n_1}}^{n_2,q_{n_2}}}(g_2) - g_2|| = 0.$$

Finally, by using Lemma 8.2.1 and definition of the operators $\overline{\mathfrak{K}_{n_1,q_{n_1}}^{n_2,q_{n_2}}}$ given by (8.14) we have

$$||\overline{\mathfrak{K}_{n_1,q_{n_1}}^{n_2,q_{n_2}}}(g_3) - g_3|| \leq 2\left(1-\eta_{n_1}^{(r_1)}\right) + \left(\frac{\eta_{n_1}^{(r_1)}}{[n_1]_{q_{n_1}}}\right) + 2\left(1-\zeta_{n_2}^{(r_2)}\right) + \left(\frac{\zeta_{n_2}^{(r_2)}}{[n_2]_{q_{n_2}}}\right)$$
$$+ x_{n_1 n_2}\left(q_{n_1}(\eta_{n_1}^{(r_1)})^2 + \frac{\eta_{n_1}^{(r_1)}}{[n_1]_{q_{n_1}}} + q_{n_2}(\zeta_{n_2}^{(r_2)})^2 + \frac{\zeta_{n_2}^{(r_2)}}{[n_2]_{q_{n_2}}}\right).$$

Since $2\left(1-\eta_{n_1}^{(r_1)}\right) + \left(\frac{\eta_{n_1}^{(r_1)}}{[n_1]_{q_{n_1}}}\right) + 2\left(1-\zeta_{n_2}^{(r_2)}\right) + \left(\frac{\zeta_{n_2}^{(r_2)}}{[n_2]_{q_{n_2}}}\right)$ is convergent to 0, as $n_1, n_2 \to \infty$, it will also converge to 0, in the sense of p.s.s.m. On the other hand since

$$0 < q_{n_1}(\eta_{n_1}^{(r_1)})^2 + \frac{\eta_{n_1}^{(r_1)}}{[n_1]_{q_{n_1}}} + q_{n_2}(\zeta_{n_2}^{(r_2)})^2 + \frac{\zeta_{n_2}^{(r_2)}}{[n_2]_{q_{n_2}}} \leq 4,$$

it follows that

$$\lim_{\varpi_1,\varpi_2\to 1^-} \frac{1}{\xi(\varpi_1,\varpi_2)} \sum_{n_1,n_2=0}^{\infty} \xi_{n_1 n_2} \varpi_1^{n_1-1} \varpi_2^{n_2-1} ||\overline{\mathfrak{K}_{n_1,q_{n_1}}^{n_2,q_{n_2}}}(g_3) - g_3|| = 0.$$

Thus, our operator defined by relation (8.14) satisfies all the conditions of Theorem 8.3.2 and hence it follows that

$$\lim_{\varpi_1,\varpi_2\to\infty} \frac{1}{\xi(\varpi_1,\varpi_2)} \sum_{n_1,n_2=1}^{\infty} \xi_{n_1 n_2} \varpi_1^{n_1-1} \varpi_2^{n_2-1} ||\overline{\mathfrak{K}_{n_1,q_{n_1}}^{n_2,q_{n_2}}}(f) - f|| = 0.$$

However, since $\langle x_{n_1 n_2}\rangle$ is not convergent to 0, we can say that the classical Korovkin theorem for the bivariate case does not work for the operators defined by (8.14).

Now, we determine the rate of convergence of the sequence $\mathfrak{K}_{n_1,q_{n_1}}^{n_2,q_{n_2}}(f)$ to f for the elements f of the Lipschitz class $Lip_L(\gamma)$, for $0 < \gamma \leq 1$ by using the p.s.s.m. For all $\mathbf{t}, \mathbf{x} \in \mathcal{I}^2$, the class $Lip_L(\gamma)$ is defined as:

$$Lip_L(\gamma) = \{f \in C(\mathbb{I}^2) : |f(\mathbf{t}) - f(\mathbf{x})| \leq L||\mathbf{t}-\mathbf{x}||^{\gamma}\},$$

where $||\mathbf{t}-\mathbf{x}|| = \Delta_{t_1,x_1}^{t_2,x_2}$ is the Euclidean norm.

Theorem 8.3.5. Let $f \in Lip_L(\gamma)$, $0 < \gamma \leq 1$. Then, for all $\varpi_1, \varpi_2 \in (0, \mathcal{R})$, $\mathbf{x} \in \mathcal{I}^2$, we have

$$\frac{\sum_{n_1,n_2=1}^{\infty} \|\mathfrak{K}_{n_1,q_{n_1}}^{n_2,q_{n_2}}(f) - (f)\| \, \xi_{n_1 n_2} \varpi_1^{n_1-1} \varpi_2^{n_2-1}}{\xi(\varpi_1, \varpi_2)}$$

$$\leq \frac{L \sum_{n_1,n_2=1}^{\infty} \delta_{n_1,n_2}^{\gamma} \xi_{n_1 n_2} \varpi_1^{n_1-1} \varpi_2^{n_2-1}}{\xi(\varpi_1, \varpi_2)},$$

where $\delta_{n_1,n_2}^2 = 4\left(1 - \eta_{n_1}^{(r_1)}\right) + \frac{\eta_{n_1}^{(r_1)}}{[n_1]_{q_{n_1}}} + 4\left(1 - \zeta_{n_2}^{(r_2)}\right) + \frac{\zeta_{n_2}^{(r_2)}}{[n_2]_{q_{n_2}}}$.

Proof. Taking into account the monotonicity and linearity of the operators (8.5), we have

$$\frac{1}{\xi(\varpi_1, \varpi_2)} \sum_{n=1}^{\infty} |\mathfrak{K}_{n_1,q_{n_1}}^{n_2,q_{n_2}}(f)(\mathbf{x}) - f(\mathbf{x})| \xi_{n_1 n_2} \varpi_1^{n_1-1} \varpi_2^{n_2-1}$$

$$\leq \frac{L}{\xi(\varpi_1, \varpi_2)} \sum_{n_1,n_2=1}^{\infty} \mathfrak{K}_{n_1,q_{n_1}}^{n_2,q_{n_2}}(\|\mathbf{t} - \mathbf{x}\|^{\gamma})(\mathbf{x}) \xi_{n_1 n_2} \varpi_1^{n_1-1} \varpi_2^{n_2-1},$$

for every $\varpi_1, \varpi_2 \in (0, \mathcal{R})$. Applying Hölder's inequality with $u_1 = \frac{2}{\gamma}$ and $v_1 = \frac{2}{2-\gamma}$, and Lemma 8.2.1, we have

$$\frac{\sum_{n_1,n_2=1}^{\infty} \|\mathfrak{K}_{n_1,q_{n_1}}^{n_2,q_{n_2}}(f) - (f)\| \, \xi_{n_1 n_2} \varpi_1^{n_1-1} \varpi_2^{n_2-1}}{\xi(\varpi_1, \varpi_2)}$$

$$\leq \frac{L \sum_{n_1,n_2=1}^{\infty} \delta_{n_1,n_2}^{\gamma} \xi_{n_1 n_2} \varpi_1^{n_1-1} \varpi_2^{n_2-1}}{\xi(\varpi_1, \varpi_2)}.$$

This completes the proof. □

8.4 $\mathcal{A}$-statistical convergence of operators $\mathfrak{K}_{n_1,q_{n_1}}^{n_2,q_{n_2}}(.)(\mathbf{x})$

We first recall the notion of $\mathcal{A}$-statistical convergence and we display an application which shows that our new result is stronger than its classical version.

A sequence $x = \langle x_{n_1,n_2} \rangle$ is said to be convergent to L in Pringsheim's sense if, for every $\epsilon > 0$, there exists $n_0 = n_0(\epsilon) \in \mathbb{N}$, such that $|x_{n_1,n_2} - L| < \epsilon$, whenever $n_1, n_2 > n_0$. The limit L is called the Pringsheim limit of x and is denoted by $P - \lim x = L$ (see [26]). Let

$$\mathcal{A} = [a_{j,k}^{(n_1,n_2)}], \ (j, k, n_1, n_2 \in \mathbb{N})$$

be a sequence of infinite matrices with non-negative real numbers. For a given sequence $x = \langle x_{n_1,n_2} \rangle$, the $\mathcal{A}$-transform of x, denoted by $\mathcal{A}x := ((\mathcal{A}x)_{j,k})$, is given by

$$(\mathcal{A}x)_{j,k} = \sum_{(n_1,n_2) \in \mathbb{N}^2} a_{j,k}^{(n_1,n_2)} x_{n_1,n_2}$$

provided the series converges in Pringsheim's sense for every $(j, k) \in \mathbb{N}^2$. Any sequence $\langle x_{n_1,n_2} \rangle$ is called $\mathcal{A}$-statistically convergent to a number l, if for every $\epsilon > 0$

$$P - \lim_{j,k} \sum_{(n_1,n_2) \in K(\epsilon)} a_{j,k}^{(n_1,n_2)} = 0,$$

where $K(\epsilon) = \{(n_1, n_2) \in \mathbb{N}^2 : |x_{n_1,n_2} - l| \geq \epsilon\}$. Clearly, a P-convergent double sequence is $\mathcal{A}$-statistically convergent to the same value but its converse is not always true. The Robinson-Hamilton conditions, or briefly, RH-regularity conditions state that a four dimensional matrix $\mathcal{A} = [a_{j,k}^{(n_1,n_2)}], \ (j, k, n_1, n_2 \in \mathbb{N})$ is said to be regular [27] if and only if

(i) for each $(n_1, n_2) \in \mathbb{N}^2$, $P - \lim_{j,k \to \infty} a_{j,k}^{(n_1,n_2)} = 0$;

(ii) $P - \lim_{j,k} \sum_{(n_1,n_2) \in \mathbb{N}^2} a_{j,k}^{(n_1,n_2)} = 1$;

(iii) $P - \lim_{j,k} \sum_{n_1 \in \mathbb{N}} |a_{j,k}^{(n_1,n_2)}| = 0$ for each $n_2 \in \mathbb{N}$;

(iv) $P - \lim_{j,k} \sum_{n_2 \in \mathbb{N}} |a_{j,k}^{(n_1,n_2)}| = 0$ for each $n_1 \in \mathbb{N}$;

(v) there exist finite positive integers M and N such that
$$\sum_{n_1,n_2 > N} |a_{j,k}^{(n_1,n_2)}| < M, \text{ holds for all } (j,k) \in \mathbb{N}^2.$$

(vi) $\sum_{(n_1,n_2) \in \mathbb{N}^2} |a_{j,k}^{(n_1,n_2)}|$ is P-convergent for every $(j, k) \in \mathbb{N}^2$;

Definition 8.4.1. *[15] Let $\langle e_{n_1,n_2} \rangle$ be a positive non-increasing double sequence and let $\mathcal{A} = [a_{j,k}^{(n_1,n_2)}] \in RH$. A real or complex valued double sequence $x = \langle x_{n_1,n_2} \rangle$ is called $\mathcal{A}$-statistically convergent to a number S with the rate of $o(e_{n_1,n_2})$, if for a given $\epsilon > 0$,*

$$P - \lim_{j,k \to \infty} \frac{1}{e_{j,k}} \sum_{(n_1,n_2) \in K(\epsilon)} a_{j,k}^{(n_1,n_2)} = 0,$$

Approximation by the double sequence of LPO   141

where $K(\epsilon) = \{(n_1, n_2) \in \mathbb{N}^2 : |x_{n_1,n_2} - S| \geq \epsilon\}$. We denote it as $x_{n_1,n_2} - S = st_{\mathcal{A}}^{(2)} - o(e_{n_1,n_2})$ as $n_1, n_2 \to \infty$.

Definition 8.4.2. *[15] Let $\langle e_{n_1,n_2} \rangle$ and $\mathcal{A} = [a_{j,k}^{(n_1,n_2)}]$ be the same as in Definition 8.4.1. A double sequence $x = \langle x_{n_1,n_2} \rangle$ is called $\mathcal{A}$-statistically bounded with the rate of $O(e_{n_1,n_2})$, if for a given $\epsilon > 0$,*

$$\sup_{j,k} \frac{1}{e_{j,k}} \sum_{(n_1,n_2) \in F(\epsilon)} a_{j,k}^{(n_1,n_2)} < \infty,$$

where $F(\epsilon) = \{(n_1, n_2) \in \mathbb{N}^2 : |x_{n_1,n_2}| \geq \epsilon\}$. It is denoted as $x_{n_1,n_2} = st_{\mathcal{A}}^{(2)} - O(e_{n_1,n_2})$ as $n_1, n_2 \to \infty$.

First, we establish a Korovkin type approximation theorem for the operators $\mathfrak{K}_{n_1,q_{n_1}}^{n_2,q_{n_2}}$ by means of the $\mathcal{A}$-statistical convergence.

Theorem 8.4.3. *For $f \in C(\mathcal{I}^2)$, the operators $\mathfrak{K}_{n_1,q_{n_1}}^{n_2,q_{n_2}}$ satisfy*

$$st_{\mathcal{A}}^{(2)} - ||\mathfrak{K}_{n_1,q_{n_1}}^{n_2,q_{n_2}}(f) - (f)|| = 0, \tag{8.16}$$

if and only if

$$st_{\mathcal{A}}^{(2)} - ||\mathfrak{K}_{n_1,q_{n_1}}^{n_2,q_{n_2}}(g_i) - (g_i)|| = 0, \text{ for } i = 1, 2, 3. \tag{8.17}$$

Proof. First, let us assume that (8.16) is true. Then (8.17) is obvious. Conversely, suppose

$$st_{\mathcal{A}}^{(2)} - ||\mathfrak{K}_{n_1,q_{n_1}}^{n_2,q_{n_2}}(g_i) - (g_i)|| = 0, \text{ for } i = 1, 2, 3,$$

is true. Since $f \in C(\mathcal{I}^2)$, it follows that for every $\epsilon > 0$, there $\exists\, \delta > 0$ such that for all $\mathbf{t}, \mathbf{x} \in \mathcal{I}^2$,

$$|f(\mathbf{t}) - f(\mathbf{x})| < \epsilon + 2M\delta^{-2}\{(\Delta_{t_1,x_1}^{t_2,x_2})^2\}, \tag{8.18}$$

where $M := \sup_{\mathbf{x} \in \mathbb{I}^2} |f(\mathbf{x})|$.

Applying the LPOs $\mathfrak{K}_{n_1,q_{n_1}}^{n_2,q_{n_2}}(.)(\mathbf{x})$ to (8.18) and taking supremum over $\mathbf{x} \in \mathcal{I}^2$, we have

$$||\mathfrak{K}_{n_1,q_{n_1}}^{n_2,q_{n_2}}(f) - (f)|| \leq \epsilon + \frac{4M}{\delta^2}\Big\{\big(||\mathfrak{K}_{n_1,q_{n_1}}^{n_2,q_{n_2}}(t_1^2 + t_2^2)(\mathbf{x}) - (x_1^2 + x_2^2)||$$

$$+ ||\mathfrak{K}_{n_1,q_{n_1}}^{n_2,q_{n_2}}(t_1)(\mathbf{x}) - x_1|| + ||\mathfrak{K}_{n_1,q_{n_1}}^{n_2,q_{n_2}}(t_2)(\mathbf{x}) - x_2||\big)\Big\}. \tag{8.19}$$

Now given $r > 0$, choose $\epsilon > 0$ such that $\epsilon < r$, and define

$$T = \{(n_1, n_2) \in \mathbb{N}^2 : \left\|\mathfrak{K}_{n_1,q_{n_1}}^{n_2,q_{n_2}}(f) - f\right\| \geq r\},$$

$$T_s = \{(n_1, n_2) \in \mathbb{N}^2 : \left\|\mathfrak{K}_{n_1,q_{n_1}}^{n_2,q_{n_2}}(g_s)(\mathbf{x}) - g_s\right\| \geq \frac{(r-\epsilon)\delta^2}{12M}\};\ s = 1, 2, 3.$$

By (8.19) it is clear that $T \subset \cup_{s=1}^{3} T_i$. Hence, we may write

$$\sum_{(n_1,n_2)\in T} a_{j,k}^{(n_1,n_2)} \leq \sum_{s=1}^{3} \sum_{(n_1,n_2)\in T_s} a_{j,k}^{(n_1,n_2)}.$$

Now, the assertion follows easily on taking the limits as $j, k \to \infty$, and the arbitrariness of ϵ. □

First, we show rate of the $\mathcal{A}$-statistical convergence in Theorem 8.4.4.

Theorem 8.4.4. *Let $\mathcal{A} = [a_{j,k}^{(n_1,n_2)}]$ be a non-negative RH-regular sumability matrix. Assume that the following condition holds:*

$$\omega_f(\delta_{n_1,n_2}) = st_{\mathcal{A}}^{(2)} - o(f_{n_1,n_2}), \quad as \ n_1, n_2 \to \infty, \tag{8.20}$$

where $\delta_{n_1,n_2} = 4\left(1 - \eta_{n_1}^{(r_1)}\right) + \frac{\eta_{n_1}^{(r_1)}}{[n_1]_{q_{n_1}}} + 4\left(1 - \zeta_{n_2}^{(r_2)}\right) + \frac{\zeta_{n_2}^{(r_2)}}{[n_2]_{q_{n_2}}}$ and $\langle f_{n_1,n_2}\rangle$ be a positive non-increasing double sequence. Then for every $f \in C(\mathcal{I}^2)$, the operator $\mathfrak{K}_{n_1,q_{n_1}}^{n_2,q_{n_2}}(f)(\mathbf{x})$ verifies the following inequality:

$$\left\|\mathfrak{K}_{n_1,q_{n_1}}^{n_2,q_{n_2}}(f) - f\right\| = st_{\mathcal{A}}^{(2)} - o(f_{n_1,n_2}), \quad as \ n_1, n_2 \to \infty.$$

Proof. Let $\mathbf{x} \in \mathcal{I}^2$ be arbitrary. By linearity of $\mathfrak{K}_{n_1,q_{n_1}}^{n_2,q_{n_2}}(.)(\mathbf{x})$ and the property (8.9), for any $\delta > 0$, we obtain

$$|\mathfrak{K}_{n_1,q_{n_1}}^{n_2,q_{n_2}}(f)(\mathbf{x}) - f(\mathbf{x})| \leq \mathfrak{K}_{n_1,q_{n_1}}^{n_2,q_{n_2}}(|f(\mathbf{t}) - f(\mathbf{x})|)(\mathbf{x})$$
$$\leq \omega_f(\delta)\left(1 + \mathfrak{K}_{n_1,q_{n_1}}^{n_2,q_{n_2}}\left(\left(\Delta_{t_1,x_1}^{t_2,x_2}\right)^2\right)(\mathbf{x})\delta^{-2}\right). \tag{8.21}$$

By using Lemma 8.2.1, we have

$$\left\|\mathfrak{K}_{n_1,q_{n_1}}^{n_2,q_{n_2}}\left(\left(\Delta_{t_1,x_1}^{t_2,x_2}\right)^2\right)\right\| \leq 4\left(1 - \eta_{n_1}^{(r_1)}\right) + \frac{\eta_{n_1}^{(r_1)}}{[n_1]_{q_{n_1}}}$$
$$+ 4\left(1 - \zeta_{n_2}^{(r_2)}\right) + \frac{\zeta_{n_2}^{(r_2)}}{[n_2]_{q_{n_2}}} = \delta_{n_1,n_2}^2,$$

as defined in Theorem 8.4.4. Thus from (8.21), we are led to

$$\left\|\mathfrak{K}_{n_1,q_{n_1}}^{n_2,q_{n_2}}(f) - f\right\| \leq \omega_f(\delta)\left\{1 + \delta^{-2}(\delta_{n_1,n_2})^2\right\}.$$

On choosing $\delta = \delta_{n_1,n_2}$, we obtain

$$\left\|\mathfrak{K}_{n_1,q_{n_1}}^{n_2,q_{n_2}}(f) - f\right\| \leq 2\,\omega_f(\delta_{n_1,n_2}). \tag{8.22}$$

Approximation by the double sequence of LPO

Now, for any $\epsilon > 0$, let us consider the sets:

$$T = \{(n_1, n_2) \in \mathbb{N}^2 : \left\|\mathfrak{K}_{n_1,q_{n_1}}^{n_2,q_{n_2}}(f) - f\right\| \geq \epsilon\},$$

$$T_1 = \{(n_1, n_2) \in \mathbb{N}^2 : \omega_f(\delta_{n_1,n_2}) \geq \frac{\epsilon}{2}\}.$$

Then in view of (8.22), we obtain $T \subset T_1$ which implies that

$$\frac{1}{f_{j,k}} \sum_{(n_1,n_2) \in T} a_{j,k}^{(n_1,n_2)} \leq \frac{1}{f_{j,k}} \sum_{(n_1,n_2) \in T_1} a_{j,k}^{(n_1,n_2)}. \tag{8.23}$$

Letting $j, k \to \infty$, we find that

$$P - \lim_{j,k \to \infty} \frac{1}{f_{j,k}} \sum_{(m,n) \in T} a_{j,k}^{(n_1,n_2)} = 0,$$

which proves the theorem. $\square$

8.4.1 Application of Theorem 8.4.4

In this section, we present an example of a sequence of LPOs, which satisfy Theorem 8.4.4 but not the classical theorem. Let $\mathcal{A} = C(1,1) = [c_{j,k}^{(n_1,n_2)}]$ the double Cesáro matrix, defined by

$$c_{j,k}^{(n_1,n_2)} = \begin{cases} \frac{1}{jk}, & \text{if } 1 \leq n_1 \leq j \text{ and } 1 \leq n_2 \leq k, \\ 0, & \text{otherwise}. \end{cases}$$

Consider the operators $\mathfrak{K}_{n_1,q_{n_1}}^{n_2,q_{n_2}}$ defined by (8.5). Using these operators, for $f \in \mathcal{C}(\mathcal{I}^2)$, we define the following LPOs on $\mathcal{C}(\mathcal{I}^2)$:

$$\overline{\mathfrak{K}_{n_1,q_{n_1}}^{n_2,q_{n_2}}}(f)(\mathbf{x}) = (1 + x_{n_1 n_2})\mathfrak{K}_{n_1,q_{n_1}}^{n_2,q_{n_2}}(f)(\mathbf{x}), \tag{8.24}$$

where the sequence $\langle x_{n_1 n_2} \rangle$ is defined by

$$\langle x_{n_1 n_2} \rangle = \begin{cases} \sqrt{n_1 n_2}, & \text{if } n_1 \text{ and } n_2 \text{ are squares of some positive integers,} \\ 0, & \text{otherwise}. \end{cases}$$

In this case, we observe that $st_{\mathcal{A}}^{(2)} - \lim_{n_1,n_2 \to \infty} x_{n_1 n_2} = 0$, although the sequence $x_{n_1 n_2}$ is not P-convergent. Now setting $\langle e_{n_1,n_2} \rangle = \langle \frac{1}{\sqrt[6]{n_1 n_2}} \rangle$, we have, for any $\epsilon > 0$

$$\frac{1}{e_{j,k}} \sum_{(n_1,n_2):|x_{n_1 n_2}| \geq \epsilon} c_{j,k}^{(n_1,n_2)} = \sqrt[6]{jk} \sum_{(n_1,n_2):|x_{n_1 n_2}| \geq \epsilon} \frac{1}{jk} \leq \frac{1}{\sqrt[3]{jk}}. \tag{8.25}$$

Taking the limit as $j, k \to \infty$ (in any manner) in (8.25), we get, for any $\epsilon > 0$,

$$P - \lim_{j,k} \frac{1}{e_{j,k}} \sum_{(n_1,n_2):|x_{n_1 n_2}| \geq \epsilon} c_{j,k}^{(n_1,n_2)} = 0$$

which gives,

$$x_{n_1 n_2} = st_{\mathcal{A}}^{(2)} - o\left(\frac{1}{\sqrt[6]{n_1 n_2}}\right) \quad \text{as } n_1, n_2 \to \infty. \tag{8.26}$$

Since $\|\overline{\mathfrak{K}_{n_1,q_{n_1}}^{n_2,q_{n_2}}}(g_0) - g_0\| = x_{n_1 n_2}$, from (8.26) we obtain

$$\|\overline{\mathfrak{K}_{n_1,q_{n_1}}^{n_2,q_{n_2}}}(g_0) - g_0\| = st_{\mathcal{A}}^{(2)} - o(e_{n_1,n_2}), \quad n_1, n_2 \to \infty.$$

From (8.24), we have

$$\overline{\mathfrak{K}_{n_1,q_{n_1}}^{n_2,q_{n_2}}}((\Delta_{t_1,x_1}^{t_2,x_2})^2)(\mathbf{x}) \leq 4\left(1 - \eta_{n_1}^{(r_1)}\right) + \frac{\eta_{n_1}^{(r_1)}}{[n_1]_{q_{n_1}}} + 4\left(1 - \zeta_{n_2}^{(r_2)}\right)$$

$$+ \frac{\zeta_{n_2}^{(r_2)}}{[n_2]_{q_{n_2}}} + x_{n_1 n_2}\left(4\left(1 - \eta_{n_1}^{(r_1)}\right) + \frac{\eta_{n_1}^{(r_1)}}{[n_1]_{q_{n_1}}} + 4\left(1 - \zeta_{n_2}^{(r_2)}\right) + \frac{\zeta_{n_2}^{(r_2)}}{[n_2]_{q_{n_2}}}\right).$$

Hence, we obtain

$$\sqrt{\|\overline{\mathfrak{K}_{n_1,q_{n_1}}^{n_2,q_{n_2}}}((\Delta_{t_1,x_1}^{t_2,x_2})^2)(\mathbf{x})\|} \leq \sqrt{\delta_{n_1 n_2}^2(1 + x_{n_1 n_2})} = x_{n_1,n_2}.$$

In this case, setting $\langle f_{n_1 n_2} \rangle = \langle \frac{1}{\sqrt[8]{n_1 n_2}} \rangle$, we have, for any $\epsilon > 0$,

$$\frac{1}{f_{j,k}} \sum_{(n_1,n_2):|x_{n_1,n_2}| \geq \epsilon} c_{j,k}^{(n_1,n_2)} = \sqrt[8]{jk} \sum_{(n_1,n_2):|x_{n_1,n_2}| \geq \epsilon} \frac{1}{jk} \leq \frac{\sqrt{jk} \sqrt[8]{jk}}{jk} = \frac{1}{\sqrt[8]{(jk)^3}},$$

which gives that

$$P - \lim_{j,k} \frac{1}{f_{j,k}} \sum_{(n_1,n_2):|x_{n_1,n_2}| \geq \epsilon} c_{j,k}^{(n_1,n_2)} = 0.$$

Thus, we get $x_{n_1,n_2} = st_{\mathcal{A}}^{(2)} - o\left(\frac{1}{\sqrt[8]{n_1 n_2}}\right)$, as $n_1, n_2 \to \infty$. By the uniform continuity of f on $\mathcal{I}^2$, we write that

$$\omega_f(x_{n_1,n_2}) = st_{\mathcal{A}}^{(2)} - o\left(\frac{1}{\sqrt[8]{n_1 n_2}}\right), \quad \text{as } n_1, n_2 \to \infty. \tag{8.27}$$

Then, the sequence of operators defined by (8.24) satisfy the condition of Theorem 8.4.4, so for all $f \in \mathcal{C}(\mathcal{I}^2)$, we have:

$$\|\overline{\mathfrak{K}_{n_1,q_{n_1}}^{n_2,q_{n_2}}}(f) - f\| = st_{\mathcal{A}}^{(2)} - o\left(\frac{1}{\sqrt[8]{n_1 n_2}}\right), \quad \text{as } n_1, n_2 \to \infty.$$

However, the Korovkin type theorem does not work for the operators given by (8.24) because $\langle x_{n_1 n_2} \rangle$ is not P-convergent.

Theorem 8.4.5. Let $\mathcal{A} = [a_{j,k}^{(n_1,n_2)}]$ be a non-negative RH-regular summability matrix. Assume that:

$$\omega_f\left(\sqrt{(1-\eta_{n_1}^{(r_1)})^2 + (1-\zeta_{n_2}^{(r_2)})^2}\right) = st_{\mathcal{A}}^{(2)} - o(e_{n_1,n_2}) \text{ as } n_1, n_2 \to \infty, \quad (8.28)$$

and

$$\omega_f^2(\Delta_{n_1,n_2}) = st_{\mathcal{A}}^{(2)} - o(f_{n_1,n_2}) \text{ as } n_1, n_2 \to \infty, \quad (8.29)$$

where $\Delta_{n_1,n_2}^2 = \frac{1}{8}\Bigg\{\left(\eta_{n_1}^{(r_1)}-1\right)^2 + \left(\zeta_{n_2}^{(r_2)}-1\right)^2 + 4(1-\eta_{n_1}^{(r_1)}) + \frac{\eta_{n_1}^{(r_1)}}{[n_1]_{q_1}}$

$$+4(1-\zeta_{n_2}^{(r_2)}) + \frac{\zeta_{n_2}^{(r_2)}}{[n_2]_{q_2}}\Bigg\} + \frac{1}{4}\Bigg\{\sqrt{\left\{4(1-\eta_{n_1}^{(r_1)}) + \frac{\eta_{n_1}^{(r_1)}}{[n_1]_{q_1}}\right\}}$$

$$\cdot\sqrt{\left\{4(1-\zeta_{n_2}^{(r_2)}) + \frac{\zeta_{n_2}^{(r_2)}}{[n_2]_{q_2}}\right\}} + \left|(\eta_{n_1}^{(r_1)}-1)\right|\left|(\zeta_{n_2}^{(r_2)}-1)\right|\Bigg\},$$

and $\omega_f^2(\Delta_{n_1,n_2})$ is the second order modulus of continuity of f.
Then for every $f \in \mathcal{C}(\mathcal{I}^2)$, the operator $\mathfrak{K}_{n_1,q_{n_1}}^{n_2,q_{n_2}}(f)(\mathbf{x})$ verifies the following inequality

$$\left\|\mathfrak{K}_{n_1,q_{n_1}}^{n_2,q_{n_2}}(f) - f\right\| = st_{\mathcal{A}}^{(2)} - o(\max\{e_{n_1,n_2}, f_{n_1,n_2}\}), \text{ as } n_1, n_2 \to \infty.$$

Proof. From Corollary 1 of [9], we have

$$\left\|\mathfrak{K}_{n_1,q_{n_1}}^{n_2,q_{n_2}}(f) - f\right\| \leq \frac{C_1}{4}\omega_f^2(\Delta_{n_1,n_2}) + \omega_f\left(\sqrt{(1-\eta_{n_1}^{(r_1)})^2 + (1-\zeta_{n_2}^{(r_2)})^2}\right),$$

where C_1 is some positive constant.
Now, for any $\epsilon > 0$, let us consider the sets:

$$T = \{(n_1, n_2) \in \mathbb{N}^2 : \left\|\mathfrak{K}_{n_1,q_{n_1}}^{n_2,q_{n_2}}(f) - f\right\| \geq \epsilon\},$$

$$T_1 = \{(n_1, n_2) \in \mathbb{N}^2 : \omega_f\left(\sqrt{(1-\eta_{n_1}^{(r_1)})^2 + (1-\zeta_{n_2}^{(r_2)})^2}\right) \geq \frac{\epsilon}{2}\}.$$

$$T_2 = \{(n_1, n_2) \in \mathbb{N}^2 : \omega_f^2(\Delta_{n_1,n_2}) \geq \frac{2\epsilon}{C_1}\}. \quad (8.30)$$

Then in view of (8.22), we obtain $T \subset T_1 \cup T_2$ which implies that

$$\frac{1}{\max\{e_{j,k}, f_{j,k}\}} \sum_{(n_1,n_2)\in T} a_{j,k}^{(n_1,n_2)} \leq \frac{1}{e_{j,k}} \sum_{(n_1,n_2)\in T_1} a_{j,k}^{(n_1,n_2)}$$

$$+ \frac{1}{f_{j,k}} \sum_{(n_1,n_2)\in T_2} a_{j,k}^{(n_1,n_2)}.$$

Considering the assumptions of Theorem 8.4.5, we obtain

$$\lim_{j,k\to\infty} \frac{1}{e_{j,k}} \sum_{(n_1,n_2)\in T_1} a_{j,k}^{(n_1,n_2)} + \lim_{j,k\to\infty} \frac{1}{f_{j,k}} \sum_{(n_1,n_2)\in T_2} a_{j,k}^{(n_1,n_2)} = 0.$$

Thus

$$\lim_{j,k\to\infty} \frac{1}{\max\{e_{j,k}, f_{j,k}\}} \sum_{(n_1,n_2)\in T} a_{j,k}^{(n_1,n_2)} = 0,$$

we reach the desired result. □

8.5 $\mathcal{A}$-statistical convergence by GBS operators

The concepts of B-continuous (Bögel continuous) and B-differentiable (Bögel differentiable) functions were introduced in [10] and [11] by Bögel. Subsequently, using these concepts of B-continuity and B-differentiability, Dobrescu and Matei [17] obtained the rate of convergence of GBS of bivariate Bernstein polynomials. Badea et al. [5] established the "Test function theorem" for B-continuous functions. Badea and Badea [4], quantified the Korovkin-type theorem for B-continuous functions. Agrawal and İspir [1] discussed the degree of approximation for the GBS operators of Chlodowsky-Szász-Charlier type for functions in a Bögel space. Baxhaku and Agrawal [8] studied the rate of approximation of B-continuous and B-differentiable functions by the GBS operators of q-Bernstein-Schurer-Stancu-Kantorovich type. Baxhaku et al. [9] obtained the degree of approximation for the bivariate positive linear operators constructed by means of q-Lagrange polynomials and the associated GBS operators. For a detailed account of the significant work in this direction, we refer the reader to [2, 9, 16]. In this section, it is shown that our statistical rates are more efficient than the classical aspects in the approximation theory for the GBS case of the operators $\mathfrak{K}_{n_1,q_{n_1}}^{n_2,q_{n_2}}(.)(\mathbf{x})$ defined in (8.5), with regard to B-continuous and B-differentiable functions.

In the following, we recall some important definitions and a lemma that are essential to study the approximation behavior of the GBS operator.

A function $f : \mathcal{I}^2 \to \mathbb{R}$ is said to be B-bounded (Bögel bounded) if there exists a positive constant M such that $|\Delta[f(\mathbf{t});(\mathbf{x})]| \leq M$, for all $\mathbf{t}, \mathbf{x} \in \mathcal{I}^2$, where $\Delta[f(\mathbf{t});(\mathbf{x})] = f(\mathbf{x}) - f(x_1,t_2) - f(t_1,x_2) + f(\mathbf{t})$ is called the mixed difference of f. We use $B_b(\mathcal{I}^2)$ to denote the space of all B-bounded functions. The function f is said to be B-continuous at a point $\mathbf{x} \in \mathcal{I}^2$ if $\lim_{\mathbf{t}\to\mathbf{x}} \Delta[f(\mathbf{t});(\mathbf{x})] = 0$. Further, f is said to be B-continuous on $\mathcal{I}^2$ if it is B-continuous at each point of $\mathcal{I}^2$. The space of all B-continuous functions is denoted by $\mathcal{C}_b(\mathcal{I}^2)$. For any

$f \in \mathcal{C}_b(\mathcal{I}^2)$ and $\delta_1, \delta_2 > 0$, the mixed modulus of smoothness is defined by

$$\omega_B(f; \delta_1, \delta_2) = \sup_{\mathbf{t}, \mathbf{x} \in \mathcal{I}^2} \{|\Delta[f(\mathbf{t}); (\mathbf{x})]| : |t_1 - x_1| < \delta_1, |t_2 - x_2| < \delta_2\},$$

and it is related to the mixed difference (cf. [4, 6]) by the inequality

$$|\Delta f[(\mathbf{t}); (\mathbf{x})]| \leq \prod_{l=1}^{2}\left(1 + \frac{|t_l - x_l|}{\delta_l}\right)\omega_B(f; \delta_1, \delta_2).$$

Lemma 8.5.1. *[5] If $f \in \mathcal{C}_b(\mathcal{I}^2)$, then, for every $\epsilon > 0$, there exists two positive numbers $C(\epsilon) = C(\epsilon, f)$ and $D(\epsilon) = D(\epsilon, f)$ that complete the following inequality:*

$$|\Delta f[(\mathbf{t}); (\mathbf{x})]| \leq C(\epsilon)(t_1 - x_1)^2 + D(\epsilon)(t_2 - x_2)^2 + \frac{\epsilon}{3}, \text{ for every } \mathbf{t}, \mathbf{x} \in \mathcal{I}^2.$$

For $f \in \mathcal{C}_b(\mathcal{I}^2)$, we define the GBS operator associated with (8.5) as;

$$\mathfrak{U}_{n_1, q_{n_1}}^{n_2, q_{n_2}}(f)(\mathbf{x}) = \mathfrak{K}_{n_1, q_{n_1}}^{n_2, q_{n_2}}(G_{x_1, x_2}(\mathbf{t}))(\mathbf{x}) \qquad (8.31)$$

where $G_{x_1, x_2}(\mathbf{t}) = f(\mathbf{x}) - \Delta f[\mathbf{t}; (\mathbf{x})]$. It is clear that the GBS operator is linear in nature.

Theorem 8.5.2. *Let $\mathcal{A} = [a_{j,k}^{(n_1, n_2)}]$ be a non-negative RH-regular summability matrix. Assume that:*

$$P - \lim_{j,k} \sum_{\mathbf{n} \in \mathbb{N}^2} a_{j,k}^{(n_1, n_2)} = 1, \qquad (8.32)$$

and

$$st_A^{(2)} - \lim_{n_1, n_2}\left(\sup_{\mathbf{x} \in \mathcal{I}^2} |\mathfrak{K}_{n_1, q_{n_1}}^{n_2, q_{n_2}}(g_i)(\mathbf{x}) - g_i(\mathbf{x})|\right) = 0, \text{ for } i=1,2,3. \qquad (8.33)$$

Then for every $f \in \mathcal{C}_b(\mathcal{I}^2)$, the operator $\mathfrak{U}_{n_1, q_{n_1}}^{n_2, q_{n_2}}(f)(\mathbf{x})$ verifies the following inequality

$$st_A^{(2)} - \left\|\mathfrak{U}_{n_1, q_{n_1}}^{n_2, q_{n_2}}(f) - f\right\| = 0.$$

Proof. Applying Lemma 8.5.1, for every $\epsilon > 0$, we have

$$|\mathfrak{U}_{n_1, q_{n_1}}^{n_2, q_{n_2}}(f)(\mathbf{x}) - f(\mathbf{x})| \leq \mathfrak{K}_{n_1, q_{n_1}}^{n_2, q_{n_2}}(|\Delta[f(\mathbf{t}); (\mathbf{x})]|)(\mathbf{x})$$

$$\leq \frac{\epsilon}{3} + C(\epsilon)\mathfrak{K}_{n_1, q_{n_1}}^{n_2, q_{n_2}}((t_1 - x_1)^2)(\mathbf{x}) + D(\epsilon)\mathfrak{K}_{n_1, q_{n_1}}^{n_2, q_{n_2}}((t_2 - x_2)^2)(\mathbf{x})$$

$$\leq \frac{\epsilon}{3} + E(\epsilon)\Big\{|\mathfrak{K}_{n_1, q_{n_1}}^{n_2, q_{n_2}}(t_1^2 + t_2^2)(\mathbf{x}) - (x_1^2 + x_2^2)|$$

$$+ 2|x_1| \, |\mathfrak{K}_{n_1, q_{n_1}}^{n_2, q_{n_2}}(t_1)(\mathbf{x}) - x_1| + 2|x_2| \, |\mathfrak{K}_{n_1, q_{n_1}}^{n_2, q_{n_2}}(t_2)(\mathbf{x}) - x_2|\Big\},$$

where $E(\epsilon) = \max\{C(\epsilon), D(\epsilon)\}$. Now, taking supremum over $\mathbf{x} \in \mathcal{I}^2$ we have

$$\|\mathfrak{U}_{n_1,q_{n_1}}^{n_2,q_{n_2}}(f) - (f)\| \leq 2E(\epsilon)\bigg\{\Big(\|\mathfrak{K}_{n_1,q_{n_1}}^{n_2,q_{n_2}}(t_1^2 + t_2^2)(\mathbf{x}) - (x_1^2 + x_2^2)\|$$
$$+ \|\mathfrak{K}_{n_1,q_{n_1}}^{n_2,q_{n_2}}(t_1)(\mathbf{x}) - x_1\| + \|\mathfrak{K}_{n_1,q_{n_1}}^{n_2,q_{n_2}}(t_2)(\mathbf{x}) - x_2\|\Big)\bigg\} + \frac{\epsilon}{3}. \quad (8.34)$$

Now given $\xi > 0$, choose $\epsilon > 0$ such that $\epsilon < 3\xi$, and define

$$T = \{(n_1, n_2) \in \mathbb{N}^2 : \left\|\mathfrak{K}_{n_1,q_{n_1}}^{n_2,q_{n_2}}(f) - f\right\| \geq \xi\},$$

$$T_s = \{(n_1, n_2) \in \mathbb{N}^2 : \|\mathfrak{K}_{n_1,q_{n_1}}^{n_2,q_{n_2}}(g_s)(\mathbf{x}) - g_s\| \geq \frac{(3\xi - \epsilon)}{18E(\epsilon)}\}; \ s = 1, 2, 3.$$

By (8.34) it is clear that $T \subset \cup_{s=1}^3 T_s$. Hence, we may write

$$\sum_{(n_1,n_2) \in T} a_{j,k}^{(n_1,n_2)} \leq \sum_{s=1}^{3} \sum_{(n_1,n_2) \in T_i} a_{j,k}^{(n_1,n_2)}. \quad (8.35)$$

Letting $j, k \to \infty$ (in any manner) in (8.35) and using (8.33), it follows that $P - \lim_{j,k\to\infty} \sum_{(n_1,n_2) \in T} a_{j,k}^{(n_1,n_2)} = 0$. Hence, we obtain

$$st_A^{(2)} - \left\|\mathfrak{U}_{n_1,q_{n_1}}^{n_2,q_{n_2}}(f) - f\right\| = 0,$$

which yields the result of Theorem 8.5.2. □

Remark 8.5.3. *We now show that our Theorem 8.5.2 is stronger than its classical version, namely, Korovkin type theorem, which was first introduced by Badea et al. [5].*

To see this first, consider the q-LPOs based multivariate Lagrange polynomials $\mathfrak{U}_{n_1,q_{n_1}}^{n_2,q_{n_2}}$ defined by (8.31). Then from Theorem Korovkin [5], we show that, for any $f \in \mathcal{C}(\mathcal{I}^2)$

$$P - \lim_{n_1,n_2} \left\|\mathfrak{U}_{n_1,q_{n_1}}^{n_2,q_{n_2}}(f) - f\right\| = 0. \quad (8.36)$$

Now let $\mathcal{A} = C(1,1)$. By using operators defined in (8.31), for $f \in \mathcal{C}(\mathcal{I}^2)$, we define the following LPOs on $C_b(\mathbb{I}^2)$:

$$\overline{\mathfrak{U}_{n_1,q_{n_1}}^{n_2,q_{n_2}}}(f)(\mathbf{x}) = (1 + x_{n_1 n_2})\mathfrak{U}_{n_1,q_{n_1}}^{n_2,q_{n_2}}(f)(\mathbf{x}), \quad (8.37)$$

where the sequence $\langle x_{n_1 n_2} \rangle$, defined by

$$\langle x_{n_1 n_2} \rangle = \begin{cases} 1, & \text{if } n_1 \text{ and } n_2 \text{ are squares of some positive integers,} \\ 0, & \text{otherwise.} \end{cases}$$

In this case, observe that

$$st_{\mathcal{A}}^{(2)} - \lim_{n_1,n_2} x_{n_1 n_2} = 0, \tag{8.38}$$

although the sequence $\langle x_{n_1 n_2} \rangle$ is not P-convergent.
Then, observe that the operators $\overline{\mathfrak{U}_{n_1,q_{n_1}}^{n_2,q_{n_2}}}$ defined by (8.37) satisfy all the hypotheses of Theorem 8.5.2. Hence, by (8.36) and (8.38), we obtain, for all $f \in C(\mathcal{I}^2)$:

$$st_{\mathcal{A}}^{(2)} - \lim_{n_1,n_2} \left\| \overline{\mathfrak{U}_{n_1,q_{n_1}}^{n_1,q_{n_2}}}(f) - f \right\| = 0.$$

Since $\langle x_{n_1 n_2} \rangle$ is not P-convergent, the sequence $\langle \overline{\mathfrak{U}_{n_1,q_{n_1}}^{n_2,q_{n_2}}} \rangle$ does not converge uniformly to the function $f \in C(\mathcal{I}^2)$, which demonstrates that Theorem 8.5.2 is a non-trivial generalization of its classical Korovkin version given in [5].

In the following theorem, we establish the order of $\mathcal{A}$-statistical convergence of the operators $\mathfrak{U}_{n_1,q_{n_1}}^{n_2,q_{n_2}}$ by using mixed modulus of smoothness (cf. [21], p. 242).

Theorem 8.5.4. *Let $\mathcal{A} = [a_{j,k}^{(n_1,n_2)}]$ be a non-negative RH-regular sumability matrix. Assume that the following conditions hold:*

$$P - \lim_{j,k} \frac{1}{e_{j,k}} \sum_{(n_1,n_2) \in \mathbb{N}^2} a_{j,k}^{(n_1,n_2)} = 1 \tag{8.39}$$

and

$$\omega_B(f; \mu_{n_1}, \nu_{n_2}) = st_{\mathcal{A}}^{(2)} - o(f_{n_1,n_2}) \text{ as } n_1, n_2 \to \infty, \tag{8.40}$$

where $\mu_{n_1} = 4(1-\eta_{n_1}^{(r_1)}) + \frac{\eta_{n_1}^{(r_1)}}{[n_1]_{q_{n_1}}}$ and $\nu_{n_2} = 4(1-\zeta_{n_2}^{(r_2)}) + \frac{\zeta_{n_2}^{(r_2)}}{[n_2]_{q_{n_2}}}$ and e_{n_1,n_2} and f_{n_1,n_2} be positive non-increasing double sequences. Then for every $f \in C_b(\mathcal{I}^2)$, the operator $\mathfrak{U}_{n_1,q_{n_1}}^{n_2,q_{n_2}}(f)(\mathbf{x})$ verifies the following inequality

$$\left\| \mathfrak{U}_{n_1,q_{n_1}}^{n_2,q_{n_2}}(f) - f \right\| = st_{\mathcal{A}}^{(2)} - o(c_{n_1,n_2}) \text{ as } n_1, n_2 \to \infty,$$

where $c_{n_1,n_2} = \max\{e_{n_1,n_2}, f_{n_1,n_2}\}$ for each $(n_1,n_2) \in \mathbb{N}^2$ and $\omega_B\left(f; \sqrt{\mu_{n_1}}, \sqrt{\nu_{n_2}}\right)$ is the mixed modulus of smoothness.

Proof. From the definition of $\Delta f[(\mathbf{t}); (\mathbf{x})]$ and the inequality (8.31), for any $\delta_1, \delta_2 > 0$, we have

$$|\mathfrak{U}_{n_1,q_{n_1}}^{n_2,q_{n_2}}(f)(\mathbf{x}) - f(\mathbf{x})| \leq \mathfrak{K}_{n_1,q_{n_1}}^{n_2,q_{n_2}}(|\Delta f[(\mathbf{t}); (\mathbf{x})]|)(\mathbf{x})$$

$$\leq \omega_B\left(f; \delta_1, \delta_2\right) \Big[\mathfrak{K}_{n_1,q_{n_1}}^{n_2,q_{n_2}}(e_0^0)(\mathbf{x}) + \delta_1^{-1}\delta_2^{-1}\mathfrak{K}_{n_1,q_{n_1}}^{n_2,q_{n_2}}(|t_1 - x_1|)(\mathbf{x})$$

$$\mathfrak{K}_{n_1,q_{n_1}}^{n_2,q_{n_2}}(|t_2 - x_2|; x_1, x_2) + \delta_2^{-1}\mathfrak{K}_{n_1,q_{n_1}}^{n_2,q_{n_2}}(|t_2 - x_2|)(\mathbf{x}) + \delta_1^{-1}\mathfrak{K}_{n_1,q_{n_1}}^{n_2,q_{n_2}}(|t_1 - x_1|)(\mathbf{x}) \Big].$$

Applying Cauchy-Schwarz inequality and Lemma 8.2.1, we obtain

$$\left\|\mathfrak{U}_{n_1,q_{n_1},\eta^{(1)},\eta^{(2)}\ldots\eta^{(r_1)}}^{n_2,q_{n_2},\zeta^{(1)},\zeta^{(2)}\ldots\zeta^{(r_2)}}(f) - f\right\| \leq \omega_B(f;\delta_1,\delta_2)\left[1 + \delta_1^{-1}\sqrt{\varkappa_{n_1,p_{n_1},q_{n_1},2}^{\eta^{(1)},\eta^{(2)}\ldots\eta^{(r_1)}}(x_1)}\right.$$

$$+ \frac{1}{\delta_1\delta_2}\sqrt{\varkappa_{n_1,p_{n_1},q_{n_1},2}^{\eta^{(1)},\eta^{(2)}\ldots\eta^{(r_1)}}(x_1)\,\varkappa_{n_2,p_{n_2},q_{n_2},2}^{\zeta^{(1)},\zeta^{(2)}\ldots\zeta^{(r_2)}}(x_2)} + \delta_2^{-1}\sqrt{\varkappa_{n_2,p_{n_2},q_{n_2},2}^{\zeta^{(1)},\zeta^{(2)}\ldots\zeta^{(r_2)}}(x_2)}\right]$$

$$\leq \omega_B(f;\delta_1,\delta_2)\left[1 + \delta_1^{-1}\sqrt{4(1-\eta_{n_1}^{(r_1)}) + \frac{\eta_{n_1}^{(r_1)}}{[n_1]_{q_1}}}\right.$$

$$+ \delta_1^{-1}\delta_2^{-1}\sqrt{4(1-\eta_{n_1}^{(r_1)}) + \frac{\eta_{n_1}^{(r_1)}}{[n_1]_{q_1}}}\sqrt{4(1-\zeta_{n_2}^{(r_2)}) + \frac{\zeta_{n_2}^{(r_2)}}{[n_2]_{q_2}}}$$

$$+ \delta_2^{-1}\sqrt{4(1-\zeta_{n_2}^{(r_2)}) + \frac{\zeta_{n_2}^{(r_2)}}{[n_2]_{q_2}}}\right].$$

For all $(n_1, n_2) \in \mathbb{N}^2$ and by choosing $\delta_1 = \sqrt{\mu_{n_1}}$ and $\delta_2 = \sqrt{\nu_{n_2}}$ which leads us to the result

$$\left\|\mathfrak{U}_{n_1,q_{n_1}}^{n_2,q_{n_2}}(f) - f\right\| \leq 4\,\omega_B\left(f;\sqrt{\mu_{n_1}},\sqrt{\nu_{n_2}}\right), \qquad (8.41)$$

Now, for any $\epsilon > 0$, let consider the sets:

$$\mathcal{T} = \{(n_1, n_2) \in \mathbb{N}^2 : \left\|\mathfrak{U}_{n_1,q_{n_1},\eta^{(1)},\eta^{(2)}\ldots\eta^{(r_1)}}^{n_2,q_{n_2},\zeta^{(1)},\zeta^{(2)}\ldots\zeta^{(r_2)}}(f) - f\right\| \geq \epsilon\},$$

$$\mathcal{T}_1 = \{(n_1, n_2) \in \mathbb{N}^2 : \omega_B\left(f;\sqrt{\mu_{n_1}},\sqrt{\nu_{n_2}}\right) \geq \frac{\epsilon}{4}\}.$$

Then in view of (8.41), we obtain $\mathcal{T} \subseteq \mathcal{T}_1$, which implies that

$$\frac{1}{c_{j,k}}\sum_{(n_1,n_2)\in\mathcal{T}} a_{j,k}^{(n_1,n_1)} \leq \frac{1}{c_{j,k}}\sum_{(n_1,n_2)\in\mathcal{T}_1} a_{j,k}^{(n_1,n_2)} \leq \frac{1}{f_{j,k}}\sum_{(n_1,n_2)\in\mathcal{T}_1} a_{j,k}^{(n_1,n_2)}, \quad (8.42)$$

where $c_{n_1,n_2} = \max\{e_{n_1,n_2}, f_{n_1,n_2}\}$. From (8.40) and letting $j, k \to \infty$ in (8.42), we find that

$$P - \lim_{j,k\to\infty} \frac{1}{c_{j,k}}\sum_{(n_1,n_2)\in T} a_{j,k}^{(n_1,n_2)} = 0.$$

This completes the requirement. □

Remark 8.5.5. *(1) Now, specializing the sequences* $\langle e_{n_1,n_2}\rangle$ *and* $\langle f_{n_1,n_2}\rangle$ *in Theorem 8.5.4, we can easily get Theorem 8.5.2 at once.*
(2) If one replaces the matrix $\mathcal{A}$, *by the double identity matrix and considers* $e_{n_1,n_2} = f_{n_1,n_2} = 1$, *for all* $n_1, n_2 \in \mathbb{N}$, *then the ordinary rate of convergence of the operators* $\mathfrak{U}_{n_1,q_{n_1}}^{n_1,q_{n_2}}$ *is obtained.*

Bibliography

[1] Agrawal, P.N. and İspir, N. 2016. Degree of approximation for bivariate Chlodowsky-Szász-Charlier type operators, *Results Math.* 69:3: 369–385.

[2] Alotaibi, A. 2022. Approximation of GBS type q-Jakimovski-Leviatan-Beta integral operators in Bögel space, *Mathematics* 10:5: 675.

[3] Altin, A., Taşdelen, F. and Erkuş, E. 2006. The q-Lagrange polynomials in several variables, *Taiwanese. J. Math.* 10:5: 1131–1137.

[4] Badea, C., Badea, I. and Gonska, H.H. 1988. Notes on the degree of approximation of B-continuous and B-differentiable functions, *J. Approx. Theory Appl.* 4:2: 95–108.

[5] Badea, C., Badea, I. and Gonska, H.H. 1986. A test function theorem and approximation by pseudo polynomials, *Bull. Austral. Math. Soc.* 34:1: 53–64.

[6] Badea, C. and Cottin, C. 1990. Korovkin-type theorems for generalized boolean sum operators, *Approximation Theory* (Kecskemét, 1990), ser. Colloq. Math. Soc. János Bolyai. Amsterdam: North-Holland, 58: 51–68.

[7] Baron, S. and Stadtmüller, U. 1997. Tauberian theorems for power series methods applied to double sequences, *J. Math. Anal. Appl.* 211:2: 574–589

[8] Baxhaku, B. and Agrawal, P.N. 2017. Degree of approximation for bivariate extension of Chlodowsky-type q-Bernstein–Stancu–Kantorovich operators, *Appl. Math. Comput.* 306: 56–72.

[9] Baxhaku, B., Agrawal, P.N. and Shukla, R. 2020. Bivariate positive linear operators constructed by means of q-Lagrange polynomials, *J. Math. Anal. Appl.* 491:2: 124337.

[10] Bögel, K. 1934. Mehrdimensionale Differentiation von Funktionen mehrerer reeller Veränderlichen, *J. Reine Angew. Math.* 170: 197–217.

[11] Bögel, K. 1935. Ubër die mehrdimensionale Differentiation Integration und beschrankte Variation, *J. Reine Angew. Math.* 173: 5–29.

[12] Boos, J. and Cass, F.P. 2000. *Classical and modern methods in summability*, Clarendon Press.

[13] Chan, W.C.C., Chyan, C. J. and Srivastava, H. M. 2001. The Lagrange polynomials in several variables, *Integral Transforms Spec. Funct.* 12: 139–148.

[14] Demirci, K., Yıldız, S. and Dirik, F. 2020. Approximation via power series method in two-dimensional weighted spaces, *Bull. Malays. Math. Sci.* 43:6: 3871–3883.

[15] Dirik, F. and Demirci, K. 2010. Four-dimensional matrix transformation and the rate of A-statistical convergence of continuous functions, *Comput. Math. Appl.* 59:8: 2976–2981.

[16] Dirik, F. and Demirci, K. 2011. Four-dimensional matrix transformation and rate of A-statistical convergence of Bögel-type continuous functions, *Stud. Univ. Babeş-Bolyai Math.* 56:3: 95–104.

[17] Dobrescu, E. Matei, I. 1966. The approximation by Bernstein type polynomials of bidimensionally continuous functions, *An. Univ. Timisoara Ser. Şti. Mat.-Fiz*, 4: 85–90.

[18] Erkuş-Duman, E. 2012. Statistical approximation by means of operators constructed by q-Lagrange polynomials, *J. Comput. Anal. Appl.* 14:1: 67–77.

[19] Duman, O. 2007. Regular matrix transformations and rates of convergence of positive linear operators, *Calcolo* 44:3: 159–164.

[20] Erkuş, E., Duman, O. and Srivastava, H.M. 2006. Statistical approximation of certain positive linear operators constructed by means of the Chan–Chyan–Srivastava polynomials, *Appl. Math. Comput.* 182:1: 213–222.

[21] Gupta, V., Rassias, Th.M., Agrawal, P.N. and Acu, A.M. 2018. *Recent Advances in Constructive Approximation Theory*, Springer.

[22] Mohiuddine, S.A., Acar, T. and Alotaibi, A. 2017. Construction of a new family of Bernstein-Kantorovich operators, *Math. Methods Appl. Sci.* 40:18: 7749–7759.

[23] Mohiuddine, S.A., Ahmad, N., Özger, F., Alotaibi, A. and Hazarika, B. 2021. Approximation by the parametric generalization of Baskakov–Kantorovich operators linking with Stancu operators, *Iran. J. Sci. Technol. Trans. A, Sci.* 45:2: 593–605.

[24] Mursaleen, M., Khan, A., Srivastava, H.M. and Nisar, K.S. 2013. Operators constructed by means of q-Lagrange polynomials and A-statistical approximation, *Appl. Math. Comput.* 219:12: 6911–6918.

[25] Özgüç, İ. and Taş, E. 2016. A Korovkin-type approximation theorem and power series method, *Results. Math.* 69:3: 497–504.

[26] Pringsheim, A. 1900. Zur theorie der zweifach unendlichen Zahlenfolgen, *Math. Ann.* 53:3: 289–321.

[27] Robison, G.M. 1926. Divergent double sequences and series, *Trans. Amer. Math. Soc.* 28:1: 50–73.

[28] Stadtmüller, U. and Tali, A.1999. On certain families of generalized Nörlund methods and power series methods, *J. Math. Anal. Appl.* 238:1: 44–66.

Chapter 9

Results on interpolative Boyd-Wong contraction in quasi-partial b-metric space

Pragati Gautam

Swapnil Verma

9.1	Introduction and preliminaries	155
9.2	Main results	159
	Bibliography	165

9.1 Introduction and preliminaries

In the early years of the twentieth century, the renowned mathematician Banach [2] commenced the concept of the Banach Contraction Principle. i.e., Let T be a self map on a non-empty set X and d is a complete metric. If there exists a constant $\rho \in [0,1)$ such that

$$d(T\xi, T\wp) \leq \rho d(\xi, \wp) \quad \text{for all } \xi, \wp \in X,$$

then it possesses a unique fixed point in X. This idea motivated researchers to explore the other forms of contractive conditions and to generalize new spaces in various directions. Also, see [1, 3, 4, 6, 22, 30, 34]. In 1994, Matthews [27] presented the notion of partial-metric space. Later on, several authors obtained generalized version of celebrated Banach contraction principle (also see [18, 21, 32, 33]). It is well known fact that a map which satisfies the prominent Banach contraction principle is continuous. Here the question arises whether a discontinuous map satisfying similar contractive conditions owns a unique fixed point. In the year 1968, Kannan [23] provided an answer to this question and defined a new type of contraction, i.e.,

$$d(T\xi, T\wp) \leq \rho[d(\xi, T\xi) + d(\wp, T\wp)] \quad \text{for all } \xi, \wp \in X,$$

DOI: 10.1201/9781003330868-9

where $\rho \in [0, \frac{1}{2})$. On evolution of new contractive mappings, Reich [31] generalized the Kannan and Banach contractions in 1972, e.g., a self map $T: X \to X$ is called a Reich-contraction mapping if there are $\alpha_1, \alpha_2, \alpha_3 \in [0,1)$ and $\alpha_1 + \alpha_2 + \alpha_3 < 1$ such that

$$d(T\xi, T\wp) \leq \alpha_1 d(\xi, T\xi) + \alpha_2 d(\wp, T\wp) + \alpha_3 d(\xi, \wp) \quad \text{for all } \xi, \wp \in X.$$

Reich–Rus–Ćirić proved that a self map $T: X \to X$ is called a Reich–Rus–Ćirić contraction map on a complete metric space (X, d) if there are $\rho \in [0, \frac{1}{3})$ such that

$$d(T\xi, T\wp) \leq \rho[d(\xi, \wp) + d(\xi, T\xi) + d(\wp, T\wp)],$$

for all $\xi, \wp \in X$, then T possesses a unique fixed point.

Very recently in 2018, Karapinar [24] adopted an interpolative approach to define the generalized Kannan-type contraction on a complete metric space.

We recall that a self-map $T: X \to X$ is said to be an interpolative Kannan type contraction for a metric space (X, d), if there are constants $\rho \in [0, 1)$ and $\alpha_1 \in (0, 1)$ such that

$$d(T\xi, T\wp) \leq \rho[d(\xi, T\xi)]^{\alpha_1} \cdot [d(\wp, T\wp)]^{1-\alpha_1} \quad \text{for all} \xi, \wp \in X \setminus Fix(T),$$

where $Fix(T) = \{z \in X : Tz = z\}$.

In continuation, Karapinar et al. [25, 26] introduced the concept of interpolative Reich–Rus–Ćirić and Hardy–Rogers type contraction and proved the following fixed point results.

Theorem 9.1.1. *[25] In the notion of partial metric space (X, d), if a mapping $T: X \to X$ is an interpolative Reich–Rus–Ćirić type contraction, i.e., there are constants $\rho \in [0, 1)$ and $\alpha_1, \alpha_2 \in (0, 1)$ such that $d(T\xi, T\wp) \leq \rho[d(\xi, \wp)]^{\alpha_2}[d(\xi, T\xi)]^{\alpha_1}; \cdot [d(\wp, T\wp)]^{1-\alpha_1-\alpha_2}$ for all $\xi, \wp \in X \setminus Fix(T)$, then T owns a fixed point.*

Theorem 9.1.2. *[26] Let (X, d) be a metric space. If the self-mapping $T: X \to X$ is an interpolative Hardy–Rogers type contraction i.e., there exists $\rho \in [0, 1)$ and $\alpha_1, \alpha_2, \alpha_3 \in (0, 1)$ with $\alpha_1 + \alpha_2 + \alpha_3 < 1$, such that*

$$d(T\xi, T\wp) \leq \rho[d(\xi, \wp)]^{\alpha_2}[d(\xi, T\xi)]^{\alpha_1} \cdot d(\wp, T\wp)]^{\alpha_3} \cdot$$
$$\left[\frac{1}{2}(d(\xi, T\wp) + d(\wp, T\xi))\right]^{1-\alpha_1-\alpha_2-\alpha_3} \quad \text{for all } \xi, \wp \in X \setminus Fix(T),$$

then T possesses a fixed point of X.

In 1974, Wong [35] generalized the Banach contraction to introduce the concept of Wong contraction by using the approach of auxiliary functions.

Definition 9.1.3. *[35] Suppose that there exist functions $f_i \colon (0, \infty) \to (0, \infty), i = 1, 2, 3, 4, 5$ such that*

(i) each f_i is upper semi-continuous from the right,

(ii) $\sum_{i=1}^{5} f_i(u) < u$ for any $u > 0$

then $\{f_i\}_{i=1}^{5}$ is said to be a set of Wong functions.

Theorem 9.1.4. *[35] Let T be a self mapping on a metric space (X, d) and $\{f_i\}_{i=1}^{5}$ be a set of Wong functions. Consider T is a Wong type contraction if*

$$d(T\xi, T\wp) \leq a_1 d(\xi, \wp) + a_2 d(\xi, T\xi) + a_3 d(\wp, T\wp) + a_4 d(\xi, T\wp) + a_5 d(T\xi, \wp)$$

for any $\xi, \wp \in X$ with $\xi \neq \wp$, where $a_i = f_i(d(\xi, \wp))/d(\xi, \wp)$. Then T has exactly one fixed point.

Gupta and Gautam [19, 20] developed a new metric space known as quasi-partial b-metric space and proved fixed point theorems on this space. Since then, especially in the last decade, several authors have contributed in the development in metric fixed point theory in the framework of quasi-partial b-metric space (see [5, 7, 9, 10, 12, 13, 15–17, 21, 28, 29]). In this paper, we proved a quasi-partial b-metric version of the interpolative Boyd Wong contraction and established the existence of fixed point in this space. Next, we will recall the basic notions of a quasi-partial b-metric space.

Definition 9.1.5. *[20] A quasi-partial b-metric on a non-empty set X is a function $qp_b : X \times X \to \mathbb{R}^+$ such that for some real number $s \geq 1$ and all $\xi, \wp, z \in X$:*

1. $qp_b(\xi, \xi) = qp_b(\xi, \wp) = qp_b(\wp, \wp)$ implies $\xi = \wp$,
2. $qp_b(\xi, \xi) \leq qp_b(\xi, \wp)$,
3. $qp_b(\xi, \xi) \leq qp_b(\wp, \xi)$,
4. $qp_b(\xi, \wp) \leq s[qp_b(\xi, z) + qp_b(z, \wp)] - qp_b(z, z)$.

(X, qp_b) is called a quasi-partial b-metric space where X is a non-empty set and qp_b defines a quasi-partial b-metric on X. The number s is called the coefficient of (X, qp_b).

Let qp_b be a quasi-partial b-metric on the set X. Then

$$d_{qp_b}(\xi, \wp) = qp_b(\xi, \wp) + qp_b(\wp, \xi) - qp_b(\xi, \xi) - qp_b(\wp, \wp) \text{ is a } b\text{-metric on } X.$$

Lemma 9.1.6. *[9] Let (X, qp_b) be a quasi-partial b-metric space. Then the following hold:*

1. *If $qp_b(\xi, \wp) = 0$ then $\xi = \wp$.*
2. *If $\xi \neq \wp$, then $qp_b(\xi, \wp) > 0$ and $qp_b(\wp, \xi) > 0$.*

Definition 9.1.7. *[10] Let (X, qp_b) be a quasi-partial b-metric. Then*

1. *A sequence $\{\xi_n\} \subset X$ converges to $\xi \in X$ if and only if*

$$qp_b(\xi, \xi) = \lim_{n \to \infty} qp_b(\xi, \xi_n) = \lim_{n \to \infty} qp_b(\xi_n, \xi).$$

2. *A sequence $\{\xi_n\} \subset X$ is called a Cauchy sequence if and only if*

$$\lim_{n,m \to \infty} qp_b(\xi_n, \xi_m) \text{ and } \lim_{m,n \to \infty} qp_b(\xi_m, \xi_n) \text{ exist (and are finite)}.$$

3. *The quasi partial b-metric space (X, qp_b) is said to be complete if every Cauchy sequence $\{\xi_n\} \subset X$ converges with respect to τ_{qp_b} to a point $\xi \in X$ such that*

$$qp_b(\xi, \xi) = \lim_{n,m \to \infty} qp_b(\xi_n, \xi_m) = \lim_{m,n \to \infty} qp_b(\xi_m, \xi_n).$$

4. *A mapping $f: X \to X$ is said to be continuous at $\xi_0 \in X$ if, for every $\varepsilon > 0$, there exists*

$$\delta > 0 \quad \text{such that } f(B(\xi_0, \delta)) \subset B(f(\xi_0), \varepsilon).$$

Lemma 9.1.8. *[28] Let (X, qp_b) be a quasi-partial b-metric space and (X, d_{qp_b}) be the corresponding b-metric space. Then (X, d_{qp_b}) is complete if (X, qp_b) is complete.*

Lemma 9.1.9. *[17] Let (X, qp_b) be a quasi-partial b-metric space and $T: X \to X$ be a given mapping. T is said to be sequentially continuous at $z \in X$ if for each sequence $\{\xi_n\}$ in X converging to z, we have: $G\xi_n \to Tz$, that is, $qp_b(G\xi_n, Tz) = qp_b(Tz, Tz)$.*

Lemma 9.1.10. *Let (X, qp_b) be a complete metric space and $\{\xi_n]\}$ be a sequence in X such that $\lim_{n \to +\infty} qp_b(\xi_n, \xi_{n+1}) = 0$. If ξ_n is not a Cauchy sequence, then there exist $\epsilon > 0$ and subsequences $\{m(k)\}$ and $\{n(k)\}$ in $\mathbb{N}$ with $n(k) > m(k) > k$ such that $qp_b(\xi_{n(k)}, \xi_{m(k)}) \geq \epsilon$ and $qp_b(\xi_{n(k)-1}, \xi_{m(k)}) < \epsilon$, so that the following holds:*

(i) $\lim_{k \to +\infty} qp_b(\xi_{n(k)-1}, \xi_{m(k)-1}) = \epsilon$;

(ii) $\lim_{k \to +\infty} qp_b(\xi_{n(k)-1}, \xi_{m(k)-1}) = \epsilon$;

Definition 9.1.11 ([35]). *Let ψ be the set of functions $\psi: [0, \infty) \to [0, \infty)$ such that*

1. $\psi(0) = 0$

2. $\psi(t) < t$ for each $t > 0$

3. ψ is upper semi-continuous from the right.

9.2 Main results

In this section, we will discuss an interpolative Boyd Wong contraction and an interpolative Reich–Rus–Ćirić type contaction to prove existence of non unique fixed point in notion of quasi-partial b-metric space.

Definition 9.2.1. *Consider (X, qp_b) be a quasi-partial b-metric space. We say the self map $T\colon X \to X$ is an interpolative Boyd-Wong type contraction, if there exist $\alpha_1, \alpha_2, \alpha_3 \in (0, \frac{1}{s})$ with $\alpha_1 + \alpha_2 + \alpha_3 < 1$ and a non-increasing function $\varphi \in \psi$ such that*

$$qp_b(T\xi, T\wp) \leq \psi[qp_b(\xi, \wp)]^{\alpha_2} \cdot [qp_b(\xi, T\wp)]^{\alpha_1} \cdot [qp_b(\wp, T\wp)]^{\alpha_3}$$
$$\cdot \left[\frac{1}{2s}(qp_b(\xi, T\wp) + qp_b(\wp, T\xi))\right]^{1-\alpha_1-\alpha_2-\alpha_3} \quad (9.1)$$

for all $\xi, \wp \in X/Fix(T)$.

Theorem 9.2.2. *Let (X, qp_b) be a complete quasi-partial b-metric space and a self map T be an interpolative Boyd-Wong type contraction. Then T owns a fixed point in X.*

Proof. Let us assume the initial point $\xi_0 \in X$, consider $\{\xi_n\}$ be a sequence given as $\xi_n = T^n(\xi_0)$ for each positive integer n. If there exist n_0 such that $\xi_{n_0} = \xi_{n_0+1}$, then ξ_{n_0} is a fixed point of T. The proof is done. Next, suppose that $\xi_n \neq \xi_{n+1}$ for all $n \geq 0$.
By substituting the values $\xi = \xi_n$ and $\wp = \xi_{n-1}$ in equation (9.1), we have
$qp_b(\xi_{n+1}, \xi_n) = qp_b(T\xi_n, T\xi_{n-1})$

$$\leq \psi([qp_b(\xi_n, \xi_{n-1})]^{\alpha_2} \cdot [qp_b(\xi_n, T\xi_n)]^{\alpha_1} \cdot [qp_b(\xi_{n-1}, T\xi_{n+1})]^{\alpha_3} \cdot$$
$$\frac{1}{2s}[(qp_b(\xi_n, \xi_n) + qp_b(\xi_{n-1}, \xi_{n+1}))]^{1-\alpha_1-\alpha_2-\alpha_3}$$

$$\leq \psi([qp_b(\xi_n, \xi_{n-1})]^{\alpha_2} \cdot [qp_b(\xi_n, \xi_{n+1})]^{\alpha_1} \cdot [qp_b(\xi_{n+1}, \xi_n)]^{\alpha_3} \cdot$$
$$\frac{1}{2s}[(qp_b(\xi_{n+1}, \xi_n) + qp_b(\xi_n, \xi_{n+1})]^{1-\alpha_1-\alpha_2-\alpha_3}) \quad (9.2)$$

Assume that $qp_b(\xi_{n-1}, \xi_n) < qp_b(\xi_n, \xi_{n+1})$ for some $n \geq 1$. Thus,

$$\frac{1}{2s}[(qp_b(\xi_{n+1}, \xi_n) + qp_b(\xi_n, \zeta_{n+1})] \leq qp_b(\xi_n, \xi_{n+1}).$$

Consequently, the inequality (9.2) yields that

$$\begin{aligned} 0 &< qp_b(\xi_n, \xi_{n+1}) \\ &\leq \psi([qp_b(\xi_{n-1}, \xi_n)]^{\alpha_2+\alpha_3} \cdot [qp_b(\xi_n, \xi_{n+1}]^{1-\alpha_2-\alpha_3}) \\ &\leq [qp_b(\xi_{n-1}, \xi_n)]^{\alpha_2+\alpha_3} \cdot [qp_b(\xi_n, \xi_{n+1}]^{1-\alpha_2-\alpha_3}. \end{aligned}$$

So,
$$[qp_b(\xi_n, \xi_{n+1})]^{\alpha_2+\alpha_3} \leq [qp_b(\xi_{n-1}, \xi_n)]^{\alpha_2+\alpha_3} \tag{9.3}$$
which is a contradiction. Thus we have
$$qp_b(\xi_n, \xi_{n+1}) \leq qp_b(\xi_{n-1}, \xi_n)$$
for all $n \geq 1$.

Hence, $\{qp_b(\xi_{n-1}, \xi_n)\}$ is a non-increasing sequence with positive terms. Let $l = \lim_{n \to \infty} qp_b(\xi_{n-1}, \xi_n)$.
$$\frac{1}{2s}(qp_b(\xi_{n-1}, \xi_n) + qp_b(\xi_n, \xi_{n+1})) \leq qp_b(\xi_{n-1}, \xi_n)$$
for all $n \geq 1$.

By using elimination, the inequality (9.2) implies that
$$qp_b(\xi_n, \xi_{n+1}) \leq \psi(qp_b(\xi_{n-1}, \xi_n)).$$
As we know that ψ is upper semi-continuous from the right, we get
$$\begin{aligned} l &= \lim_{n\to\infty} qp_b(\xi_n, \xi_{n+1}) \\ &\leq \lim_{n\to\infty} Sup\psi(qp_b(\xi_{n-1}, \xi_n)) \\ &= \psi(l) \\ &< l \end{aligned}$$
which is a contradiction. Therefore, we have
$$l = \lim_{n\to\infty} qp_b(\xi_n, \xi_{n+1}) = 0 \tag{9.4}$$

We prove that $\{\xi_n\}$ is a Cauchy sequence.

We argue by contradiction, that is, $\{\xi_n\}$ is not a Cauchy sequence. This means that there exists $\epsilon > 0$ for which we can find subsequences of integers $\{m_k\}$ and $\{n_k\}$ with $n_k > m_k > k$ such that
$$qp_b(\xi_{m_k}, \xi_{n_k}) \geq \epsilon. \tag{9.5}$$
Further, corresponding to m_k, we can choose n_k such aa way that it is the smallest integer with $n_k > m_k$ and satisfying (9.5). Then
$$qp_b(\xi_{m_k}, \xi_{n_k-1}) < \epsilon \tag{9.6}$$
By substituting the values $\xi = \xi_{n_k-1}$ and $\wp = \wp_{m_k-1}$ in (9.1), we find that
$$qp_b(\xi_{n_k}, \xi_{m_k}) \tag{9.7}$$
$$= qp_b(T\xi_{n_k-1}, T\xi_{m_k-1})$$
$$\leq \psi([qp_b(\xi_{n_k-1}, \xi_{m_k-1})]^{\alpha_2}[qp_b(\xi_{n_k-1}, T\xi_{n_k-1})]^{\alpha_1} \cdot [qp_b(\xi_{m_k-1}, T\xi_{m_k-1})]^{\alpha_3}) \cdot$$
$$[\frac{1}{2s}(qp_b(\xi_{n_k-1}, T\xi_{m_k-1}) + qp_b(\xi_{m_k-1}, T\xi_{n_k-1})))]^{1-\alpha_1-\alpha_2-\alpha_3} \tag{9.8}$$
$$= \psi([qp_b(\xi_{n_k-1}, \xi_{m_k-1})]^{\alpha_2}[qp_b(\xi_{n_k-1}, \xi_{n_k})]^{\alpha_1} \cdot [qp_b(\xi_{m_k-1}, \xi_{m_k})]^{\alpha_3}) \cdot$$
$$[\frac{1}{2s}(qp_b(\xi_{n_k-1}, T\xi_{m_k}) + qp_b(\xi_{m_k-1}, T\xi_{n_k})))]^{1-\alpha_1-\alpha_2-\alpha_3}.$$

Using the upper semi-continuity of ψ, (9.4) and Lemma 9.1.10, we obtain that
$$\epsilon \leq \psi(0) = 0$$
which is a contradiction. Therefore, $\{\xi_n\}$ is a Cauchy sequence in the complete quasi-partial b-metric space (X, qp_b), so there exists $u \in X$ such that $\lim_{n \to +\infty} qp_b(\xi_n, u) = 0$. Assume that $\xi \neq T\xi$. Since $\xi_n \neq T\xi_n$ for each $n \geq 0$, by letting $\xi = \xi_n$ and $\wp = u$ in (9.1), we get

$$\begin{aligned} qp_b(\xi_{n+1}, Tu) &= qp_b(T\xi_n, Tu) \\ &\leq \psi([qp_b(\xi_n, u)]^{\alpha_2} \cdot [qp_b(\xi_n, T\xi_n)]^{\alpha_1} \cdot [qp_b(u, Tu)]^{\alpha_3} \cdot \\ &\quad [\frac{1}{2s}(qp_b(\xi_n, Tu) + qp_b(u, \xi_{n+1}))]^{1-\alpha_1-\alpha_2-\alpha_3}). \quad (9.9) \end{aligned}$$

Letting $n \to \infty$, in the inequality (9.9) and using the upper semi-continuity of ψ, we find out
$$qp_b(u, Tu) \leq \psi(0) = 0,$$
which is a contradiction. Thus $Tu = u$. Hence, T owns a fixed point. □

Corollary 9.2.3. *Let us consider a complete quasi-partial b-metric space (X, qp_b) and a self map $T \colon X \to X$. Assume there exist $\alpha_1, \alpha_2 \in (0, \frac{1}{s})$ with $\alpha_1 + \alpha_2 < 1$ and a non-decreasing function $\psi \in \Psi$ such that*

$$qp_b(T\xi, T\wp) \leq \psi([qp_b(\xi, \wp)]^{\alpha_2} \cdot [qp_b(\xi, T\xi)]^{\alpha_1} \cdot [\frac{1}{s}qp_b(\wp, T\wp)]^{1-\alpha_1-\alpha_2}), \quad (9.10)$$

for all $\xi, \wp \in X/Fix(T)$. Then T has a fixed point in X.

Corollary 9.2.4. *Let us consider a complete quasi-partial b-metric space (X, qp_b) and a self map $T \colon X \to X$. Suppose there exist $\alpha_1, \alpha_2, \alpha_3 \in (0, \frac{1}{s})$ with $\alpha_1 + \alpha_2 + \alpha_3 < 1$ and $\lambda \in [0, 1)$ such that*

$$\begin{aligned} qp_b(T\xi, T\wp) &\leq \lambda([qp_b(\xi, \wp)]^{\alpha_2} \cdot [qp_b(\xi, T\xi)]^{\alpha_1} \cdot [qp_b(\wp, T\wp)]^{\alpha_3} \\ &\quad \cdot [\frac{1}{2s}(qp_b(\xi, T\wp) + qp_p(\wp, T\xi)]^{1-\alpha_1-\alpha_2-\alpha_3}), \quad (9.11) \end{aligned}$$

for all $\xi, \wp \in X/Fix(T)$. Then T posses a fixed point in X.

Proof. It is sufficient to take $\psi(t) = \lambda t$ in Theorem 9.2.2. □

In our next result, we will prove the existence of fixed point in the realm of interpolative Reich–Rus–Ćirić contraction in the framework of quasi-partial b-metric space.

Definition 9.2.5. *Let (X, qp_h) be a complete quasi-partial b-metric space. Let a self map $T \colon X \to X$ is said to be interpolative Reich–Rus–Ćirić contraction if there are some $\lambda \in [0, 1)$, $\alpha_2, \alpha_3 \in (0, 1)$, $\alpha_2 + \alpha_3 < 1$, and $s \geq 1$ such that the condition*

$$\frac{1}{s} qp_b(T\xi, T\wp) \leq \lambda [qp_b(\xi, \wp)]^{\alpha_2} [qp_b(\xi, T\xi)]^{\alpha_3} [qp_b(\wp, T\wp)]^{1-\alpha_2-\alpha_3} \quad (9.12)$$

is satisfied for all $\xi, \wp \in X$ such that $T\xi \neq \xi$ whenever $T\wp \neq \wp$.

Theorem 9.2.6. Let a self map $T: X \to X$ is an interpolative Reich–Rus–Ćirić type contr (X, qp_b, s). Then T owns a fixed point in X.

Proof. Let us consider an arbitrary point $\xi_0 \in X$. Now suppose a sequence ξ_n such that $\xi_n = T^n(\xi_0), n \geq 0$.
If, $\xi_n \neq \xi_{n+1}$, For each $n \geq 0$. We have

$$qp_b(\xi_{n+1}, \xi_n) = qp_b(T\xi_n, T\xi_{n-1})$$
$$\leq \lambda[qp_b(\xi_n, \xi_{n-1})]^{\alpha_2} \cdot [qp_b(\xi_n, T\xi_n)]^{\alpha_3} \cdot [qp_b(\xi_{n-1}, T\xi_{n-1})]^{1-\alpha_3-\alpha_2}$$
$$= \lambda[qp_b(\xi_n, \xi_{n-1})]^{\alpha_2} \cdot [qp_b(\xi_n, \xi_{n+1})]^{\alpha_3} \cdot [qp_b(\xi_{n-1}, \xi_n)]^{1-\alpha_3-\alpha_2}$$
$$= [qp_b(\xi_n, \xi_{n+1})]^{\alpha_3} \cdot [qp_b(\xi_{n-1}, \xi_n)]^{1-\alpha_3}$$
$$qp_b(\xi_{n+1}, \xi_n)^{1-\alpha_3} \leq \lambda[qp_b(\xi_{n-1}, \xi_n)]^{1-\alpha_3}.$$

By using the above inequality, we get $qp_b(\xi_{n+1}, \xi_n)$ is a non increasing sequence. Next, we shall prove that $\lim_{n \to \infty} qp_b(\xi_{n-1}, \xi_n) = 0$.
Let us assume that $\lim_{n \to \infty} qp_b(\xi_{n-1}, \xi_n) = l$, where $l \geq 1$ Consider,

$$qp_b(\xi_n, \xi_{n+1}) \leq \lambda qp_b(\xi_{n-1}, \xi_n)$$
$$\leq \lambda^n qp_b(\xi_0, \xi_1)$$

Thus $qp_b(\xi_n, \xi_{n+1}) = 0$ when $n \to \infty$ because $\lambda < 1$.

$$\lim_{n \to \infty} qp_b(\xi_n, \xi_{n+1}) = 0 \tag{9.13}$$

Now, we will prove that ξ_n is a Cauchy sequence in (X, qp_b, s).
We have,
$\lim_{n \to \infty} qp_b(\xi_n, \xi_{n+1}) = s[qp_b(\xi_n, \wp) + qp_b(\wp, \xi_{n+1})] - qp_b(\wp, \wp) = 0 \longrightarrow \lim_{n \to \infty \wp} qp_b(\xi_n, \xi_n) = 0$.
Let k be the smallest integer which satisfies above equation such that

$$qp_b(\xi_{l_k-1}, \xi_{n_k}) < \epsilon.$$

By the definition of a quasi-partial b-metric space,

$$\epsilon \leq qp_b(\xi_{l_k}, \xi_{n_k})$$
$$\leq s[qp_b(\xi_{l_k}, \xi_{l_k-1}) + qp_b(\xi_{l_k-1}, \xi_{n_k})] - qp_b(\xi_{l_k-1}, \xi_{l_k-1})$$
$$\leq s[qp_b(\xi_{l_k}, \xi_{l_k-1}) + qp_b(\xi_{l_k-1}, \xi_{n_k})]$$
$$< sqp_b(\xi_{l_k}, \xi_{l_k-1}) + s\epsilon.$$

Thus,
$$\lim_{n \to \infty} qp_b(\xi_{l_k}, \xi_{n_k}) = \epsilon.$$

That is,
$$\lim_{n \to \infty} (qp_b(\xi_{l_k}, \xi_{n_k}) - qp_b(\xi_{l_k-1}, \xi_{n_k-1})) = s\epsilon.$$

Since
$$qp_b(\xi_{l_k}, \xi_{n_k}) \leq s[qp_b(\xi_{l_k}, \xi_{l_k+1}) + qp_b(\xi_{l_k+1}, \xi_{n_k})] + qp_b(\xi_{n_k+1}, \xi_{n_k}),$$
and
$$qp_b(\xi_{l_k+1}, \xi_{n_k+1}) \leq qp_b(\xi_{l_k}, \xi_{l_k+1}) + qp_b(\xi_{l_k}, \xi_{n_k}) + qp_b(\xi_{n_k+1}, \xi_{n_k}).$$
Consequently, we conclude that
$$qp_b(\xi_{n_k+1}, \xi_{n_k}) \leq qp_b(\xi_{l_k}, \xi_{l_k+1}). \tag{9.14}$$
On taking the limit as $k \to \infty$, together with (9.13) and (9.14) we have
$$\lim_{k \to \infty} qp_b(\xi_{l_k+1}, \xi_{n_k+1}) = \epsilon.$$
Therefore, there exists $n_1 \in \mathbb{N}$ such that for all $k \geq n_1$ we have
$$qp_b(\xi_{l_k}, \xi_{n_k}) > \frac{\epsilon}{2} \text{ and } qp_b(\xi_{l_k+1}, \xi_{n_k+1}) > \frac{\epsilon}{2} > 0. \tag{9.15}$$
Since T is continuous, we have
$qp_b(T\xi_{l_k}, T\xi_{n_k}) \leq ([qp_b(\xi_{l_k}, \xi_{n_k})]^{\alpha_2}[qp_b(\xi_{l_k}, T\xi_{l_k})]^{\alpha_3}[qp_b(\xi_{n_k}, T\xi_{n_k})]^{1-\alpha_3-\alpha_2}),$
when $k \to \infty$,
$$qp_b(T\xi_{l_k}, T\xi_{n_k}) \leq \lambda(\epsilon) < \epsilon.$$
This implies that $qp_b(T\xi_{l_k}, T\xi_{n_k}) < \epsilon$ which is a contradiction, and therefore, ξ_n is a Cauchy sequence. Regarding the completeness of the quasi-partial b-metric space (X, qp_b, s), we deduce that there is some $\xi \in X$ so that
$$\lim_{n \to \infty} qp_b(\xi_n, \xi) = 0.$$
Since T is continuous, we have $\xi = \lim_{n \to \infty} \xi_{n+1} = \lim_{n \to \infty} T\xi_n = T(\lim_{n \to \infty} \xi_n) = T\xi$. Hence T has a fixed point. □

Example 9.2.7. Let $X = \{3, 4, 6, 12\}$. Define a quasi-partial b-metric space $qp_b(\xi, \wp) = |\xi - \wp| + \xi$ as given in Table 9.1,
$$\psi(t) = \begin{cases} t & ,0 \leq t \leq 1 \\ \frac{t}{3} & ,t > 1 \end{cases}$$

TABLE 9.1:

$qp_b(\xi, \wp)$	3	4	6	12
3	3	4	6	12
4	5	4	6	12
6	9	8	6	12
12	21	20	18	12

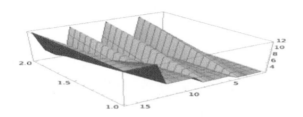

FIGURE 9.1: T has fixed points 3 and 4.

We define a self-mapping T on X by $T\colon \begin{pmatrix} 3 & 4 & 6 & 12 \\ 3 & 4 & 3 & 4 \end{pmatrix}$ as shown in Figure 9.1.
Let us consider $(\xi, \wp \in XFixT)$, then $(\xi, \wp) \in \{(6,6), (12,12), (6,12), (12,6)\}$.
Choose $\alpha_1 = \alpha_2 = \alpha_3 = \frac{1}{5}$.
Case 1: Let $(\xi, \wp) = (6, 12)$. We have

$$qp_b(T\xi, T\wp) = qp_b(3, 4) = 4 \leq \psi(13.46)$$
$$= \psi(qp_b(\xi, \wp))^{\frac{1}{5}}(qp_b(\xi, T\xi))^{\frac{1}{5}}(qp_b(\wp, T\wp))^{\frac{1}{5}} \cdot \frac{1}{2s}(qp_b(\xi, T\wp) + qp_b(\wp, T\xi))^{\frac{2}{5}}.$$

Case 2: Let $(\xi, \wp) = (6, 6)$, we have

$$qp_b(T\xi, T\wp) = qp_b(3, 3) = 3 \leq \psi(8.28)$$
$$= \psi(qp_b(\xi, \wp))^{\frac{1}{5}}(qp_b(\xi, T\xi))^{\frac{1}{5}}(qp_b(\wp, T\wp))^{\frac{1}{5}} \cdot \frac{1}{2s}(qp_b(\xi, T\wp) + qp_b(\wp, T\xi))^{\frac{2}{5}}.$$

Thus 3 and 4 are the fixed point of T. Hence, T is an interpolative Boyd Wong contraction.

Example 9.2.8. Let $X = \{x, y, z, w\}$. Define a complete quasi partial b-metric space as:
$qp_b(x,x) = qp_b(y,y) = qp_b(z,z) = qp_b(w,w) = 1$; $qp_b(x,x) = qp_b(y,y) = 3$; $qp_b(x,x) = qp_b(y,y) = 1/4$; $qp_b(x,x) = qp_b(y,y) = 5$; $qp_b(x,x) = qp_b(y,y) = 6$; $qp_b(x,x) = qp_b(y,y) = 1/2$; $qp_b(x,x) = qp_b(y,y) = 1/4$
We define a self-map T on X by $T\colon \begin{pmatrix} 3 & 4 & 6 & 12 \\ 3 & 4 & 3 & 4 \end{pmatrix}$ Choose $\alpha_1 = \alpha_2 = \alpha_3 = \frac{1}{4}$. We can conclude that

$$qp_b(Tx, Ty) = qp_b(y, x) = 3 \leq \psi(3^{3/4})$$
$$= \psi(qp_b(\xi, \wp))^{\frac{1}{4}}(qp_b(\xi, T\xi))^{\frac{1}{4}}(qp_b(\wp, T\wp))^{\frac{1}{5}} \cdot \frac{1}{2s}(qp_b(\xi, T\wp) + qp_b(\wp, T\xi))^{\frac{1}{4}}.$$

Thus, y is the fixed point of T in the setting of interpolative Boyd Wong contraction. Many more fixed points can be obtained in similar manner. Hence, a fixed point exists but is not unique.

Bibliography

[1] Aydi, H., Chen, C. M. and Karapınar, E. 2019. Interpolative Reich-Rus-Ćirić Type Contractions via the Branciari Distance. *Mathematics* 7: 84, doi:10.3390/math7010084.

[2] Banach, S. 1922. Sur les opérationsdans les ensembles abstraits et leur application aux équationsintégrales. *Fund Math.* 3: 133–181.

[3] Bakhtin, I.A. 1989. The contraction principle in quasi-metric spaces. *Int. Funct. Anal.* 30: 26–37.

[4] Branciari, A. 2000. A fixed point theorem of Banach-Caccioppoli type on a class of generalized metric spaces. *Publ. Math. Debrecen* 57: 31–37.

[5] Ćirić, L.J., Samet, B., Aydi, H. and Vetro, C. 2011. Common fixed points of generalized contractions on partial-metric spaces and an application. *Appl. Math. Comput.* 218: 2398–2406.

[6] Czerwik, S. 1993. Contraction mappings in b-metric spaces. *Acta Math. Inform. Univ. Ostrav.* 1: 5–11.

[7] Debnath, P. and de La Sen, M. 2019. Set-valued interpolative Hardy-Rogers and set-valued Reich-Rus-Ćirić-type contractions in b-metric spaces. *Mathematics* 7: 849, doi:10.3390/math7090849.

[8] Gaba, Y. U. and Karapınar, E. 2019. A new approach to the interpolation contraction. *Axioms* 8: 1–4.

[9] Gautam, P. and Verma, S. 2021. Fixed point via implicit contraction mapping on quasi-partial b-metric space. *J. Anal.* 2021: https://doi.org/10.1007/s41478-021-00309-6.

[10] Gautam, P., Sánchez Ruiz, L. M., Verma, S. and Gupta, G. 2021. Common Fixed point results on generalized weak compatible mapping in quasi-partial b-metric space. *J. Math.* 2021: Article ID 5526801.

[11] Gupta, A.; Gautam, P. 2015. A version of coupled fixed point theorems on quasi-partial b-metric spaces. *Adv. Fixed Point Theory*, 5: 407–419.

[12] Gautam, P., Verma, S., De La Sen, M. and Marwaha, P.R. 2021. On ω−interpolative berinde weak contraction in quasi-partial b-metric spaces. *Int. J. Anal. Appl.* 19:4: 619–632.

[13] Gautam, P., Verma, S., De La Sen, M. and Sundriyal, S. 2021. Fixed point results for ω−interpolative Chatterjea type contraction in quasi-partial b-metric space. *Int. J. Anal. Appl.* 19:2: 280–287.

[14] Gupta, A. and Gautam, P. 2016. A new common tripled fixed point result in two quasi-partial b-metric spaces. *J. Adv. Math. Comput.* 1–20.

[15] Gupta, A. and Gautam, P. 2015. Some coupled fixed point theorems on quasi–partial b–metric spaces. *Int. J. Math. Anal.* 9: 293–306.

[16] Gautam, P., Mishra, V. N., Ali, R. and Verma, S. 2021. Interpolative Chatterjea and cyclic Chatterjea contraction on quasi–partial b–metric space. *AIMS Math.* 6: 1727–1742.

[17] Gautam, P., Sánchez Ruiz, L.M. and Verma, S. 2021. Fixed Point of Interpolative Rus–Reich–Ćirić Contraction Mapping on Rectangular Quasi-Partial b-Metric Space. *Symmetry* 13:32: 2–16.

[18] Geraghty, M. 1973. On contractive mappings. *Proc. Amer. Math. Soc.* 40: 604–608.

[19] Gupta, A. and Gautam, P. 2016. Topological structure of quasi-partial b-metric spaces. *Int. J. Pure Math. Sci.* 17: 8–18.

[20] Gupta, A. and Gautam, P. 2015. Quasi partial b-metric spaces and some related fixed point theorems. *Fixed Point Theory Appl.* 18: doi:10.1186/s13663-015-0260-2.

[21] Hardy, G.E. and Rogers, T.D. 1973. A generalization of a fixed point theorem of Reich. *Can. Math. Bull.* 16: 201–206.

[22] Janos, L. 1978. A converse of Banach's contraction theorem. *Proc. Amer. Math. Soc.* 68: 121–124.

[23] Kannan, R. 1968. Some results on fixed points. *Bull. Calcutta Math. Soc.* 60: 71–76.

[24] Karapınar, E. 2018. Revisiting the Kannan type contractions via interpolation. *Adv. Theory Nonlinear Anal. Appl.* 2: 85–87.

[25] Karapınar, E., Agarwal, R.P. and Aydi, H. 2018. Interpolative Reich-Rus-Ćirić type contractions on partial-metric spaces. *Mathematics* 6:256: doi:10.3390/math6110256.

[26] Karapınar, E., Alqahtani, O. and Aydi, H. 2018. On interpolative Hardy-Rogers type contractions. *Symmetry* 11:8: doi:10.3390/sym11010008.

[27] Matthews, S.G. 1994. Partial metric topology. *Ann. N. Y. Acad. Sci.* 728: 183–197.

[28] Mishra, V.N., Sánchez Ruiz, L.M., Gautam, P. and Verma, S. 2020. Interpolative Reich–Rus–Ćirić and Hardy–Rogers Contraction on Quasi-Partial b-Metric Space and Related Fixed Point Results. *Mathematics* 8: 1598.

[29] Mohammadi, B., Parvaneh, V. and Aydi, H. 2019. On extended interpolative Ćirić–Reich–Rus type F-contractions and an application. *J. Inequal. Appl.* 290: https://doi.org/10.1186/s13660-019-2227-z

[30] Rakotch, E. 1962. A note on contractive mappings. *Amer. Math. Soc.* 13: 459–465.

[31] Reich, S. 1972. Fixed point of contractive functions. *Boll. Un. Mat. Ital.* 4: 26–42

[32] Rhoades, B.E. 1977. A comparison of various definitions of contractive mappings. *Trans. Amer. Math. Soc.* 226: 257–290.

[33] Shukla, S. 2014. Partial b-metric spaces and fixed point theorems. *Mediterr. J. Math.* 11: 703–711.

[34] Wilson, W. A. 1931. On quasi-metric spaces. *Amer. J. Math.* 53: 675–684.

[35] Wong, C.S. 1974. Generalized contractions and fixed point theorems. *Proc. Amer. Math. Soc.* 42: 409–417.

Chapter 10

Applications of differential transform method on some functional differential equations

Anil Kumar

Giriraj Methi

Sanket Tikare

10.1	Introduction	169
10.2	Preliminaries	170
	10.2.1 Definition of differential transform	171
	10.2.2 Faà di Bruno's formula and Bell polynomials	171
	10.2.3 Description of the method	173
	10.2.4 Convergence results	174
	10.2.5 Error estimate	175
10.3	Applications	175
	10.3.1 Example 1	176
	10.3.2 Example 2	178
	10.3.3 Example 3	181
	10.3.4 Example 4	184
	10.3.5 Example 5	186
	Bibliography	188

10.1 Introduction

The functional differential equations with proportional delay were first studied by Ockendon and Taylor [18] in their work of collecting electric current for the pantograph of an electric locomotive, hence named pantograph equations. The investigation of these equations is important since they find applications in economic activities involving production, planning and decision

DOI: 10.1201/9781003330868-10

making, number theory, biological phenomena, probability concepts applied on algebraic structures, electrodynamics, quantum mechanics, nautical science, and astrophysics, among others [2, 6, 14, 15]. Further, time delays are significant in engineering problems such as feedback loops equipped with sensors and actuators, the transmission of signals to remote center, in predictions and control systems, etc. [1, 9, 11, 16, 28].

Several analytical and numerical methods have been proposed by many researchers to study solutions of proportional delay differential equations (PDDEs) which include Adomian decomposition method (ADM) [5], Homotopy perturbation method (HPM) [26], Homotopy analysis method (HAM) [12], Variational iteration method (VIM) [8, 29], Taylor series [25], Rungakutta method [30] and Collocation method [4]. Due to calculation of Adomian polynomials in ADM, evaluation of integrals in HPM, finding Lagrangian multipliers in VIM and discretization of variables and complex calculations in numerical methods make them unsuitable. We propose a simple approach involving the differential transformation in this chapter. The differential transformation has been introduced by G. Pukhov as the "Taylor transform" in 1976 and applied to the study of electrical circuits [19]. The differential transformation is closely related to Taylor expansion of real analytic functions. It has applications in solving different types of problems for all classes of differential equations (ordinary, partial, delayed, fractional, fuzzy etc.). The recent development and applications of DT are discussed in [3, 7, 13, 17, 20, 21, 23, 24] and references therein.

In the present chapter, the differential transformation is applied to solve proportional delay differential equations. The nonlinearity in the problems is addressed by using the partial ordinary Bell polynomials in the Faà di Bruno's formula. The results obtained by this technique are compared with analytical solutions. Detailed error analysis is provided. However, to the best of our knowledge, no researcher has applied the DTM using Bell polynomials on the practical problems discussed in the Section 10.3.

The chapter is organized as follows. In Section 10.2, we introduce the main idea and basic formulae of the differential transformation and provide necessary results to handle nonlinearity using partial ordinary Bell polynomials and discuss convergence results. Applications of the method are presented in Section 10.3.

10.2 Preliminaries

In this section, we discuss the main idea and basic formulae of the differential transformation as well as notation and results related to transformation of general nonlinear terms.

10.2.1 Definition of differential transform

Let $w(v)$ be a real analytical function in a domain Ω and $v = v_0$ be an arbitrary point in Ω. Then, $w(v)$ can be expanded in a Taylor series in a neighborhood of the point $v = v_0$.

Definition 10.2.1. *[22] The differential transformation of a real function $w(v)$ at a point $v_0 \in \mathbb{R}$ is $\mathcal{D}\{w(v)\}[v_0] = \{W(k)[v_0]\}_{k=0}^{\infty}$, where $W(k)[v_0]$, the differential transform of the kth derivative of the function $w(v)$ at v_0, is defined as*

$$W(k)[v_0] = \frac{1}{k!}\left[\frac{d^k w(v)}{dv^k}\right]_{v=v_0}. \tag{10.1}$$

Definition 10.2.2. *[22] The inverse differential transformation is given by*

$$w(v) = \mathcal{D}^{-1}\left\{\{W(k)[v_0]\}_{k=0}^{\infty}\right\}[v_0] = \sum_{k=0}^{\infty} W(k)[v_0](v - v_0)^k. \tag{10.2}$$

Combining Definition 10.2.1 and Definition 10.2.2 give an expression of the function w in the form of the Taylor series:

$$w(v) = \sum_{k=0}^{\infty} \frac{1}{k!}\left[\frac{d^k w(v)}{dv^k}\right]_{v=v_0} (v - v_0)^k. \tag{10.3}$$

In practical applications, the function $w(v)$ is expressed by a finite sum

$$w(v) = \sum_{k=0}^{N} W(k)[v_0](v - v_0)^k, \tag{10.4}$$

since N can be chosen large enough to ensure that the effect of the remainder $\sum_{k=N+1}^{\infty} W(k)[v_0](v - v_0)^k$ is arbitrarily small. The results which are used in this chapter are listed in Table 10.1 without proofs.

10.2.2 Faà di Bruno's formula and Bell polynomials

In the literature, it has been observed that differential transformation is not applied directly to nonlinear terms like $w^n, n \in \mathbb{N}$ or e^w. Authors [23] used Adomian polynomials to compute the differential transform of nonlinear terms. However, the differential transformation of nonlinear terms can be determined without calculating and evaluating symbolic derivatives by applying Faà di Bruno's formula to non-linear terms.

TABLE 10.1: Formulae of the differential transform method

	Original function	Transformed function
1	$\dfrac{d^n w(v)}{dv^n}$	$(k+1)(k+2)(k+3)\ldots(k+n)W(k+n)$
2	$w(v) = v^n$	$\delta(k-n) = \begin{cases} 1, k=n \\ 0, k \neq n \end{cases}$
3	$e^{\alpha v}$	$\dfrac{\alpha^k}{k!}$
4	$w_1(v) w_2(v)$	$\sum_{i=0}^{k} W_1(i) W_2(k-i)$
5	$w(\alpha v)$	$\alpha^k W(k)$
6	$w_1(\alpha_1 v) w_2(\alpha_2 v)$	$\sum_{i=0}^{k} (\alpha_1)^i (\alpha_2)^{k-i} W_1(i) W_2(k-i)$

Definition 10.2.3. *[10] The partial ordinary Bell polynomials are the polynomials $\hat{B}_{k,l}(\hat{x}_1, \ldots, \hat{x}_{k-l+1})$ in an infinite number of variables $\hat{x}_1, \hat{x}_2, \ldots$ defined by the series expansion*

$$\sum_{k \geq l} \hat{B}_{k,l}(\hat{x}_1, \ldots, \hat{x}_{k-l+1}) v^k = \left(\sum_{m \geq 1} \hat{x}_m v^m \right)^l, l = 0, 1, 2, \ldots. \qquad (10.5)$$

Lemma 10.2.4. *[22] The partial ordinary Bell polynomials $\hat{B}_{k,l}(\hat{x}_1, \ldots, \hat{x}_{k-l+1}), l = 0, 1, 2, \ldots, k \geq l$ satisfy the recurrence relation*

$$\hat{B}_{k,l}(\hat{x}_1, \ldots, \hat{x}_{k-l+1}) = \sum_{l=1}^{k-l+1} \frac{i.l}{k} \hat{x}_i \hat{B}_{k-i,l-1}(\hat{x}_1, \ldots, \hat{x}_{k-i-l+2}), \qquad (10.6)$$

where $\hat{B}_{0,0} = 1$ and $\hat{B}_{k,0} = 0$ for $k \geq 1$.

Theorem 10.2.5. *[22] Let g and f be real functions analytic near v_0 and $g(v_0)$, respectively, and let h be the composition $h(v) = (f \circ g)(v) = f(g(v))$. Denote $D\{g(v)\}[v_0] = \{G(k)\}_{k=0}^{\infty}$, $D\{f(v)\}[g(v_0)] = \{F(k)\}_{k=0}^{\infty}$ and $D\{(f \circ g)(v)\}[v_0] = \{H(k)\}_{k=0}^{\infty}$ the differential transformations of functions g, f, and h at v_0, $g(v_0)$, and v_0, respectively. Then the numbers $H(k)$ in the sequence $\{H(k)\}_{k=0}^{\infty}$ satisfy the relations*

$$H(0) = F(0)$$
$$H(k) = \sum_{l=1}^{k} F(l) \hat{B}_{k,l}(G(1), \ldots, G(k-l+1)) \text{ for } k \geq 1. \qquad (10.7)$$

Applications of differential transform method 173

10.2.3 Description of the method

Consider the proportional delay differential equation defined by

$$w^n(v) = f\left(v, w(v), w'(v), \ldots, w^{(n-1)}(v), w\left(\frac{v}{\alpha_1}\right),\right.$$
$$\left. w\left(\frac{v}{\alpha_2}\right), \ldots, w\left(\frac{v}{\alpha_r}\right)\right), \quad (10.8)$$

where $\alpha_i \geq 1$ and $w^{(n)}$ is the nth derivative of w with respect to v, for $n, r \in \mathbb{N}$.

Consider equation (10.8) subject to initial function $\phi(v) \in C^n([v^*, 0], \mathbb{R})$ where $v^* < 0$ such that

$$\phi(v_0) = w(v_0), \phi'(v_0) = w'(v_0), \ldots, \phi^{(n-1)}(v_0) = w^{(n-1)}(v_0), \quad (10.9)$$

and subject to initial conditions

$$w(v_0) = w_0, w'(v_0) = w_1, \ldots, w^{(n-1)}(v_0) = w_{n-1}. \quad (10.10)$$

Now equation (10.8) can be written in the form

$$L(w) + R(w) + M(v) = N(w). \quad (10.11)$$

The linear terms are split into L and R, where L is the highest order bounded linear operator and R is the remaining of the linear operators which are also bounded, M are continuous known functions satisfy the Lipschitz condition, and N are nonlinear terms.

Apply DT with Bell polynomial on equation (10.10)–(10.11),

$$\mathcal{D}(L(w)) + \mathcal{D}(R(w)) + \mathcal{D}(M(v)) = \mathcal{B}(N(w)), \quad (10.12)$$

$$\mathcal{D}(w(v_0)) = w_0, \mathcal{D}(w'(v_0)) = w_1, \ldots, \mathcal{D}\left(w^{(n-1)}(v_0)\right) = w_{n-1}, \quad (10.13)$$

where $\mathcal{D}$ is DT operator and $\mathcal{B}$ is Bell polynomial operator.

From equation (10.12)–(10.13) we obtain following recursive relations,

$$\frac{(k+n)!}{k!}W(k+n) + W(k) + M(k) = \mathcal{B}(N(w)), \quad (10.14)$$

$$W(0) = w_0, W(1) = w_1, \ldots, W(n-1) = \frac{1}{(n-1)!}w_{(n-1)}. \quad (10.15)$$

If $N(w) = H(v) = f(g(v))$ then nonlinear Bell polynomial operator $\mathcal{B}$ are defined by Theorem (10.2.5) as

$$H(0) = F(0),$$
$$H(k) = \sum_{l=1}^{k} F(l) \cdot \hat{B}_{k,l}(G(1), \ldots, G(k-l+1)) \text{ for } k \geq 1. \quad (10.16)$$

We can easily obtain different components using equations (10.14)–(10.16), and then using inverse transformation, we obtain approximate solution in the form of Taylor series

$$w(v) = \sum_{k=0}^{\infty} W(k)(v - v_0)^k. \tag{10.17}$$

10.2.4 Convergence results

In this section, we discuss the convergence results used in this chapter. The proof is taken from [23, 27].

Theorem 10.2.6. *Let f be an analytical function in $[0, T] \times \mathbb{R}^n \times \mathbb{R}^n$. Assume that problem (10.8) has unique solution in some interval $[0, T]$. Let $B_k = W(k)v^k$. If there exist a constant δ, $0 < \delta < 1$, $k_0 \in \mathbb{N}$ such that $\|(B_{k+1}(v))\| \leq \alpha \|(B_k(v))\|$ for all, then the series converges to the unique solution on the interval $J = [0, \alpha]$, $\alpha \leq T$.*

Proof. Let $C^n(J)$ be a Banach space of vector-valued functions $h(v) = (h_1(v), \ldots, h_p(v))^T$ with continuous derivatives up to order n and norm $\|h(v)\| = \max_{i=1,\ldots,p} \max_{j=0,\ldots,n} \max_{v \in J} \left| h_i^{(j)}(v) \right|$.

Assume $S_l = \sum_{k=0}^{l} B_k(v)$. Now it is sufficient to prove that sequence $\{S_l\}$ is a Cauchy sequence in $C^n(J)$.

Due to

$$\|S_{l+1} - S_l\| = \|B_{l+1}(v)\| \leq \delta \|B_l(v)\| \leq \ldots \delta^{l-n_0+1} \|B_{n_0}(v)\|$$

for every $l, m \in \mathbb{N}$, $l \geq m > n_0$, we get

$$\|S_l - S_m\| = \left\| \sum_{j=m}^{l-1} (S_{j+1} - S_j) \right\| \leq \sum_{j=m}^{l-1} \|(S_{j+1} - S_j)\| \leq \sum_{j=m}^{l-1} \delta^{j-n_0+1} \|B_{n_0}(v)\|$$

$$= \delta^{m-n_0+1}(1 + \delta + \delta^2 + \cdots + \delta^{l-m+1}) \|B_{n_0}(v)\|$$

$$= \frac{1 - \delta^{l-m}}{1 - \delta} \delta^{m-n_0+1} \|B_{n_0}(v)\|. \tag{10.18}$$

Since $0 < \delta < 1$, it follows that

$$\lim_{l,m \to \infty} \|S_l - S_m\| = 0.$$

Therefore $\{S_l\}$ is a Cauchy sequence in $C^n(J)$ and the proof is complete. □

Theorem 10.2.7. *Suppose that the assumptions of Theorem (10.2.6) are valid. Then for the truncated series $\sum_{k=0}^{m} B_k(v)$, the following error estimate holds*

$$\left\| w(v) - \sum_{k=0}^{m} B_k(v) \right\| \leq \frac{1}{1-\delta} \delta^{m-m_0+1} \max_{i=1,\ldots,p} \max_{j=0,\ldots,n} \left| \frac{m_0!}{(m_0-j)!} W_i(m_0) \alpha^{m_0-j} \right|$$

for any $m \geq 0$, $m \geq m_0$.

Proof. Without loss of generality, we can choose $m_0 \geq n$, where n is the order of the system (10.8). From inequality (10.18) we have

$$\|S_l - S_m\| \leq \frac{1-\delta^{l-m}}{1-\delta} \delta^{m-m_0+1} \|B_{m_0}(v)\|$$

$$= \frac{1-\delta^{l-m}}{1-\delta} \delta^{m-m_0+1} \max_{i=1,\ldots,p} \max_{j=0,\ldots,n} \left| \frac{m_0!}{(m_0-j)!} W_i(m_0) \alpha^{m_0-j} \right| \tag{10.19}$$

for $l \geq m \geq m_0$.
From $0 < \delta < 1$ it follows $(1-\delta^{l-m}) < 1$. Hence, inequality (10.19) can be reduced to

$$\frac{1}{1-\delta} \delta^{m-m_0+1} \max_{i=1,\ldots,p} \max_{j=0,\ldots,n} \left| \frac{m_0!}{(m_0-j)!} W_i(m_0) \alpha^{m_0-j} \right|$$

Hence, we use the fact that $l \to \infty$, $S_l \to w(v)$, and so proof is complete. □

10.2.5 Error estimate

For comparison, absolute error and maximum absolute error are computed and defined as

$$E_N(v) := |w(v) - w_N(v)|,$$
$$E_{N,\infty} := \max_{0 \leq v \leq 1} E_N(v),$$

where $w(v)$ is the exact solution and $w_N(v)$ is the truncated series solution with degree N. Furthermore, the relative error between exact and approximate solution is defined by $R_N(v) := \frac{E_N(v)}{|w(v)|}$.

10.3 Applications

Five examples are discussed to show reliability and accuracy of the presented method. The MATHEMATICA software version 11 has been used for numerical computations.

10.3.1 Example 1

Consider the following linear proportional DDE

$$w''(v) = 2e^{\frac{2v}{3}} w\left(\frac{v}{3}\right) - w'(v) - w\left(\frac{v}{2}\right) + e^{v/2}, \ 0 \leq v \leq 1, \qquad (10.20)$$

with initial conditions

$$w(0) = w'(0) = 1. \qquad (10.21)$$

The exact solution is given by

$$w(v) = e^v. \qquad (10.22)$$

Applying differential transform to equations (10.20)–(10.21), we obtain the following recursive relation

$$W(k+2) = \frac{1}{(k+1)(k+2)} \left(2 \sum_{r=0}^{k} \left(\frac{1}{3}\right)^k \frac{2^{k-r}}{(k-r)!} W(r) - (k+1)W(k+1) \right.$$
$$\left. - \left(\frac{1}{2}\right)^k W(k) + \left(\frac{1}{2}\right)^k \frac{1}{k!} \right), \qquad (10.23)$$

$$W(0) = W(1) = 1. \qquad (10.24)$$

Using equations (10.23)–(10.24), we obtain the following components,

$$k = 0, \ W(2) = \frac{1}{2}(2W(0) - W(1) - W(0) + 1) = \frac{1}{2!},$$
$$k = 1, \ W(3) = \frac{1}{6} \left(2\left(\frac{2}{3}W(0) + \frac{1}{3}W(1)\right) - 2W(2) - \frac{1}{2}W(1) + \frac{1}{2}\right) = \frac{1}{3!},$$
$$k = 2, \ W(4) = \frac{1}{4!}, \dots, \text{ and so on.} \qquad (10.25)$$

Now, using equation (10.4), the series solution is given by

$$w(v) = 1 + v + \frac{1}{2!}v^2 + \frac{1}{3!}v^3 + \dots, \qquad (10.26)$$

which converges to the exact solution given by equation (10.22). The approximate solution for $N = 12$ is compared with the analytical solution in Table 10.2, where N represents a number of terms considered. Table 10.3 lists the maximal absolute error of approximate results obtained by the present method for $N = 5, 10$, and 15. Figure 10.1 depicts absolute errors for the numerical solutions for $N = 5, 10$, and 15. From these results, it is clear that absolute errors and maximal absolute errors all decline systematically with the increase in N.

Applications of differential transform method 177

TABLE 10.2: Comparison of numerical solution $w(v)$ with exact solution when $N = 12$ for Example 1

v	$w(v)$	$w_N(v)$	$R_N(v)$
0.1	1.105170918	1.105170918	2.0E-16
0.2	1.221402758	1.221402758	2.0E-16
0.3	1.349858807	1.349858807	1.6E-16
0.4	1.491824697	1.491824697	7.4E-16
0.5	1.648721270	1.648721270	1.2E-14
0.6	1.822118800	1.822118800	1.2E-13
0.7	2.013752707	2.013752707	8.1E-13
0.8	2.225540928	2.225540928	4.2E-12
0.9	2.459603111	2.459603111	1.7E-11
1.0	2.718281828	2.718281828	6.3E-11

TABLE 10.3: Maximum absolute errors for w of Example 1

N	$E_{N,\infty}$
5	9.9E-03
10	3.0E-07
15	8.1E-13

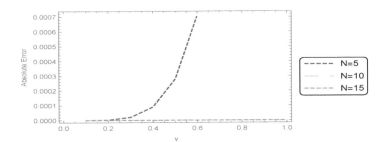

FIGURE 10.1: Absolute errors for $N = 5$, 10, and 15 of Example 1.

10.3.2 Example 2

Consider the following linear system of proportional DDEs

$$u'(v) = u(v) + 2e^v w\left(\frac{v}{3}\right) - e^{2v}, \qquad (10.27)$$

$$w'(v) = 2w(v) + e^{2v} u\left(\frac{v}{2}\right), \quad 0 \le v \le 1, \qquad (10.28)$$

with initial conditions

$$u(0) = w(0) = 1. \qquad (10.29)$$

The exact solution is given by

$$u(v) = e^{2v}, \quad w(v) = e^{3v}. \qquad (10.30)$$

Employing the differential transform to equations (10.27)–(10.29), we obtain the following recursive relation

$$U(k+1) = \frac{1}{(k+1)}\left(U(k) + 2\sum_{r=0}^{k}\frac{1}{r!}\left(\frac{1}{3}\right)^{k-r} W(k-r) - \frac{2^k}{k!}\right), \qquad (10.31)$$

$$W(k+1) = \frac{1}{(k+1)}\left(2W(k) + \sum_{r=0}^{k}\frac{2^r}{r!}\left(\frac{1}{2}\right)^{k-r} U(k-r)\right), \qquad (10.32)$$

$$U(0) = W(0) = 1. \qquad (10.33)$$

Using equations (10.31)–(10.33), we obtain the following components,

$$k = 0, \ U(1) = U(0) + 2W(0) - 1 = 2,$$
$$W(1) = 2W(0) + U(0) = 3,$$
$$k = 1, \ U(2) = \frac{1}{2}\left(U(1) + 2\left(\frac{1}{3}W(1) + W(0)\right) - 2\right) = \frac{2^2}{2!},$$
$$W(2) = \frac{1}{2}\left(2W(1) + \frac{1}{2}U(1) + 2U(0)\right) = \frac{3^2}{2!},$$
$$k = 2, \ U(3) = \frac{2^3}{3!},$$
$$W(3) = \frac{3^3}{3!}, \ldots, \text{ and so on.} \qquad (10.34)$$

Now, with the help of equation (10.4), the series solution is given by

$$u(v) = 1 + 2v + \frac{2^2}{2!}v^2 + \frac{2^3}{3!}v^3 + \ldots, \qquad (10.35)$$

$$w(v) = 1 + 3v + \frac{3^2}{2!}v^2 + \frac{3^3}{3!}v^3 + \ldots, \qquad (10.36)$$

which converges to the exact solution given by equation (10.30).

Applications of differential transform method

The approximate solution for $N = 12$ is compared with the analytical solution in Tables (10.4)–(10.5), where N represents a number of terms considered. Table (10.6) lists the maximal absolute error of approximate results obtained by the present method for $N = 5, 10$, and 15. Figure 10.2 depicts absolute errors for the numerical solutions for $N = 5$, 10, and 15. From these results, it is clear that absolute errors and maximal absolute errors all decline systematically with the increase in N.

TABLE 10.4: Comparison of numerical solution $u(v)$ with exact solution when $N = 12$ for Example 2

v	$u(v)$	$u_N(v)$	$R_N(v)$
0.1	1.221402758	1.221402758	0
0.2	1.491824698	1.491824698	2.4E-14
0.3	1.822118800	1.822118800	2.6E-12
0.4	2.225540928	2.225540928	6.8E-11
0.5	2.718281828	2.718281826	8.3E-10
0.6	3.320116923	3.320116902	6.1E-09
0.7	4.055199967	4.055199834	3.2E-08
0.8	4.953032424	4.953031755	1.3E-07
0.9	6.049647464	6.049644666	4.6E-07
1.0	7.389056099	7.389046016	1.3E-06

TABLE 10.5: Comparison of numerical solution $w(v)$ with exact solution when $N = 12$ for Example 2

v	$w(v)$	$w_N(v)$	$R_N(v)$
0.1	1.349858808	1.349858808	9.8E-16
0.2	1.8221188	1.8221188	2.6E-12
0.3	2.459603111	2.459603111	2.5E-10
0.4	3.320116923	3.320116902	6.1E-09
0.5	4.48168907	4.481688764	6.8E-08
0.6	6.049647464	6.049644666	4.6E-07
0.7	8.166169913	8.166151643	2.2E-06
0.8	11.02317638	11.0230832	8.4E-06
0.9	14.87973172	14.87933803	2.6E-05
1.0	20.08553692	20.08410308	7.1E-05

TABLE 10.6: Maximum absolute errors for u and w of example 2

N	$E_{N,\infty}$ for u	$E_{N,\infty}$ for w
5	3.8E-01	3.7E-00
10	3.4E-04	2.2E-02
15	2.8E-08	1.3E-05

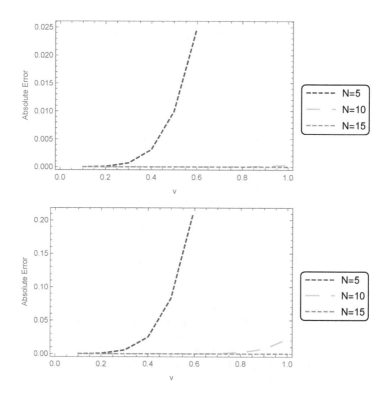

FIGURE 10.2: Absolute errors for u and w for $N = 5$, 10, and 15 of Example 2.

10.3.3 Example 3

Consider the following nonlinear system of proportional DDEs

$$u'(v) = u\left(\frac{v}{2}\right) w\left(\frac{v}{4}\right) - u(v) - u\left(\frac{v}{4}\right), \tag{10.37}$$

$$w'(v) = u\left(\frac{v}{2}\right) w(v) - w(v) - w\left(\frac{v}{2}\right), \quad 0 \leq v \leq 1, \tag{10.38}$$

with initial conditions

$$u(0) = w(0) = 1. \tag{10.39}$$

The exact solution is given by

$$u(v) = e^{-v}, \quad w(v) = e^{v}. \tag{10.40}$$

Applying differential transform to equations (10.37)–(10.39), we obtain the following recursive relation

$$U(k+1) = \frac{1}{(k+1)} \left(\sum_{r=0}^{k} \left(\frac{1}{2}\right)^{2k-r} U(r) W(k-r) - U(k) - \left(\frac{1}{4}\right)^{k} U(k) \right), \tag{10.41}$$

$$W(k+1) = \frac{1}{(k+1)} \left(\sum_{r=0}^{k} \left(\frac{1}{2}\right)^{r} U(r) W(k-r) + W(k) - \left(\frac{1}{2}\right)^{k} W(k) \right), \tag{10.42}$$

$$U(0) = W(0) = 1. \tag{10.43}$$

Using equations (10.41)–(10.43), we obtain the following components,

$k = 0$, $U(1) = U(0) W(0) - U(0) - U(0) = -1$,
$W(1) = U(0) W(0) + W(0) - W(0) = 1$,

$k = 1$, $U(2) = \frac{1}{2}\left(\frac{1}{4}U(0)W(1) + \frac{1}{2}U(1)W(0) - U(1) - \frac{1}{4}U(1)\right) = -\frac{1}{2!}$,

$W(2) = \frac{1}{2}\left(U(0)W(1) + \frac{1}{2}U(1)W(0) + W(1) - \frac{1}{2}U(1)\right) = \frac{1}{2!}$,

$k = 2$, $U(3) = -\frac{1}{3!}$,

$W(3) = \frac{1}{3!}, \ldots$, and so on. $\tag{10.44}$

Now, with the help of equation (10.4), the series solution is given by

$$u(v) = 1 - v + \frac{1}{2!}v^2 - \frac{1}{3!}v^3 + \ldots, \tag{10.45}$$

$$w(v) = 1 + v + \frac{1}{2!}v^2 + \frac{1}{3!}v^3 + \ldots, \tag{10.46}$$

which converges to the exact solution given by equation (10.40).

The approximate solution for $N = 12$ is compared with the analytical solution in Tables (10.7)–(10.8), where N represents a number of terms considered. Table 10.9 lists the maximal absolute error of approximate results obtained by the present method for $N = 5, 10$, and 15. Figure 10.3 depicts absolute errors for the numerical solutions for $N = 5, 10$, and 15. From these results, it is clear that absolute errors and maximal absolute errors all decline systematically with the increase in N.

TABLE 10.7: Comparison of numerical solution $u(v)$ with exact solution when $N = 12$ for Example 3

v	$u(v)$	$u_N(v)$	$R_N(v)$
0.0	1.000000000	1.000000000	0
0.1	0.904837418	0.904837418	1.2E-16
0.2	0.818730753	0.818730753	1.3E-16
0.3	0.740818220	0.740818220	1.3E-16
0.4	0.670320046	0.670320046	1.4E-15
0.5	0.606530659	0.606530659	3.1E-14
0.6	0.548811636	0.548811636	3.6E-13
0.7	0.496585303	0.496585303	2.9E-12
0.8	0.449328964	0.449328964	1.8E-11
0.9	0.406569659	0.406569659	9.4E-11
1.0	0.367879441	0.367879441	4.0E-10

TABLE 10.8: Comparison of numerical solution $w(v)$ with exact solution when $N = 12$ for Example 3

v	$w(v)$	$w_N(v)$	$R_N(v)$
0.0	1.000000000	1.000000000	0
0.1	1.105170918	1.105170918	2.0E-16
0.2	1.221402758	1.221402758	2.0E-16
0.3	1.349858807	1.349858807	1.6E-16
0.4	1.491824697	1.491824697	7.4E-16
0.5	1.648721270	1.648721270	1.2E-14
0.6	1.822118800	1.822118800	1.2E-13
0.7	2.013752707	2.013752707	8.1E-13
0.8	2.225540928	2.225540928	4.2E-12
0.9	2.459603111	2.459603111	1.7E-11
1.0	2.718281828	2.718281828	6.3E-11

TABLE 10.9: Maximum absolute errors for u and w of Example 3

N	$E_{N,\infty}$ **for** u	$E_{N,\infty}$ **for** w
5	1.2E-03	9.9E-03
10	2.3E-08	3.0E-07
15	7.1E-13	8.1E-13

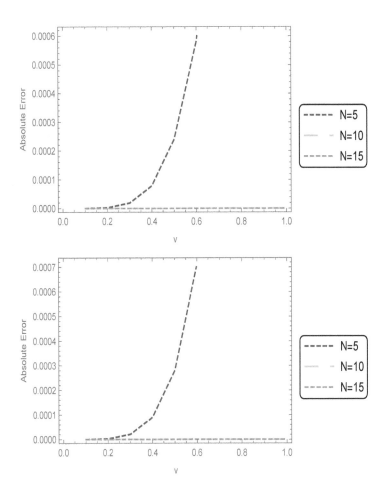

FIGURE 10.3: Absolute errors for u and w for $N = 5$, 10, and 15 of Example 3.

10.3.4 Example 4

Consider the following nonlinear proportional DDE

$$(1+v)w'(v) = (1+v)e^{w\left(\frac{v}{2}\right)} - \frac{1}{2}v^2 - \frac{3}{2}v, \ 0 \leq v \leq 1, \tag{10.47}$$

with initial condition

$$w(0) = 0. \tag{10.48}$$

The exact solution is given by

$$w(v) = \log(1+v). \tag{10.49}$$

Denote $h(v) = f(g(v))$, where $g(v) = w\left(\frac{v}{2}\right)$ and $f(x) = e^x$. Differential transform of $f(x)$ is represented by $F(k)$ then

$$F(k) = \frac{1}{k!}, \tag{10.50}$$

and differential transform of $h(v)$ is represented by $H(k)$ then using theorem (10.2.5), we have $H(0) = 1$ and

$$H(k) = \left(\frac{1}{2}\right)^k \sum_{l=1}^{k} F(l) \hat{B}_{k,l}(W(1), \ldots, W(k-l+1)) \text{ for } k \geq 1.$$

Applying differential transform to equations (10.47)–(10.48), we obtain the following recursive relation

$$W(k+1) = \frac{1}{(k+1)} \left(-\sum_{r=0}^{k} \delta(r-1)(k-r+1)W(k-r+1) + H(k) \right.$$
$$\left. + \sum_{r=0}^{k} \delta(r-1) H(k-r) - \frac{1}{2}\delta(k-2) - \frac{3}{2}\delta(k-1) \right) \tag{10.51}$$
$$W(0) = 0. \tag{10.52}$$

Using equations (10.50)–(10.52) we obtain following components,

$$k = 0, \ W(1) = H(0) = 1,$$
$$k = 1, \ H(1) = \left(\frac{1}{2}\right) F(1) \hat{B}_{1,1}(W(1)) = \frac{1}{2} F(1) W(1) = \frac{1}{2},$$
$$W(2) = \frac{1}{2}\left(-W(1) + H(1) + H(0) - \frac{3}{2}\right) = -\frac{1}{2},$$
$$k = 2, \ H(2) = \left(\frac{1}{2}\right)^2 \left(\sum_{l=1}^{k} F(l) \hat{B}_{2,1}(W(1), W(2))\right)$$

Applications of differential transform method 185

$$= \frac{1}{4}\left(F(1)W(2) + F(2)W^2(1)\right) = 0,$$

$$W(3) = \frac{1}{3}\left(-2W(2) + H(2) + H(1) - \frac{1}{2}\right)$$

$$= \frac{1}{3}, \ldots \text{ and so on.} \quad (10.53)$$

Now, with the help of equation (10.4), the series solution is given by

$$w(v) = v - \frac{1}{2}v^2 + \frac{1}{3}v^3 - \cdots, \quad (10.54)$$

which converges to the exact solution given by equation (10.49).

The approximate solution for $N = 15$ is compared with the analytical solution in Table 10.10, where N represents a number of terms considered. Table 10.11 lists the maximal absolute error of approximate results obtained by the present method for $N = 5$, 10, and 15. Figure 10.4 depicts absolute errors for the numerical solutions for $N = 5$, 10, and 15.

TABLE 10.10: Comparison of numerical solution $w(v)$ with exact solution when $N = 15$ for Example 4

v	$w(v)$	$w_N(v)$	$R_N(v)$
0.1	0.0953101798	0.0953101798	7.2E-16
0.2	0.1823215568	0.1823215568	1.8E-12
0.3	0.2623642645	0.2623642647	7.9E-10
0.4	0.3364722366	0.3364722561	5.7E-08
0.5	0.4054651081	0.4054657568	1.5E-06
0.6	0.4700036292	0.4700149031	3.9E-05
0.7	0.5306282511	0.5307535353	2.3E-04
0.8	0.5877866649	0.5887909192	1.7E-03
0.9	0.6418538862	0.6481288691	9.7E-03
1.0	0.6931471806	0.7253718504	4.6E-02

TABLE 10.11: Maximum absolute errors for w of Example 4

N	$E_{N,\infty}$
5	9.0E-02
10	4.7E-02
15	3.2E-02

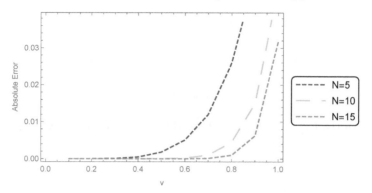

FIGURE 10.4: Absolute errors for $N = 5$, 10, and 15 of Example 4.

10.3.5 Example 5

Consider the following nonlinear proportional DDE

$$w'(v) = 4w'\left(\frac{v}{2}\right)\sqrt{1 - w^2\left(\frac{v}{2}\right)} - 2, \ 0 \le v \le 1, \quad (10.55)$$

with initial condition

$$w(0) = 0. \quad (10.56)$$

The exact solution is given by

$$w(v) = \sin(2v). \quad (10.57)$$

Denote $h(v) = f(g(v))$, where $g(v) = w\left(\frac{v}{2}\right)$ and $f(x) = \sqrt{1 - x^2}$. Differential transform of $f(x)$ is represented by $F(k)$ then

$$F(k) = \begin{cases} \binom{1/2}{k}(-1)^k, & k \text{ is even} \\ 0, & k \text{ is odd}, \end{cases} \quad (10.58)$$

and differential transform of $h(t)$ is represented by $H(k)$ then using theorem (10.2.5), we have $H(0) = 1$ and

$$H(k) = \left(\frac{1}{2}\right)^k \sum_{l=1}^{k} F(l)\,\hat{B}_{k,l}(W(1), \ldots, W(k - l + 1)) \ \text{ for } k \ge 1.$$

Applying differential transform to equations (10.55)–(10.56), we obtain the following recursive relation

$$W(k+1) = \frac{4}{(k+1)}\left(\sum_{r=0}^{k}\left(\frac{1}{2}\right)^{k-r+1}(k - r + 1)\,H(r)\,W(k - r + 1) - 2\delta(k)\right)$$
$$(10.59)$$
$$W(0) = 0. \quad (10.60)$$

Using equations (10.58)–(10.60) we obtain following components,

$$k = 0, \ W(1) = \frac{4}{2} H(0) W(1) - 2 = 2,$$

$$k = 1, \ H(1) = \left(\frac{1}{2}\right) F(1) \hat{B}_{1,1}(W(1)) = \frac{1}{2} F(1) W(1) = 0,$$

$$W(2) = \frac{4}{2}\left(\frac{2}{4} H(0) W(2) + \frac{1}{2} H(1) W(1)\right) = 0,$$

$$k = 2, \ H(2) = \left(\frac{1}{2}\right)^2 \left(\sum_{l=1}^{k} F(l) \hat{B}_{2,1}(W(1), W(2))\right)$$

$$= \frac{1}{4}\left(F(1) W(2) + F(2) W^2(1)\right) = -\frac{1}{2},$$

$$W(3) = \frac{4}{3}\left(\frac{3}{8} H(0) W(3) + \frac{2}{4} H(1) W(2) + \frac{1}{2} H(2) W(1)\right)$$

$$= -\frac{4}{3}, \ldots \text{ and so on.} \quad (10.61)$$

Now, with the help of equation (10.4), the series solution is given by

$$w(v) = 2v - \frac{4}{3}v^3 + \ldots, \quad (10.62)$$

which converges to the exact solution given by equation (10.57).

The approximate solution for $N = 12$ is compared with the analytical solution in Table (10.12), where N represents a number of terms considered. Table (10.13) lists the maximal absolute error of approximate results obtained by the present method for $N = 5, 10,$ and 15. Figure 10.5 depicts absolute errors for the numerical solutions for $N = 5, 10,$ and 15. From these results, it

TABLE 10.12: Comparison of numerical solution $w(v)$ with exact solution when $N = 12$ for Example 5

v	$w(v)$	$w_N(v)$	$R_N(v)$
0.1	0.1986693308	0.1986693308	7.2E-16
0.2	0.3894183423	0.3894183423	1.3E-16
0.3	0.5646424734	0.5646424734	2.8E-15
0.4	0.7173560909	0.7173560909	3.7E-13
0.5	0.8414709848	0.8414709846	1.2E-11
0.6	0.9320390860	0.9320390843	1.8E-09
0.7	0.9854497300	0.9854497174	1.2E-08
0.8	0.9995736030	0.9995735316	7.1E-08
0.9	0.9738476309	0.9738473016	3.3E-07
1.0	0.9092974268	0.9092961360	1.4E-02

TABLE 10.13: Maximum absolute errors for w of Example 5

N	$E_{N,\infty}$
5	2.4E-02
10	5.0E-05
15	3.6E-10

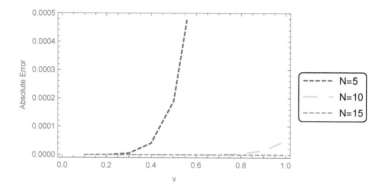

FIGURE 10.5: Absolute errors for $N = 5, 10$, and 15 of Example 5.

is clear that absolute errors and maximal absolute errors all decline systematically with the increase in N.

Bibliography

[1] Adak, D., Bairagi, N. and Hakl, R. 2020. Chaos in delay-induced Leslie-Gower prey-predator-parasite model and its control through prey harvesting, *Nonlinear Analysis: Real World Appl.* 5: 102998–103020.

[2] Ajello, W.G., Freedman, H.I. and Wu, J. 1992. A model of stage structured population growth with density depended time delay, SIAM J. Appl. Math. 52: 855–869.

[3] Benhammouda, B. and Hector, V.L. 2016. A new multi step technique with differential transform method for analytical solution of some nonlinear variable delay differential equations, *Springer Plus* 5: 1–17.

[4] Bellour, A. and Bousselsal, M. 2014. Numerical solution of delay integro-differential equations by using Taylor collocation method, *Math. Methods Appl. Sci.* 37: 1491–1506.

[5] Blanco-Cocom, L., Estrella, A.G. and Avila-Vales, E. 2013. Solving delay differential systems with history functions by the Adomian decomposition method, *Appl. Math. Comput.* 218: 5994–6011.

[6] Buhmann, M.D. and Iserles, A. 1993. Stability of the discretized pantograph differential equation, *Math. Comput.* 60: 575–589.

[7] Chen, C.K. and Ho, S.H. 1999. Solving partial differential equations by two-dimensional differential transform method, *Appl. Math. Comput.* 106: 171–179.

[8] Chen, X. and Wang, L. 2010. The variational iteration method for solving a neutral functional-differential equation with proportional delays, *Comput. Math. Appl.* 59: 2696–2702.

[9] Cherepennikov, V. 2013. Numerical analytical method of studying some linear functional differential equations, *Numer. Anal. Appl.* 6: 236–246.

[10] Comtet, L. 1974. *Advanced Combinatorics:The Art of Finite and Infinite Expansions*, Springer Science & Business Media.

[11] Dugard, L. and Ei, V. 1997. *Stability and Control of Time-delay Systems*, Verlag Berlin Heidelberg.

[12] Duarte, J., Januario, C. and Martins, N. 2016. Analytical solutions of an economic model by the homotopy analysis method, *App. Math Sci.*, 10:49: 2483–2490.

[13] Ayaz, F. 2004. Solution of the systems of differential equations by differential transform method, *Appl. Math. Comput.* 147: 547–567.

[14] Fox, L., Mayers, D.F., Ockendon, J.A. and Tayler, A.B. 1971. On a functional differential equation, *J. Inst. Math. Appl.* 8: 271–307.

[15] Herrera, A.R. 2013. Chaos in delay differential equations with applications in population dynamics, *Discrete Contin. Dyn. Syst.* 33: 1633–1644.

[16] Kuang, Y. 1993. *Delay Differential Equations with Applications in Population Dynamics*, Academic Press.

[17] Methi, G. and Kumar, A. 2019. Numerical Solution of Linear and Higher-order Delay Differential Equations using the Coded Differential Transform Method, *Comput. Res. Model.*, 11:6: 1091–1099.

[18] Ockendon, J.R. and Tayler, A.B. 1971. The dynamics of a current collection system for an electric locomotive, *Proc. Roy. Soc. London Ser. A*, 6: 447–468.

[19] Pukhov, G.E. 1982. Differential transforms and circuit theory, *Circuit Theory Appl.* 10: 265–276.

[20] Ravi Kanth, A.S.V. and Aruna, K. 2008. Solution of singular two-point boundary value problems using differential transformation method, *Physics Letters A*, 372: 4671–4673.

[21] Rebenda, J. and Šmarda, Z. 2017. A differential transformation approach for solving functional differential equations with multiple delays, *Commun. Nonlinear Sci. Numer. Simul.* 48: 246–257.

[22] Rebenda, J. 2018. An application of Bell polynomials in numerical solving of nonlinear differential equations, *Aplimat Proceedings* 2018: 1–10.

[23] Rebenda, J. and Šmarda, Z. 2019. Numerical algorithm for nonlinear delayed differential systems of nth order, *Adv. Diff. Equ.* 26: 1–13.

[24] Rebenda, J. and Pátíková, Z. 2020. Differential Transform Algorithm for Functional Differential Equations with Time-Dependent Delays, *Complexity* 2020: 1–12.

[25] Sezer, M., Yalcinbas, S. and Sahin, N. 2008. Approximate solution of multi-pantograph equation with variable coefficients, *J. Comput. Appl. Math.* 214: 406–416.

[26] Shakeri, F. and Dehghan, M. 2008. Solution of delay differential equations via a homotopy perturbation method, *Math. Comput. Model.* 48: 486–498.

[27] Warne, P., Warne, D., Sochacki, J., Parker G. and Carothers, D. 2006. Explicit A-Priori error bounds and adaptive error control for approximation of nonlinear initial value differential systems, *Comput. Math. Appl.* 52:12: 1695–1710.

[28] Widatalla, S and Koroma, M. 2012. Approximation algorithm for a system of pantograph equations, *J. Appl. Math.* 2012: 1–9.

[29] Yu, Z.H. 2008. Variational iteration method for solving the multi-pantograph delay equation, *Phys. Lett. A,* 372: 6475–6479.

[30] Zhao, J.J., Xu, Y., Wang, H.X. and Liu, M.Z. 2006. Stability of a class of Runge-Kutta methods for a family of pantograph equations of neutral type, *Appl. Math. Comput.* 181:2: 1170–1181.

Chapter 11

Solvability of fractional integral equation via measure of noncompactness and shifting distance functions

Bhuban Chandra Deuri

Anupam Das

11.1	Introduction	191
	11.1.1 Some notations	192
	11.1.2 Measure of noncompactness	192
11.2	Main result	193
11.3	Application	196
	Bibliography	201

11.1 Introduction

Integral equations are extremely useful in solving actual problems. Measures of noncompactness (MNC) and fixed point theory are valuable tools for solving many sorts of integral equations encountered in real world settings. Darbo and Schauder fixed point theorems are important in solving functional integral problems. We mention (see [4, 5, 8–18, 20]) for solving integral and differential equations using fixed point theorems and MNC.
We have shown several new fixed point theorems in this article by applying the idea of shifting distance functions and a measure of noncompactness. We've also used this theorem to investigate the existence of solutions to functional integral equations in Banach algebra, and we've confirmed our findings using an example.

DOI: 10.1201/9781003330868-11

11.1.1 Some notations

Let $(P, \|\cdot\|)$ be a real Banach space and $C(t,q) = \{a \in P : \|a - t\| \leq q\}$. If $\mathcal{E}(\neq \emptyset) \subseteq P$. Also, $\bar{\mathcal{E}}$ and $Conv\mathcal{E}$ represent closure and convex closure of $\mathcal{E}$. Additional, let

I. $\mathbb{R} = (-\infty, \infty)$,

II. $\mathbb{R}_+ = [0, \infty)$.

III. $\mathfrak{M}_P :=$ All nonempty and bounded subsets of P are collected here,

IV. $\mathfrak{N}_P :=$ The collection of all relatively compact sets.

11.1.2 Measure of noncompactness

The measure of noncompactness is a key technique to formulate fixed point in Banach spaces. It can be applied to solved differential equations, integral equations, integro-differential equations, and functional equations, among other.

The next definition of a MNC is provided in [1].

Definition 11.1.1. Let $O : \mathfrak{M}_P \to [0, \infty)$ be a mapping, which is called a MNC in P if the conditions given below hold :

(i) Family $\ker O = \{\mathcal{E} \in \mathfrak{M}_P : O(\mathcal{E}) = 0\} \neq \emptyset$ with $\ker O \subset \mathfrak{N}_P$.

(ii) $\mathcal{E} \subseteq \mathcal{F} \implies O(\mathcal{E}) \leq O(\mathcal{F})$.

(iii) $O(\mathcal{E}) = O(\bar{\mathcal{E}})$.

(iv) $O(Conv\mathcal{E}) = O(\mathcal{E})$.

(v) $O(\sigma\mathcal{E} + (1-\sigma)\mathcal{F}) \leq \sigma O(\mathcal{E}) + (1-\sigma)O(\mathcal{F})$ for $\sigma \in [0,1]$.

(vi) if $\mathcal{E}_v \in \mathfrak{M}_P$, $\mathcal{E}_v = \bar{\mathcal{E}}_v$, $\mathcal{E}_{v+1} \subset \mathcal{E}_v$ for $v = 1, 2, 3, \ldots$ and $\lim_{v \to \infty} O(\mathcal{E}_v) = 0$ then $\bigcap_{v=1}^{\infty} \mathcal{E}_v \neq \emptyset$.

Recall that, the intersection set $\mathcal{E}_\infty$ from (vi) belongs to $\ker O$ family. Indeed, since $O(\mathcal{E}_\infty) \leq O(\mathcal{E}_n)$ for any n, we've accepted the fact that $O(\mathcal{E}_\infty) = 0$. This is giving $\mathcal{E}_\infty \in \ker O$.

The Banach space of real functions defined and continuous on $I \times I$ is represented by the space $C(I \times I)$ with the norm

$$\|r\| = \sup |r(u,s)| : u, s \in I,$$

where $r \in C(I \times I)$. This space has the structure of Banach algebra in relation to the usual product of functions.

Solvability of fractional integral equation

For arbitrary fixed $\delta > 0$, fix $P \in \mathfrak{M}_C(I \times I)$ and $r \in P$, we denote by $\omega(r, \delta)$ the continuity modulus of r, i.e.,

$$\omega(r,\delta) = \sup\{|r(u,s) - r(\beta,\gamma)| : u, s, \beta, \gamma \in I, |u - \beta| \leq \delta, |s - \gamma| \leq \delta\}.$$

Also, let
$$\omega(P,\delta) = \sup\{\omega(r,\delta) : r \in P\}$$

and
$$\omega_0(P) = \lim_{\delta \to 0} \omega(P, \delta).$$

It can be shown that $\omega_0(P)$ is a MNC in $C(I \times I)$.
The following are some of the most important theorems that we recall:

Theorem 11.1.2. (Shauder [2]) Let K be a nonempty, closed and convex subset of a Banach space P. Then $\mathfrak{J} : K \to K$ continuous and compact mapping has at least one fixed point.

Theorem 11.1.3. (Darbo [3]) Let K be a nonempty, bounded, closed and convex (Nbcc) subset of a Banach space P. Let $\mathfrak{J} : K \to K$ be a continuous map. Suppose that, $\exists \lambda \in [0, 1)$ such that

$$O(\mathfrak{J}\mathcal{E}) \leq \lambda O(\mathcal{E}), \ \mathcal{E} \subseteq K.$$

Then $\mathfrak{J}$ has a fixed point.

11.2 Main result

We shall use some of the related concepts shown below to prove our fixed point theorem.

Definition 11.2.1. ([6]) Let $\Psi_\mathcal{A}$ be the family of all functions $\mathcal{A} : \mathbb{R}_+ \times \mathbb{R}_+ \to \mathbb{R}$ st

1. $\mathcal{A}$ is strictly increasing and continuous;

2. For each sequences $\{s_n\}, \{\alpha_n\} \subseteq \mathbb{R}_+$, $\lim_{n \to \infty} s_n = \lim_{n \to \infty} \alpha_n = 0 \Leftrightarrow \lim_{n \to \infty} \mathcal{A}(s_n, \alpha_n) = -\infty$.

Definition 11.2.2. ([6]) Let $\Omega_{\mathcal{B},\gamma}$ denotes the set of pairs $(\mathcal{B}, \gamma)$, where $\mathcal{B} : \mathbb{R}_+ \to \mathbb{R}$ and $\gamma : \mathbb{R}_+ \times \mathbb{R}_+ \to [0, 1)$ st

1. For each sequences $\{s_n\} \subseteq \mathbb{R}_+$, so $\limsup_{n \to \infty} \mathcal{B}(s_n) \geq 0 \Leftrightarrow \limsup_{n \to \infty}(s_n) \geq 1$;

2. For each sequences $\{s_n\}, \{\alpha_n\} \subseteq \mathbb{R}_+$, $\limsup_{n\to\infty} \gamma(s_n, \alpha_n) = 1 \Rightarrow \lim_{n\to\infty} s_n = \lim_{n\to\infty} \alpha_n = 0$;

3. For each sequences $\{s_n\}, \{\alpha_n\} \subseteq \mathbb{R}_+$, $\sum_{n=1}^{\infty} \mathcal{B}(\gamma(s_n, \alpha_n)) = -\infty$.

Definition 11.2.3. ([7]) Let the two functions be $\Delta_1, \Delta_2 : \mathbb{R}_+ \to \mathbb{R}$, so the pair of maps (Δ_1, Δ_2) is called a pair of shifting distance function, if it satisfies the given conditions:

1. For $x, y \in \mathbb{R}_+$ if $\Delta_1(x) \leq \Delta_2(y)$, then $x \leq y$.

2. For $x_n, y_n \in \mathbb{R}_+$ and $\lim_{n\to\infty} x_n = \lim_{n\to\infty} y_n = z$, if $\Delta_1(x_n) \leq \Delta_2(y_n)\ \forall\ n$, then $z = 0$.

We denote by Δ a pair (Δ_1, Δ_2) of shifting distance functions.

e.g., we take $\Delta_1(\alpha) = \alpha$, $\Delta_2(\alpha) = \epsilon\alpha$, $\alpha \geq 0$ and $\epsilon \in [0, 1)$. They are obviously a pair of shifting distance functions.

Definition 11.2.4. ([19]) Let G be the class of all maps $\mathcal{B} : \mathbb{R}_+ \times \mathbb{R}_+ \to \mathbb{R}_+$ fulfilling the below conditions:
(1) $\max\{h, l\} \leq \mathcal{B}(h, l)$ for $h, l \geq 0$.
(2) $\mathcal{B}$ is non-decreasing and continuous.
(3) $\mathcal{B}(h + l, s + f) \leq \mathcal{B}(h, s) + \mathcal{B}(l, f)$.

For example $\mathcal{B}(w, l) = w + l$.

Theorem 11.2.5. Let K be a Nbcc subset of a Banach space P. Also, let $\mathfrak{J} : K \to K$ be continuous and $\tau : \mathbb{R}_+ \to \mathbb{R}_+$ be non-decreasing and continuous maps. Suppose that for any $0 < w < l < \infty\ \exists\ 0 < \gamma(w, l) < 1$ such that $\forall\ \mathcal{E} \subseteq K$,

$$w \leq \mathcal{B}(O(\mathcal{E}), \tau(O(\mathcal{E}))) \leq l$$
$$\implies \Delta_1[\mathcal{B}(O(\mathfrak{J}\mathcal{E}), \tau(O(\mathfrak{J}\mathcal{E})))] \leq \Delta_2[\gamma(w, l)\mathcal{B}(O(\mathcal{E}), \tau(O(\mathcal{E})))],$$

where O is an arbitrary MNC and $(\Delta_1, \Delta_2) \in \Delta$ and $\mathcal{B} \in G$. So $\mathfrak{J}$ has at least one fixed point in K.

Proof. Let us create a sequence (K_σ) st $K_0 = K$ and $K_{\sigma+1} = Conv(\mathfrak{J}K_\sigma)$ for $\sigma \geq 0$. We observe that $\mathfrak{J}K_0 = \mathfrak{J}K \subseteq K = K_0$, $K_1 = Conv(\mathfrak{J}K_0) \subseteq K = K_0$, therefore by continuing this process, we have $K_0 \supseteq K_1 \supseteq K_2 \supseteq \ldots \supseteq K_\sigma \supseteq K_{\sigma+1} \supseteq \ldots$. Let $\exists$ a natural number σ st $O(K_\sigma) = 0$. So, K_σ is compact. By Schauder's fixed point theorem, we reached the conclusion that $\mathfrak{J}$ has a fixed point. Suppose $O(K_\sigma) > 0$ for some $\sigma \geq 0$ that is $\mathcal{B}(O(K_\sigma), \tau(O(K_\sigma))) > 0\ \forall\ \sigma \geq 0$. Let $\mathcal{E} = K_\sigma$ for some $\sigma \in \mathbb{N}$.
For $w \leq \mathcal{B}(O(K_\sigma), \tau(O(K_\sigma))) \leq l$ gives

$$\Delta_1[\mathcal{B}(O(K_{\sigma+1}), \tau(O(K_{\sigma+1})))] = \Delta_1[\mathcal{B}(O(Conv\mathfrak{J}K_\sigma), \tau(O(Conv\mathfrak{J}K_\sigma)))]$$
$$= \Delta_1[\mathcal{B}(O(\mathfrak{J}K_\sigma), \tau(O(\mathfrak{J}K_\sigma)))]$$
$$\leq \Delta_2[\gamma(w, l)\mathcal{B}(O(K_\sigma), \tau(O(K_\sigma)))]$$

Using part (1) of Definition 11.2.3, we get

$$\mathfrak{B}(O(K_{\sigma+1}), \tau(O(K_{\sigma+1}))) \leq \gamma(w,l)\mathfrak{B}(O(K_\sigma), \tau(O(K_\sigma)))$$
$$< \mathfrak{B}(O(K_\sigma), \tau(O(K_\sigma))); \quad since \quad 0 < \gamma(w,l) < 1.$$

Hence $\{\mathfrak{B}(O(K_\sigma), \tau(O(K_\sigma)))\}$ is a non-negative decreasing sequence, thus $\exists$ $\alpha \geq 0$ st $\lim_{\sigma \to \infty} \mathfrak{B}(O(K_\sigma), \tau(O(K_\sigma))) = \lim_{\sigma \to \infty} \mathfrak{B}(O(K_{\sigma+1}), \tau(O(K_{\sigma+1}))) = \alpha$. Suppose $\alpha > 0$. Then, $0 < \alpha = w \leq \mathfrak{B}(O(K_\sigma), \tau(O(K_\sigma))) \leq \mathfrak{B}(O(K_0), \tau(O(K_0))) = l$ for all $\sigma \geq 0$.

Then by (2) of Definition 11.2.3, we get $\alpha = 0$.
This implies that
$$\lim_{\sigma \to \infty} \mathfrak{B}(O(K_\sigma), \tau(O(K_\sigma))) = 0. \tag{11.1}$$

Using (11.1) and (1) of property **G**, we obtain
$$\lim_{\sigma \to \infty} O(K_\sigma) = 0 = \lim_{n \to \infty} \tau(O(K_\sigma)).$$

Since $K_\sigma \supseteq K_{\sigma+1}$ according to the Definition 11.1.1, we come to a conclusion that $K_\infty = \bigcap_{\sigma=1}^{\infty} K_\sigma$ is non-empty, closed and convex subset of K and K_∞ is invariant under $\mathfrak{J}$. Therefore, the Theorem 11.1.2 makes it clear that $\mathfrak{J}$ has a fixed point in $K_\infty \subseteq K$. □

The next theorems are the consequences that can be generated by theorem 11.2.5.

Theorem 11.2.6. Let K be a Nbcc subset of a Banach space P. Also $\mathfrak{J} : K \to K$ be continuous and $\tau : \mathbb{R}_+ \to \mathbb{R}_+$ be non-decreasing and continuous maps. Suppose that for any $0 < w < l < \infty$ $\exists$ $0 < \gamma(w,l) < 1$ such that $\forall \mathcal{E} \subseteq K$,

$$w \leq O(\mathcal{E}) + \tau(O(\mathcal{E})) \leq l$$
$$\implies \Delta_1[O(\mathfrak{J}\mathcal{E}) + \tau(O(\mathfrak{J}\mathcal{E}))] \leq \Delta_2[\gamma(w,l)(O(\mathcal{E}) + \tau(O(\mathcal{E})))],$$

where O is an arbitrary MNC and $(\Delta_1, \Delta_2) \in \Delta$. So $\mathfrak{J}$ has at least one fixed point in K.

Proof. Taking $\mathfrak{B}(w,l) = w + l$ in Theorem 11.2.5. we get the above result. □

Theorem 11.2.7. Let K be a Nbcc subset of a Banach space P. Also $\mathfrak{J} : K \to K$ be continuous mapping. Suppose for any $0 < w < l < \infty$ $\exists$ $0 < \gamma(w,l) < 1$ st $\forall \mathcal{E} \subseteq K$,

$$w \leq O(\mathcal{E}) \leq l \implies \Delta_1[O(\mathfrak{J}\mathcal{E})] \leq \Delta_2[\gamma(w,l)O(\mathcal{E})]$$

where O is an arbitrary MNC and $(\Delta_1, \Delta_2) \in \Delta$. So $\mathfrak{J}$ has at least one fixed point in K.

Proof. Putting $\mathfrak{B}(w,l) = w + l$ and $\tau \equiv 0$ in Theorem 11.2.5, we obtain the result. □

Theorem 11.2.8. Let K be a Nbcc subset of a Banach space P. Also $\mathfrak{J} : K \to K$ be continuous mapping. Suppose that for any $0 < w < \mathfrak{l} < \infty \; \exists \; 0 < \gamma(w, \mathfrak{l}) < 1$ st $\forall \; \mathcal{E} \subseteq K$,
$$w \leq O(\mathcal{E}) \leq \mathfrak{l} \implies O(\mathfrak{J}\mathcal{E}) \leq \gamma(w, \mathfrak{l}) O(\mathcal{E})$$
where O is an arbitrary MNC. So $\mathfrak{J}$ has at least one fixed point in K.

Proof. Taking $\Delta_1(\varsigma) = \varsigma = \Delta_2(\varsigma)$ for all $\varsigma \geq 0$ in Theorem 11.2.7, we obtain DFPT. □

11.3 Application

In this part, we define some notations. Let

1. $(P, \| \cdot \|) =$ A Real Banach algebra,

2. $C(I, P) :=$ The space consisting of all continuous functions $\nu : I = [0, 1] \to P$,

3. $C_+(I) :=$ The space of positive real valued continuous function defined on I,

4. $C_+^1(I) :=$ The space of positive real valued continuous differentiable mapping defined on I.

We will establish the existence of a solution $\nu \in C(I, P)$ to the integral equation shown below:
$$\nu(s) = G(s, \nu(s)) + \frac{1}{\Gamma(\varpi)} \int_0^s \frac{J'(\iota)}{(J(s) - J(\iota))^{1-\varpi}} h(s, \iota, \nu(\iota)) d\iota, \quad (11.2)$$
where $\iota, s \in [0, 1] = I, 0 < \varpi < 1$.

Assumptions:

(B1) The continuous mapping $G : I \times P \to P$ such that $\exists \; \mathcal{A} \in \Psi_\mathcal{A}, (\mathcal{B}, \gamma) \in \Omega_{\mathcal{B}, \gamma}$ and $\Lambda : \mathbb{R}_+ \to \mathbb{R}_+$ be a nondecreasing mapping st

$$\| G(s, \nu) - G(s, \pi) \| > 0$$
$$\implies \mathcal{A}(\| G(s, \nu) - G(s, \pi) \|, \Lambda(\| G(s, \nu) - G(s, \pi) \|))$$
$$\leq \mathcal{A}(\| \nu - \pi \|, \Lambda(\| \nu - \pi \|)) + \mathcal{B}(\gamma(\| \nu - \pi \|, \Lambda(\| \nu - \pi \|))). \quad (11.3)$$

Also, $\exists$ a mapping $\mu_1 : \mathbb{R}_+ \to \mathbb{R}_+$ st
$$\| g(s, \nu) \| \leq \mu_1(\| \nu \|)$$
and $\mathcal{M}_0 = \sup\{|\mu_1(s)| : s \in \mathbb{R}_+\} < \infty$.

(B2) Let $h : \mathtt{I} \times \mathtt{I} \times \mathbb{R} \to \mathbb{R}$ be a continuous mapping st $h : \mathtt{I} \times \mathtt{I} \times \mathbb{R}_+ \to \mathbb{R}_+$ and $\mathcal{H}_0 = \sup \{|h(s,\iota,\nu(\iota))| : \iota, s \in \mathtt{I}, \nu \in C_+(\mathtt{I})\} < \infty.$

(B3) The mapping $J : \mathtt{I} \to \mathbb{R}_+$ is nondecreasing and C_+^1.

(B4) $\liminf_{\alpha \to \infty} \frac{\mathcal{H}_0 (J(1)-J(0))^\varpi}{\alpha \Gamma(\varpi+1)} < 1.$

We denote a closed ball, $B_r(\theta) = \{\nu \in C(\mathtt{I},\mathtt{P}) : \| \nu \| \leq r\}$, where center is θ and radius is r.

Theorem 11.3.1. Under assumptions $(B1)\check{\ }(B4)$, Eq. (11.2) has at least one solution $\nu \in C(\mathtt{I},\mathtt{P})$.

Proof. The integral operator defined as $\mathfrak{J} : C(\mathtt{I},\mathtt{P}) \to C(\mathtt{I},\mathtt{P})$ by

$$\mathfrak{J}\nu(s) = G(s,\nu(s)) + \mathfrak{L}\nu(s),$$

where

$$\mathfrak{L}\nu(s) = \frac{1}{\Gamma(\varpi)} \int_0^s \frac{J'(\iota)}{(J(s)-J(\iota))^{1-\varpi}} h(s,\iota,\nu(\iota)) d\iota,$$

where $\iota, s \in I$. We prove $fix(\mathfrak{J}) \neq \emptyset$.

Consider the two mappings $\mathfrak{J}_1, \mathfrak{J}_2 : C(\mathtt{I},\mathtt{P}) \to C(\mathtt{I},\mathtt{P})$ define by

$$\mathfrak{J}_1\nu(s) = G(s,\nu(s)), \text{ and } \mathfrak{J}_2\nu(s) = \mathfrak{L}\nu(s),$$

where $\mathfrak{J} = \mathfrak{J}_1 + \mathfrak{J}_2$. So $\mathfrak{J}_1$ is well defined. Now, we prove that $\mathfrak{J}_2$ is well defined. Let $\nu \in C(\mathtt{I},\mathtt{P})$ be given and fixed. Also, let $\epsilon > 0$ be arbitrary, let $\iota_1, \iota_2 \in I$ (assume that $\iota_2 \geq \iota_1$) with $|\iota_2 - \iota_1| \leq \epsilon$ and $r_0 = \| \nu \|$. Then we get

$$\Gamma(\varpi) |(\mathfrak{L}\nu)(\iota_2) - (\mathfrak{L}\nu)(\iota_1)|$$

$$= \left| \int_0^{\iota_2} \frac{J'(\iota)}{(J(\iota_2)-J(\iota))^{1-\varpi}} h(\iota_2,\iota,\nu(\iota)) d\iota \right.$$

$$\left. - \int_0^{\iota_1} \frac{J'(\iota)}{(J(\iota_1)-J(\iota))^{1-\varpi}} h(\iota_1,\iota,\nu(\iota)) d\iota \right|$$

$$\leq \left| \int_0^{\iota_2} \frac{J'(\iota)}{(J(\iota_2)-J(\iota))^{1-\varpi}} h(\iota_2,\iota,\nu(\iota)) d\iota \right.$$

$$\left. - \int_0^{\iota_2} \frac{J'(\iota)}{(J(\iota_2)-J(\iota))^{1-\varpi}} h(\iota_1,\iota,\nu(\iota)) d\iota \right|$$

$$+ \left| \int_0^{\iota_2} \frac{J'(\iota)}{(J(\iota_2)-J(\iota))^{1-\varpi}} h(\iota_1,\iota,\nu(\iota)) d\iota \right.$$

$$\left. - \int_0^{\iota_1} \frac{J'(\iota)}{(J(\iota_2)-J(\iota))^{1-\varpi}} h(\iota_1,\iota,\nu(\iota)) d\iota \right|$$

$$+ \left| \int_0^{\iota_1} \frac{J'(\iota)}{(J(\iota_2) - J(\iota))^{1-\varpi}} h(\iota_1, \iota, \nu(\iota)) d\iota \right.$$

$$\left. - \int_0^{\iota_1} \frac{J'(\iota)}{(J(\iota_1) - J(\iota))^{1-\varpi}} h(\iota_1, \iota, \nu(\iota)) d\iota \right|$$

$$\leq \int_0^{\iota_2} \frac{J'(\iota)}{(J(\iota_2) - J(\iota))^{1-\varpi}} |h(\iota_2, \iota, \nu(\iota)) d\iota - h(\iota_1, \iota, \nu(\iota))| \, d\iota$$

$$+ \int_{\iota_1}^{\iota_2} \frac{J'(\iota)}{(J(\iota_2) - J(\iota))^{1-\varpi}} |h(\iota_1, \iota, \nu(\iota))| \, d\iota$$

$$+ \int_0^{\iota_1} \left| \frac{J'(\iota)}{(J(\iota_2) - J(\iota))^{1-\varpi}} - \frac{J'(\iota)}{(J(\iota_1) - J(\iota))^{1-\varpi}} \right| |h(\iota_1, \iota, \nu(\iota))| \, d\iota.$$

Denote

$$\omega(h, \epsilon) = \sup \left\{ |h(s, \iota, \nu) - h(s_1, \iota, \nu)| : \iota, s, s_1 \in \mathbf{I}; \nu \in [-r_0, r_0]; |s - s_1| \leq \epsilon \right\}.$$

Then,

$$\Gamma(\varpi) |(\mathfrak{L}\nu)(\iota_2) - (\mathfrak{L}\nu)(\iota_1)|$$

$$\leq \frac{\omega(h, \epsilon)}{\varpi} (J(\iota_2) - J(0))^\varpi + \frac{\mathcal{H}_0}{\varpi} (J(\iota_2) - J(\iota_1))^\varpi$$

$$+ \frac{\mathcal{H}_0}{\varpi} [(J(\iota_2) - J(0))^\varpi + (J(\iota_2) - J(\iota_1))^\varpi - (J(\iota_1) - J(0))^\varpi]$$

$$\leq \frac{\omega(h, \epsilon)}{\varpi} (J(\iota_2) - J(0))^\varpi + \frac{2\mathcal{H}_0}{\varpi} (J(\iota_2) - J(\iota_1))^\varpi$$

$$\leq \frac{\omega(h, \epsilon)}{\varpi} (J(\iota_2) - J(0))^\varpi + \frac{2\mathcal{H}_0}{\varpi} \omega(J, \epsilon)^\varpi$$

i.e., $\| (\mathfrak{L}\nu)(\iota_2) - (\mathfrak{L}\nu)(\iota_1) \| \leq \frac{\omega(h, \epsilon)}{\Gamma(\varpi + 1)} (J(\iota_2) - J(0))^\varpi + \frac{2\mathcal{H}_0}{\Gamma(\varpi + 1)} \omega(J, \epsilon)^\varpi.$

By referring to the concept of uniform continuity of the mapping h on the set $\mathbf{I}^2 \times [\check{\ } r_0, r_0]$ also f on $\mathbf{I}$, we get $\omega(h, \epsilon) \to 0$ with $\omega(J, \epsilon) \to 0$ as $\epsilon \to 0$, therefore $\mathfrak{L}\nu \in C(\mathbf{I}, \mathbf{P})$ and hence, $\mathfrak{J}_2\nu \in C(\mathbf{I}, \mathbf{P})$.

We show that $\mathfrak{J}_2$ is a continuous operator. Let $\varepsilon > 0$ be given and Fix $\pi \in C(\mathbf{I}, \mathbf{P})$.

For each $s \in \mathbf{I}$, we have

$$\Gamma(\varpi) |(\mathfrak{L}\nu)(s) - (\mathfrak{L}\pi)(s)|$$

$$= \left| \int_0^s \frac{J'(\iota)}{(J(s) - J(\iota))^{1-\varpi}} h(s, \iota, \nu(\iota)) d\iota - \int_0^s \frac{J'(\iota)}{(J(s) - J(\iota))^{1-\varpi}} h(s, \iota, \pi(\iota)) d\iota \right|$$

$$\leq \int_0^s \frac{J'(\iota)}{(J(s) - J(\iota))^{1-\varpi}} |h(s, \iota, \nu(\iota)) - h(s, \iota, \pi(\iota))| \, d\iota$$

$$\leq \frac{(J(1) - J(0))^\varpi}{\varpi} \mathcal{H}_{\delta_2},$$

where $\mathcal{H}_{\delta_2} = \sup\{|h(s,\iota,\nu) - h(s,\iota,\pi)| : \iota, s \in \mathtt{I}; \|\nu - \pi\| \leq \delta_2\}$.
So,
$$\|(\mathfrak{L}\nu) - (\mathfrak{L}\pi)\| \leq \frac{(J(1) - J(0))^\varpi}{\Gamma(\varpi + 1)} \mathcal{H}_{\delta_2}.$$

Also, we have
$$|(\mathfrak{L}\nu)(s)| \leq \frac{1}{\Gamma(\varpi)} \int_0^s \frac{J'(\iota)}{(J(s) - J(\iota))^{1-\varpi}} |h(s,\iota,\nu(\iota))|\, d\iota$$
$$\leq \frac{\mathcal{H}_0}{\Gamma(\varpi)} \int_0^s \frac{J'(\iota)}{(J(s) - J(\iota))^{1-\varpi}}\, d\iota$$
$$\leq \frac{\mathcal{H}_0 (J(1) - J(0))^\varpi}{\Gamma(\varpi + 1)}, \forall\, s \in \mathtt{I} \qquad (11.4)$$

Now, suppose we setting $\delta = \min\{\delta_2, \delta_1\}$, so for any $\nu \in C(\mathtt{I}, \mathtt{P})$ st $\|\nu - \pi\| < \delta$, using triangle inequality, we get

$$\|\mathfrak{J}_2\nu(s) - \mathfrak{J}_2\pi(s)\| = \|\mathfrak{L}\nu(s) - \mathfrak{L}\pi(s)\|$$
$$\leq \frac{(J(1) - J(0))^\varpi}{\Gamma(\varpi + 1)} \mathcal{H}_{\delta_2} \leq \frac{(J(1) - J(0))^\varpi}{\Gamma(\varpi + 1)} \varepsilon_2 \leq \varepsilon,$$

where $\varepsilon_2 = \frac{\Gamma(\varpi+1)\varepsilon}{1+(J(1)-J(0))^\varpi}$.
We need to prove $\mathfrak{J}_2$ is a compact operator. Let $B = \{\nu \in C(\mathtt{I}, \mathtt{P}) : \|\nu\| \leq 1\}$ be an open unit ball of $C(\mathtt{I}, \mathtt{P})$. Then, we show that a compact subset of $C(\mathtt{I}, \mathtt{P})$ is $\overline{\mathfrak{J}_2(B)}$. To see this, we have to only prove that $\mathfrak{J}_2(B)$ is an equicontinuous and an uniformly bounded subset of $C(\mathtt{I}, \mathtt{P})$. *First*, we claim that $\mathfrak{J}_2(B) = \{\mathfrak{J}_2\nu : \nu \in B\}$ is uniformly bounded.
For any $\nu \in B$,

$$\|\mathfrak{J}_2\nu(s)\| = \|\mathfrak{L}\nu(s)\| \leq \frac{\mathcal{H}_0 (J(1) - J(0))^\varpi}{\Gamma(\varpi + 1)}.$$

Hence, putting $\mathcal{M} := \frac{\mathcal{H}_0(J(1)-J(0))^\varpi}{\Gamma(\varpi+1)}$, concluding that $\mathfrak{J}_2(B)$ is uniformly bounded. Next, we prove that $\mathfrak{J}_2(B)$ is an equi-continuous and uniformly subset of $C(\mathtt{I}, \mathtt{P})$. Seeing this, let $\varepsilon > 0$ and let $\nu \in B$ be arbitrary. Since $\mathfrak{L}\nu$ is uniformly continuous, $\exists$ some $\delta_2(\varepsilon) > 0$ such that

$$\forall \iota_1, \iota_2 \in \mathtt{I}, |\iota_2 - \iota_1| < \delta_2(\varepsilon) \Rightarrow \|\mathfrak{L}\nu(\iota_2) - \mathfrak{L}\nu(\iota_1)\| < \varepsilon_2.$$

Let $\delta(\varepsilon) = \min\{\delta_2(\varepsilon), \varepsilon_2\}$, where the given ε_2 depend on ε. Therefore, if $\iota_1, \iota_2 \in \mathtt{I}$ satisfies $0 < \iota_2 - \iota_1 \in \mathtt{I} < \delta(\varepsilon)$ and $\nu \in B$,

$$\|\mathfrak{J}_2\nu(\iota_2) - \mathfrak{J}_2\nu(\iota_1)\| = \|\mathfrak{L}\nu(\iota_2) - \mathfrak{L}\nu(\iota_1)\|$$
$$\leq \varepsilon_2 \leq \varepsilon.$$

Therefore, $\mathfrak{J}_2$ is a compact operator. Now, we claim that $\mathfrak{J}_1$ satisfies (11.3).

Let $\nu, \pi \in C(\mathbf{I},\mathbf{P})$ and $\|\mathfrak{J}_1\nu\check{\ }\mathfrak{J}_1\pi\| > 0$. Using the fact that every continuous function reaches its maximum value on a compact set, $\exists\ s \in \mathbf{I}$ st $0 < \|\mathfrak{J}_1\nu\check{\ }\mathfrak{J}_1\pi\| = \|G(s,\nu(s))\check{\ }G(s,\pi(s))\|$. By $(B1)$ and applying the fact that $\mathcal{A}$ and Λ are strictly increasing maps, we get

$$\mathcal{A}(\|\mathfrak{J}_1\nu\check{\ }\mathfrak{J}_1\pi\|, \Lambda(\|\mathfrak{J}_1\nu\check{\ }\mathfrak{J}_1\pi\|))$$
$$= \mathcal{A}(\|G(s,\nu(s)) - G(s,\pi(s))\|, \Lambda(\|G(s,\nu(s)) - G(s,\pi(s))\|))$$
$$\leq \mathcal{A}(\|\nu - \pi\|, \Lambda(\|\nu - \pi\|)) + \mathcal{B}(\gamma(\|\nu - \pi\|, \Lambda(\|\nu - \pi\|))).$$

Hence $\mathfrak{J}_1$ satisfies (11.3). Now, we show that $\exists$ some $\mathcal{M}_1 > 0$ st $\|\mathfrak{J}_1\nu\| \leq \mathcal{M}_1$ holds for each $\nu \in C(\mathbf{I},\mathbf{P})$. By $(B1)$

$$\|\mathfrak{J}_1\nu(s)\| = \|G(s,\nu)\| \leq \mu_1(\|\nu\|) \leq \mathcal{M}_0.$$

Therefore, there exists $\mathcal{M}_0 > 0$ such that

$$\nu \in C(\mathbf{I},\mathbf{P}) \Rightarrow \|\mathfrak{J}_1\nu\| \leq \mathcal{M}_0\ \forall\ \nu \in C(\mathbf{I},\mathbf{P}).$$

Lastly, we show that $\exists$ some $r > 0$, st $\mathfrak{J}(B_r(\theta)) \subseteq B_r(\theta)$ with $B_r(\theta)$. On the contrary, for any $\alpha > 0\ \exists$ some $\nu_\alpha \in B_r(\theta)$ st $\mathfrak{J}(\nu_\alpha) > \alpha$. This gives that $\liminf_{\alpha \to \infty} \frac{1}{\alpha}\|\mathfrak{J}(\nu_\alpha)\| \geq 1$. On the other hand, we obtain

$$\|\mathfrak{J}\nu_\alpha(s)\| \leq \|G(s,\nu_\alpha(s))\| + \|\mathfrak{L}\nu_\alpha(s)\|$$
$$\leq \|\mathfrak{J}_1\nu_\alpha\| + \|\mathfrak{L}\nu_\alpha(s)\|$$
$$\leq \mathcal{M}_0 + \|\mathfrak{L}\nu_\alpha\|$$
$$\leq \mathcal{M}_0 + \frac{\mathcal{H}_0(J(1) - J(0))^\varpi}{\Gamma(\varpi + 1)}.$$

Thus, based on the above estimation and condition $(B4)$, we have

$$\liminf_{\alpha \to \infty} \frac{1}{\alpha}\|\mathfrak{J}(\nu_\alpha)\| \leq \liminf_{\alpha \to \infty} \frac{\mathcal{H}_0(J(1) - J(0))^\varpi}{\alpha\Gamma(\varpi + 1)} < 1$$

which is a contradiction. So, according to Theorem 11.1.3, we get $\mathfrak{J}$ has at least one fixed point in B_r. Thus, the equation (11.2) has at least one solution in $B_r(\theta) \subseteq C(\mathbf{I},\mathbf{P})$. $\square$

Example 11.3.2. Consider the integral equation of fractional order $0 < \sigma$,

$$\nu(s) = \frac{2se^{-\sigma(s+4)}}{s^2+1}\sin(|\nu(s)|) + \frac{1}{8\Gamma(\frac{1}{2})}\int_0^s \frac{4t^3}{(s^4-t^4)^{\frac{1}{2}}}\frac{t^2}{(1+s^4)(1+\nu^4(s))}dt. \tag{11.5}$$

Here

$$G(s,\nu(s)) = \frac{2se^{-\sigma(s+4)}}{s^2+1}\sin(|\nu(s)|),\ J(t) = t^4,$$

$$h(s,t,\nu(t)) = \frac{t^2}{8(1+s^4)(1+\nu^4(s))}\ \text{and}\ \varpi = \frac{1}{2}.$$

Define the mapping $G : \mathrm{I} \times \mathbb{R} \to \mathbb{R}$ given by $G(s, \nu(s)) = \frac{2se^{-\sigma(s+4)}}{s^2+1} \sin(v(s))$. G is continuous and

$$|G(s, \nu(s)) - G(s, \pi(s))| \leq \frac{2se^{-\sigma(s+4)}}{s^2+1} |\sin(v) - \sin(\pi)| \leq e^{-4\sigma} |v - \pi|. \tag{11.6}$$

Also, define $\mu_1 : \mathbb{R}_+ \to \mathbb{R}_+$ by $\mu_1(s) = \sin(s)$ with $\mathcal{M}_0 = 1 < \infty$ such that

$$|G(s, \nu(s))| = \frac{2se^{-\sigma(s+4)}}{s^2+1} |\sin(|v(s)|)| \leq |\sin(|v(s)|)| = \mu_1(|v(s)|).$$

Next, the mapping $\mathcal{A} : \mathbb{R}_+ \times \mathbb{R}_+ \to \mathbb{R}_+$ defined by $\mathcal{A}(s,t) = \ln(s+t)$, $\mathcal{B} : \mathbb{R}_+ \to \mathbb{R}_+$ given by $\mathcal{B}(s) = \ln(s)$, $\gamma : \mathbb{R}_+ \times \mathbb{R}_+ \to [0,1)$ given by $\gamma(s,t) = e^{\check{}4\sigma}$ and the mapping $\Lambda : \mathbb{R}_+ \to \mathbb{R}_+$ defined by $\Lambda(s) = s^2$. Obviously the inequality (11.6) gives that the condition (11.3) holds.
Evidently, if $|G(s, \nu(s)) - G(s, \pi(s))| > 0$, then we obtain

$$\begin{aligned}
&\mathcal{A}(|G(s,\nu(s)) - G(s,\pi(s))|, \Lambda(|G(s,\nu(s)) - G(s,\pi(s))|)) \\
&= \mathcal{A}(|G(s,\nu(s)) - G(s,\pi(s))|, |G(s,\nu(s)) - G(s,\pi(s))|^2) \\
&= \ln[|G(s,\nu(s)) - G(s,\pi(s))| + |G(s,\nu(s)) - G(s,\pi(s))|^2] \\
&\leq \ln[e^{\check{}4\sigma}(|\nu - \pi| + |\nu - \pi|^2)] \\
&= \ln(|\nu - \pi| + |\nu - \pi|^2) + \ln(e^{\check{}4\sigma}) \\
&= \mathcal{A}(|\nu - \pi|, |\nu - \pi|^2) + \mathcal{B}(e^{\check{}4\sigma}) \\
&= \mathcal{A}(|\nu - \pi|, \Lambda(|\nu - \pi|)) + \mathcal{B}(\gamma(|\nu - \pi|, \Lambda(|\nu - \pi|))).
\end{aligned}$$

Finally, $h : \mathrm{I} \times \mathrm{I} \times \mathbb{R} \to \mathbb{R}$ is a continuous mapping such that $h(s,t,\nu(t)) = \frac{t^2}{8(1+s^4)(1+\nu^4(s))}$ with $\mathcal{H}_0 = \frac{1}{8}$. Then, we have

$$\liminf_{\alpha \to \infty} \frac{\mathcal{H}_0(J(1) - J(0))^{\varpi}}{\alpha \Gamma(\varpi + 1)} = \liminf_{\alpha \to \infty} \frac{1}{4\Gamma(\frac{1}{2})\alpha} = 0 < 1.$$

This satisfied the assumption $(B4)$. As a result of the above, it is obvious that Equation (11.5) satisfies all of Theorem 11.3.1's assumptions. Hence, the equation (11.2) has a solution in $C(\mathrm{I}, \mathrm{P})$.

Bibliography

[1] Banaś, J. and Goebel, K. 1980. *Measure of Noncompactness in Banach Spaces*, Lecture Notes in Pure and Applied Mathematics, Marcel Dekker, New York, 60.

[2] Agarwal, R.P. and O'Regan, D. 2004. *Fixed point theory and applications*, Cambridge University Press.

[3] Darbo, G. 1955. Punti uniti in trasformazioni a codominio non compatto (Italian), *Rend. Sem. Mat. Univ. Padova*. 24: 84–92.

[4] Aghajani, A., Allahyari, R. and Mursaleen, M. 2014. A generalization of Darbo's theorem with application to the solvability of systems of integral equations, *J. Comput. Appl. Math*. 260: 68–77.

[5] Arab, R., Allahyari, R. and Haghighi, A.S. 2014. Existence of solutions of infinite systems of integral equations in two variables via measure of noncompactness, *Appl. Math. Comput*. 246: 283–291.

[6] Parvaneh, V., Hussain, N. and Kadelburg, Z. 2016. Generalized Wardowski type fixed point theorems via α-admissible FG-contractions in b-metric spaces, *Acta Math. Sci*. 36: 1445–1456.

[7] Nashine, H.K., Arab, R., Agarwal, R.P. and Sen, M. 2017. Positive solutions of fractional integral equations by the technique of measure of noncompactness, *J. Inequa. Appl*. 2017:225.

[8] Parvaneh, V., Khorshidi, M., De La Sen, M., Işık, H. and Mursaleen, M. 2020. Measure of noncompactness and a generalized Darbo's fixed point theorem and its applications to a system of integral equations, *Adv. Diff. Equ*. 1: 1–13.

[9] Parvaneh, V., Banaei, S., Roshan, J. R. and Mursaleen, M. 2021. On tripled fixed point theorems via measure of noncompactness with applications to a system of fractional integral equations, *Filomat* 35: 4897–4915.

[10] Das, A., Suwan, I., Deuri, B.C. and Abdeljawad, T. 2021. On solution of generalized proportional fractional integral via a new fixed point theorem, *Adv. Diff. Equ*. 2021, Article number: 427 (2021).

[11] Bazgir, H., and Ghazanfari, B. 2018. Existence of Solutions for Fractional Integro-Differential Equations with Non-Local Boundary Conditions, *Comput. Appl. Math*. 23, 36; doi:10.3390/mca23030036.

[12] Banaei, S., Mursaleen, M. and Parvaneh, V. 2020. Some fixed point theorems via measure of noncompactness with applications to differential equations, *Comput. Appl. Math*. 39: 1–12.

[13] Kuratowski, K. 1930. Sur les espaces complets, *Fund. Math*. 15: 301–309.

[14] Mursaleen, M. and Rizvi, Syed M. H. 2016. Solvability of infinite systems of second order differential equations in c_0 and ℓ_1 by Meir-Keeler condensing operators, *Proc. Amer. Math. Soc*. 144: 4279–4289.

[15] Banaei, S., Parvaneh, V. and Mursaleen, M. 2021. Measures of noncompactness and infinite systems of integral equations of Urysohn type in $L_\infty(G)$, *Carpathian J. Math.* 37: 407–416.

[16] Sanhan, S., Sanhan, W. and Mongkolkeha, C. 2018. New Existence of Fixed Point Results in Generalized Pseudodistance Functions with Its Application to Differential Equations, *Mathematics* 2018, 6(12), 324.

[17] Roshan, J.R. 2017. Existence of solutions for a class of system of functional integral equation via measure of noncompactness, *J. Comput. Appl. Math.* 313: 129–141.

[18] Das, A., Parvaneh, V., Deuri, B.C. and Bagherid, Z. 2022. Application of a generalization of Darbo's fixed point theorem via Mizogochi-Takahashi mappings on mixed fractional integral equations involving (k, s)-Riemann–Liouville and Erdélyi–Kober fractional integrals, *Int. J. Nonlinear Anal. Appl.* 13: 859–869.

[19] Das, A., Hazarika, B. and Kumam, P. 2019. Some new generalization of Darbo's fixed point theorem and its application on integral equations, *Mathematics* 2019, 7(3), 214.

[20] Arab, R., Nashine, H.K., Can, N.H. and Binh, T.T. 2020. Solvability of functional-integral equations (fractional order) using measure of noncompactness, *Adv. Diff. Equ.* 2020, Article number: 12 (2020).

Chapter 12

Generalized fractional operators and inequalities integrals

Juan E. Nápoles Valdés

Florencia Rabossi

12.1	Introduction	205
12.2	Integral inequalities with some integral operators	208
	12.2.1 Generalized integral operators	208
	12.2.2 Generalized fractional integral operators	212
	12.2.3 Weighted integral operators	220
12.3	A general formulation of the notion of convex function	222
12.4	Integral inequalities and fractional derivatives	223
	Bibliography	225

12.1 Introduction

Throughout the History of Mathematics, we find concepts that are the center of various theories and theoretical developments, it may even happen that this initial concept has received innumerable extensions and generalizations that make a researcher who begins in a certain area that it involves, you may be overwhelmed by such theoretical sublimation.

One of these concepts is that of convex function, present today in multiple mathematical disciplines ranging from Optimization to Function Theory and center of, possibly, the most fruitful nucleus in the study of integral inequalities, mainly linked to the estimation of the integral mean value of a certain function over a given interval. This concept has spread in various directions (the interested reader can consult [36], where a fairly complete overview of the generalizations and extensions of the convex function concept is presented). Let's start by presenting the concept of a convex function as follows.

Definition 12.1.1. *A function* $\phi : I \to \mathbb{R}$, $I := [a_1, a_2]$ *is said to be* **convex** *if the inequality*

$$\phi(\lambda x + (1-\lambda)y) \leq \lambda\phi(x) + (1-\lambda)\phi(y) \tag{12.1}$$

holds for all $x, y \in I$ *and* $\lambda \in [0, 1]$.

If the above inequality is reversed, then the function ϕ will be the concave on $[a_1, a_2]$.

The main extensions and generalizations of the classical notion of convexity, which will be used in our chapter, are presented below.

In [47], the author defined the h-convex functions:

Definition 12.1.2. *Let* $I \subset \mathbb{R} \setminus \{0\}$ *be a real interval and* $h : I \to \mathbb{R}$ *be a positive function. A function* $\phi : I \to \mathbb{R}$ *is said to be* **h-convex**, *if* ϕ *is nonnegative and*

$$\phi(tx + (1-t)y) \leq h(t)\phi(x) + h(1-t)\phi(y) \tag{12.2}$$

for all $x, y \in I$ *and* $t \in [0, 1]$. *If the inequality in (12.2) is reversed, then* ϕ *is said to be h-concave.*

In 1984, G. Toader defines the class of m-convex functions as follows:

Definition 12.1.3. *Let* ϕ *be a real valued function on* $[0, b]$. *We will say that it is* **m-convex**, *where* $m \in [0, 1]$, *if we have*

$$\phi(tx + m(1-t)y) \leq t\phi(x) + m(1-t)\phi(y) \tag{12.3}$$

for any $x, y \in [a, b]$ *and* $t \in [0, 1]$. *Also,* ϕ *is m-concave if* $-\phi$ *is m-convex. With* $K_m(b)$ *will denote the class of all m-convex functions over* $[0, b]$ *which* $\phi(0) \leq 0$.

Remark 12.1.4. *Clearly, 1-convex functions are the classical convex functions, and 0-convex functions are the "starshaped" functions, that is, those functions* ϕ *that satisfies the inequality* $\phi(tx) \leq t\phi(x)$, *with* $t \in [0, 1]$.

Geometrically a function $\phi : [0, b] \longrightarrow \mathbb{R}$ is m-convex if for any $x, y \in [0, b]$, say $x \leq y$, the segment between the points $(x, \phi(x))$ and $(m, m\phi(y))$ is above the graph of ϕ in $[x, my]$.

Now we consider that X will be a non-empty convex subset of a real vector space and Y will be a real vector space (see [44]).

Definition 12.1.5. *A function* $\phi : X \to \mathbb{R}$ *is said to be* **quasi-convex** *if*

$$\phi((1-t)x + ty) \leq \max(\phi(x), \phi(y)), \tag{12.4}$$

for any $x, y \in X$ *and* $t \in (0, 1)$.

In [20], the authors introduced the idea of η-convex functions (and η-quasi convex functions) as generalizations of ordinary convex functions and gave the following definition:

Definition 12.1.6. *A function $\phi : [a_1, a_2] \to \mathbb{R}$ is said to be η-**convex** if the inequality*

$$\phi(tx + (1-t)y) \leq \phi(x) + t\eta(\phi(x), \phi(y)) \qquad (12.5)$$

holds for all $x, y \in [a_1, a_2]$ and $t \in [0, 1]$, and η is defined by $\eta : \phi([a_1, a_2]) \times \phi([a_1, a_2]) \to \mathbb{R}$.

Obviously, if $\eta(x, y) = x - y$ we obtain the classic convex notion.
Also f is called η-**quasi-convex**, if

$$\phi(tx + (1-t)y) \leq max\{\phi(y), \phi(y) + \eta(\phi(x), \phi(y))\} \qquad (12.6)$$

with the same conditions.

One of the most important inequalities, that has attracted many inequality experts in the last few decades, is the famous Hermite–Hadamard inequality:

$$\phi\left(\frac{a_1 + a_2}{2}\right) \leq \frac{1}{a_2 - a_1}\int_{a_1}^{a_2} \phi(x)\, dx \leq \frac{\phi(a_1) + \phi(a_2)}{2} \qquad (12.7)$$

holds for any function ϕ convex on the interval $[a_1, a_2]$. This inequality was published by Hermite in 1883 and, independently, by Hadamard in 1893. It gives an estimation of the mean value of a convex function, and it is important to note that it also provides a refinement to the Jensen inequality.

They also play a fundamental role in convex analysis, their applications and generalizations (see [19, 39]). This direction of work has become, in recent years, one of the most fruitful areas of work in Mathematics, this development has led to performed in four basic directions:

1. Using various notions of convexity, to get a pretty good idea from the multitude of concepts that have been derived from the classical notion of convex function, we recommend [36]. In [4] the notion of generalized convex and, within the framework of a non-conforming integral, they obtain new inequalities of the Hermite-Hadamard type.

2. Refinement of the mesh used, that is, instead of working directly at the extreme points of the interval $(a; b)$, the points $\frac{3a+b}{4}$, $\frac{a+b}{2}$, and $\frac{a+3b}{4}$, which provides advantages in integration formulas two numerical, although it has the disadvantage that in the case of generalized operators, additional conditions must be imposed on the core, in those points of "union" (see [9]).

3. Improvement of the estimates of the left and right members of (12.7), by introducing auxiliary functions, see for example [14, 48, 49].

4. Obtaining new inequalities of type (12.7) with new operators generalized and fractional integrals (cf. [4, 22, 23, 38]).

To encourage comprehension of the subject, we present the definition of Riemann-Liouville fractional integral (with $0 \leq a_1 < t < a_2 \leq \infty$).

Definition 12.1.7. *Let $\phi \in L_1[a_1, a_2]$. Then the Riemann-Liouville fractional integrals of order $\alpha \in \mathbb{C}$, $\Re(\alpha) > 0$ are defined by (right and left respectively):*

$$^{\alpha}I_{a_1^+}\phi(x) = \frac{1}{\Gamma(\alpha)} \int_{a_1}^{x} (x-t)^{\alpha-1}\phi(t)\,dt, \quad x > a_1 \qquad (12.8)$$

$$^{\alpha}I_{a_2^-}\phi(x) = \frac{1}{\Gamma(\alpha)} \int_{x}^{a_2} (t-x)^{\alpha-1}\phi(t)\,dt, \quad x < a_2. \qquad (12.9)$$

12.2 Integral inequalities with some integral operators

12.2.1 Generalized integral operators

Next we present the generalized integral operators for locally integrable functions.

Definition 12.2.1. *Let I be an interval $I \subseteq \mathbb{R}$, $a, t \in I$ and $\alpha \in \mathbb{R}$. The integral operator J_T^{α}, right and left, is defined for every locally integrable function ϕ on I as*

$$J_{T,a_1^+}^{\alpha}(\phi)(x) = \int_{a_1}^{x} \frac{\phi(t)}{T(x-t,\alpha)}\,dt, x > a_1. \qquad (12.10)$$

$$J_{T,a_2^-}^{\alpha}(\phi)(x) = \int_{x}^{a_2} \frac{\phi(t)}{T(t-x,\alpha)}\,dt, a_2 > x. \qquad (12.11)$$

Remark 12.2.2. *To have a clearer idea of the amplitude of the Definition 12.2.1, let's consider some particular cases:*

1. *Putting $T \equiv 1$, $n = 0$, we obtain the classical Riemann integral.*

2. *If $\frac{1}{T(x-t,\alpha)} = \frac{(x-t)^{(\alpha-1)}}{\Gamma(\alpha)}$ and $\frac{1}{T(t-x,\alpha)} = \frac{(t-x)^{(\alpha-1)}}{\Gamma(\alpha)}$, $n = 0$, then we obtain the Riemann-Liouville fractional integral, right and left.*

3. *If $\frac{1}{T(x-t,\alpha)} = \frac{(x-t)^{(\frac{\alpha}{k}-1)}}{k\Gamma_k(\alpha)}$, and $\frac{1}{T(t-x,\alpha)} = \frac{(t-x)^{(\frac{\alpha}{k}-1)}}{k\Gamma_k(\alpha)}$, $n = 0$, then we obtain the k-Riemann-Liouville fractional integral, right and left [33].*

4. *If $\frac{1}{T(x-t,\alpha)} = \frac{(h(x)-h(t))^{(\frac{\alpha}{k}-1)}h'(t)}{k\Gamma_k(\alpha)}$, and $w\frac{1}{T(t-x,\alpha)} = \frac{(h(t)-h(x))^{(\frac{\alpha}{k}-1)}h'(t)}{k\Gamma_k(\alpha)}$, $n = 0$, then we obtain the right and left-sided fractional integrals of a function f with respect to another function h on $[\nu_1, \nu_2]$ (see [3]).*

Let's see an extension of the Hermite-Hadamard Inequality for h-convex functions.

Theorem 12.2.3. *(see [42]) Let $\phi : I \subset \mathbb{R} \longrightarrow \mathbb{R}$ be a h-convex function on I°. If $\phi \in L[a_1, a_2]$, then we have*

$$\frac{1}{2h\left(\frac{1}{2}\right)} \phi\left(\frac{a_1 + a_2}{2}\right) \leq \frac{1}{a_2 - a_1} \int_{a_1}^{a_2} \phi(x)\, dx \leq (\phi(a_1) + \phi(a_2)) \int_0^1 h(t)\, dt.$$

Theorem 12.2.4. *([35]) Let ϕ be a continuous real function defined on some interval I, $I \subset \mathbb{R}$, $a_1, a_2 \in \mathring{I}$, $a_1 < a_2$. If ϕ is h-convex function we have the following inequality:*

$$\frac{\Pi}{h\left(\frac{1}{2}\right)} \phi\left(\frac{a_1 + a_2}{2}\right) \leq \frac{1}{a_2 - a_1} \left[J^\alpha_{T, a_1^+} \phi(a_2) + J^\alpha_{T, a_2^-} \phi(a_1)\right]$$
$$\leq (H(t) + H(1-t))(\phi(a_1) + \phi(a_2)), \tag{12.12}$$

with $\Pi = \int_0^1 \frac{dt}{T((a_2 - a_1)t, \alpha)}$, $H(t) = \int_0^1 \frac{h(t)dt}{T((a_2 - a_1)t, \alpha)}$ and $H(1-t) = \int_0^1 \frac{h(1-t)dt}{T((a_2-a_1)t, \alpha)}$.

Proof. From (12.2) we obtain, with $x = ta_1 + (1-t)a_2$, $y = ta_2 + (1-t)a_1$ and $t = \frac{1}{2}$ we have

$$\phi\left(\frac{a_1 + a_2}{2}\right) \leq h\left(\frac{1}{2}\right) \phi(ta_1 + (1-t)a_2) + h\left(\frac{1}{2}\right) \phi(ta_2 + (1-t)a_1)$$

$$\phi\left(\frac{a_1 + a_2}{2}\right) \leq h\left(\frac{1}{2}\right) (\phi(ta_1 + (1-t)a_2) + \phi(ta_2 + (1-t)a_1)).$$

Multiplying by $\frac{1}{T((a_2 - a_1)t, \alpha)}$, integrating between 0 and 1, and making the change of variables $u = ta_1 + (1-t)a_2$ in the first integral and $u = ta_2 + (1-t)a_1$ in the second, we easily obtain the first inequality.

To obtain the second inequality, from (12.2) we successively obtain

$$\phi(ta_1 + (1-t)a_2) \leq h(t)\phi(a_1) + h(1-t)\phi(a_2)$$

and

$$\phi(ta_2 + (1-t)a_1) \leq h(t)\phi(a_2) + h(1-t)\phi(a_1),$$

adding term to term and integrating we obtain

$$\phi(ta_1 + (1-t)a_2) + \phi(ta_2 + (1-t)a_1) \leq (h(t) + h(1-t))(\phi(a_1) + \phi(a_2))$$

$$\frac{1}{a_2 - a_1} \left[J^\alpha_{T, a_1^+} \phi(a_2) + J^\alpha_{T, a_2^-} \phi(a_1)\right] \leq (H(t) + H(1-t))(\phi(a_1) + \phi(a_2)),$$

which is the required inequality. This completes the proof. □

Remark 12.2.5. *Let's take $T \equiv 1$, that is, let's work with the classic Riemann integral, in this case $\Pi \equiv 1$, if we consider $h(t) = t$, that is, in the case of a function convex, the classic Hermite-Hadamard inequality (12.7) is obtained. If we consider the notion of s-convex (in the second sense, see [14]), i.e., $h(t) = t^s$, then from the previous theorem we obtain Theorem 1 of [16]. If the function is h-convex, and $T \equiv 1$, then this result covers Theorem 6 of [42].*

Remark 12.2.6. *The right member of the Equation (12.12) coincides with the left member of (2.1) in [46], if we consider the kernel $t^{1-\alpha}$, that is, if we consider integrals fractional Riemann-Liouville type.*

Our next result will be basic from now on. Let us denote $\Theta(u) = \int_0^u \frac{ds}{T((b-a)s,\alpha)}$ so

Lemma 12.2.7. *Let ϕ a real function such that $\phi' \in L([a_1,a_2])$, $I \subset \mathbb{R}$, $a_1, a_2 \in \mathring{I}$, $a_1 < a_2$. If ϕ is h-convex the inequality*

$$-\Theta(1)(\phi(a_1)+\phi(a_2)) + \frac{1}{a_2-a_1}\left[J_{T,a_1+}^\alpha(\phi)(a_2) + J_{T,a_2-}^\alpha(\phi)(a_1)\right]$$
$$= (a_2-a_1)\int_0^1 [\Theta(1-t)-\Theta(t)]\phi'(ta_2+(1-t)a_1)dt \quad (12.13)$$

is fulfilled.

Proof. Let's put

$$I = \int_0^1 \Theta(1-t)\phi'(ta_2+(1-t)a_1)dt - \int_0^1 \Theta(t)\phi'(ta_2+(1-t)a_1)dt = I_1 - I_2.$$

We must integrate by parts I_1 and I_2, then we make a change of variables. By subtracting $I_2 - I_1$, the required equality is obtained. This completes the proof. □

Remark 12.2.8. *If we take the kernel $T \equiv 1$, and $\Theta(s) = s$, then this result is equivalent to Lemma 2.1 of [15].*

Let's call

$$LHH = -\Theta(1)(\phi(a_1)+\phi(a_2)) + \frac{1}{a_2-a_1}\left[J_{T,a_1+}^\alpha(\phi)(a_2) + J_{T,a_2-}^\alpha(\phi)(a_1)\right].$$

Theorem 12.2.9. *Let ϕ a real function such that $\phi' \in L([a_1,a_2])$, $I \subset \mathbb{R}$, $a_1, a_2 \in \mathring{I}$, $a_1 < a_2$. If $|\phi'|$ is h-convex then the following inequality*

$$|LHH| \leq (a_2-a_1)\left(A|\phi'(a_2)| + B|\phi'(a_1)|\right) \quad (12.14)$$

holds, with $A = \int_0^1 |\Theta(1-t)-\Theta(t)|h(t)dt$ and $B = \int_0^1 |\Theta(1-t)-\Theta(t)|h(1-t)dt$.

Proof. From Lemma 12.2.7, using the h-convexity of $|\phi'|$ and using the Definition of A and B, it is very easy to conclude the proof. □

Remark 12.2.10. *Under the conditions of the previous Remark, this Theorem contains as a particular case Theorem 2.2 of [15].*

Theorem 12.2.11. Let ϕ a real function such that $\phi' \in L([a_1, a_2])$, $I \subset \mathbb{R}$, $a_1, a_2 \in \mathring{I}$, $a_1 < a_2$. If $|\phi'|^q$ is h-convex, with $q \geq 1$, then we have the following inequality

$$|LHH| \leq (a_2 - a_1) C^{1-\frac{1}{q}} \left(A|\phi'(a_2)|^q + B|\phi'(a_1)|^q \right)^{\frac{1}{q}} \qquad (12.15)$$

with A and B as before and $C = \int_0^1 |\Theta(1-t) - \Theta(t)| dt$.

Proof. From Lemma 12.2.7, the well-known power mean inequality and using the h-convexity of $|\phi'|^q$ we have the inequality sought. □

A variant of the previous result is the one presented below.

Theorem 12.2.12. Let ϕ a real function such that $\phi' \in L([a_1, a_2])$, $I \subset \mathbb{R}$, $a_1, a_2 \in \mathring{I}$, $a_1 < a_2$. If $|\phi'|^q$ is h-convex, with $q > 1$, then we have the following inequality

$$|LHH| \leq (a_2 - a_1) \left(D^{\frac{1}{p}} + E^{\frac{1}{p}} \right).$$

$$\cdot \left(|\phi'(a_2)|^q \int_0^1 h(t) dt + |\phi'(a_1)|^q \int_0^1 h(1-t) dt \right)^{\frac{1}{q}} \qquad (12.16)$$

with $D = \int_0^1 \Theta(1-t)^p dt$, $E = \int_0^1 \Theta(t)^p dt$ and $\frac{1}{p} + \frac{1}{q} = 1$.

Proof. From Lemma 12.2.7, using the Hölder inequality and the h-convexity of $|\phi'|^q$ we obtain the required inequality. That completes the Theorem's proof. □

Remark 12.2.13. *Under the conditions of the previous Remarks, the Theorem 2.3 of [15] is contained in this result.*

Taking into account the Definition (12.2.1) (see [21]) seen above and, considering that the nucleus of the integral operator, T, is an absolutely continuous function, we have the following lemma.

Lemma 12.2.14. Let $\phi : [a_1, a_2] \to \mathbb{R}$ be a differentiable function on (a_1, a_2) with $a_1 < a_2$. If $\phi' \in L[a_1, a_2]$, then the following equality holds

$$\frac{\Pi}{2}(\phi(a_1) + \phi(a_2)) - \frac{1}{2(a_2 - a_1)} \left[J_{\mathbb{T}, a_1^+}^{\alpha, [a_1, a_2]} \phi(a_2) - J_{\mathbb{T}, a_2^-}^{\alpha, [a_1, a_2]} \phi(a_1) \right]$$

$$= \frac{a_2 - a_1}{2} \int_0^1 [\mathbb{T}(1-t) - \mathbb{T}(t)] \phi'(t a_1 + (1-t) a_2) dt, \qquad (12.17)$$

where $\Pi = \int_0^1 \frac{dt}{T(t,\alpha)}$ and $\mathbb{T}(t) = \int_0^t \frac{ds}{T(s,\alpha)}$.

Proof. Using integration by parts in the right member of (12.17), making the change of variables $s = ta_1 + (1-t)a_2$, rearranging and a simple algebraic work, leads us to the desired result. □

Remark 12.2.15. *If $T(t, \alpha) = t^\alpha$ then the previous result become the Lemma 2 of [43].*

Theorem 12.2.16. *Let $\phi : [0, a_2] \to \mathbb{R}$ be a twice differentiable mapping, such that $|\phi''|$ is an m-convex function in $[a_1, a_2]$, then the following inequalities holds:*

$$\left| \frac{\phi(a_1) + m\phi(a_2)}{2} - \frac{\Pi}{2(a_2 - a_1)} \left[J_{T,a_1^+}^\alpha \phi(a_2) - J_{T,a_2^-}^\alpha \phi(a_1) \right] \right|$$
$$\leq \frac{\alpha(a_2 - a_1)^2}{2(\alpha+1)(\alpha+2)} \left[\frac{|\phi''(a_1)| + m|\phi''(a_2)|}{2} \right]. \tag{12.18}$$

12.2.2 Generalized fractional integral operators

We are now in a position to define the generalized integral operators.

Definition 12.2.17. *([24]) Let $a_1 < a_2$ and $\phi \in L^1((a_1, a_2); \mathbb{R})$. The right and left side Hadamard fractional integrals of order α, with $Re(\alpha) > 0$, are defined, respectively, by*

$$H_{a_1^+}^\alpha \phi(x) = \frac{1}{\Gamma(\alpha)} \int_{a_1}^x \left(\log \frac{x}{t} \right)^{\alpha-1} \frac{\phi(t)}{t} dt \tag{12.19}$$

and

$$H_{a_2^-}^\alpha \phi(x) = \frac{1}{\Gamma(\alpha)} \int_x^{a_2} \left(\log \frac{t}{x} \right)^{\alpha-1} \frac{\phi(t)}{t} dt \tag{12.20}$$

with $x \in (a_1, a_2)$.

When $\alpha \in (0, 1)$, Hadamard differential operators are given by the following expressions:

$$(^H D_{a_1^+}^\alpha \phi)(x) = x \frac{d}{dx} \left(H_{a_1^+}^{1-\alpha} \phi(x) \right) = \frac{1}{(1-\alpha)} x \frac{d}{dx} \int_{a_1}^x \left(\log \frac{x}{t} \right)^{-\alpha} \frac{\phi(t)}{t} dt,$$

$$(^H D_{b^-}^\alpha \phi)(x) = -x \frac{d}{dx} \left(H_{a_2^-}^{1-\alpha} \phi(x) \right) = \frac{-1}{\Gamma(1-\alpha)} x \frac{d}{dx} \int_x^{a_2} \left(\log \frac{t}{x} \right)^{-\alpha} \frac{\phi(t)}{t} dt,$$

with $x \in (a_1, a_2)$.

In [28], the author introduced new fractional integral operators, called the Katugampola fractional integrals, in the following way.

Definition 12.2.18. *Let $0 < a_1 < a_2$, $\phi : [a_1, a_2] \to \mathbb{R}$ an integrable function, and $\alpha \in (0, 1), \rho > 0$ two fixed real numbers. The right and left side Katugampola fractional integrals of order α are defined, respectively, by*

$$K_{a_1^+}^{\alpha,\rho} \phi(x) = \frac{\rho^{1-\alpha}}{\Gamma(\alpha)} \int_{a_1}^x \frac{t^{\rho-1}}{(x^\rho - t^\rho)^{1-\alpha}} \phi(t) dt, \tag{12.21}$$

and
$$K_{a_2-}^{\alpha,\rho}\phi(x) = \frac{\rho^{1-\alpha}}{\Gamma(\alpha)} \int_x^{a_2} \frac{x^{\rho-1}}{(t^\rho - x^\rho)^{1-\alpha}} \phi(t)dt, \qquad (12.22)$$

with $x \in (a_1, a_2)$.

An interesting definition is the following: the Generalized Proportional Fractional (see [41]).

Definition 12.2.19. *Let $U \in X_\Psi^q(0, +\infty)$, $0 < a_1 < a_2$, there is an increasing, positive monotone function Ψ defined on $[0, +\infty)$ having continuous derivative Ψ' on $[0, +\infty)$ with $\Psi(0) = 0$. Then the left-sided and right-sided GPF-integral operator of a function U in the sense of another function Ψ of order $\eta > 0$ are stated as:*

$$^\Psi T_{v_1+}^{\eta,\varsigma} U(\varsigma) = \frac{1}{\varsigma^\eta \Gamma(\eta)} \int_{v_1}^\varsigma \frac{exp\left[\frac{\varsigma-1}{\varsigma}(\Psi(\varsigma) - \Psi(\xi))\right]}{(\Psi(\varsigma) - \Psi(\xi))^{1-\eta}} U(\xi)\Psi'(\xi)d\xi, \quad v_1 < \varsigma, \qquad (12.23)$$

and

$$^\Psi T_{v_2-}^{\eta,\varsigma} U(\varsigma) = \frac{1}{\varsigma^\eta \Gamma(\eta)} \int_{v_1}^\varsigma \frac{exp\left[\frac{\varsigma-1}{\varsigma}(\Psi(\xi) - \Psi(\varsigma))\right]}{(\Psi(\xi) - \Psi(\varsigma))^{1-\eta}} U(\xi)\Psi'(\xi)d\xi, \quad \varsigma < v_2, \qquad (12.24)$$

where the proportionality index $\varsigma \in (0,1]$, $\eta \in \mathbb{C}$, $Re(\eta) > 0$, and Γ is the Gamma function.

Definition 12.2.20. *([34]) Let $h \in X_T^q(0, +\infty)$, T a continuous, positive function on $[0, +\infty)$ with $T(0) = 0$. The right and left side Generalized k-Proportional Fractional Integral Operators with General Kernel of order γ of h are defined, respectively by*

$$J_{T,a_1+}^{\frac{\gamma}{k},\lambda} h(\chi) = \frac{1}{\lambda^\gamma k \Gamma_k(\gamma)} \int_{a_1}^\chi \frac{G(\mathbb{T}_+(\chi,s), \lambda) T(s) h(s)}{(\mathbb{T}_+(\chi,s))^{1-\frac{\gamma}{k}}} ds, \qquad (12.25)$$

$$J_{T,a_2-}^{\frac{\gamma}{k},\lambda} h(\chi) = \frac{1}{\lambda^\gamma k \Gamma_k(\gamma)} \int_\chi^{a_2} \frac{G(\mathbb{F}_-(s,\chi), \lambda) T(s) h(s)}{(\mathbb{T}_-(s,\chi))^{1-\frac{\gamma}{k}}} ds, \qquad (12.26)$$

with $\chi \in (a,b)$, $\mathbb{T}_+(\chi,s) = \int_s^\chi T(r)dr$, $\mathbb{T}_-(s,\chi) = \int_\chi^s T(r)dr$ and $G(\mathbb{T}_+(\chi,s),1) = G(\mathbb{F}_-(\chi,s),1) = 1$.

Of course there are other integral fractional operators, variations of the previous ones can be considered, so we will omit them.

Remark 12.2.21. *Next, we will show how many integral operators are particular cases of (12.25) and (12.26).*

1. *If in Definition 12.2.20, we make $k = 1$, $T = 1$, and $\lambda = 1$, we obtain the Riemann-Liouville operators of Definition 12.1.7.*

2. Under the above conditions, if $k \neq 1$ then, from Definition 12.2.20, the k-fractionals operators of [33] are obtained.

3. If $T(s) = \frac{1}{s}$, $\lambda = 1$ and $k = 1$, then the Hadamard fractional operator is reproduced, see Definition 12.2.17, and [24, 30, 45].

4. If $T(s) = \frac{1}{s^\rho}$, $\lambda = 1$, and $k = 1$, then we obtain the Katugampola's fractional operator of Definition 12.2.18, see [28, 29].

5. Choosing $\lambda = 1$, $T(s) = g'(s)$, and $k = 1$, then we get the integral operator of [31].

6. Taking $T(s) = \frac{1}{s}$, $k \neq 1$ and $G(\mathbb{T}_+(\chi, s), \lambda) = exp\left[\frac{\lambda-1}{\lambda}\left(ln\frac{\chi}{s}\right)\right]$, we obtain the integral operator of [40].

7. We can obtain an integral operator with a non-singular nucleus, of the Riemann-Liouville type, by putting $\gamma = k = 1$, $T(t) = 1$, and $G(\mathbb{T}_+(x,s), \alpha) = exp\left[-\frac{1-\alpha}{\alpha}(x-s)\right]$. A slight modification of the operator of [1].

8. Choosing $\lambda \neq 1$, $T(s) = g'(s)$, $k = 1$ and $G(\mathbb{T}_+(\chi, s), \lambda) = exp\left[\frac{\lambda-1}{\lambda}(g(\chi) - g(s))\right]$, then we obtain the integral operator of [41], called GFP, presented in Definition 12.2.19.

Here are some integral inequalities in the context of the generalized fractional operator of the Definition 12.2.20.

Theorem 12.2.22. *Under the conditions $a_1 < \varsigma \leq a_2$, $\theta > 0, \sigma > 0$ and $\gamma > 0$, suppose that ω is a continuous positive non-decreasing function on $[a_1, a_2]$ and $h : [a_1, a_2] \to \mathbb{R}^+$ is also a positive continuous function. Then the following inequality is valid*

$$J_{T,a_1+}^{\frac{\gamma}{k},\lambda}[(\varsigma-a_1)^\sigma h(\varsigma)\omega^\theta(\varsigma)]J_{T,a_1+}^{\frac{\gamma}{k},\lambda}h(\varsigma) \geq J_{T,a_1+}^{\frac{\gamma}{k},\lambda}[(\varsigma-a_1)^\sigma h(\varsigma)]J_{T,a_1+}^{\frac{\gamma}{k},\lambda}[h(\varsigma)\omega^\theta(\varsigma)]. \tag{12.27}$$

Proof. Taking into account the positivity and not decreasing of the function ω on $[a_1, a_2]$ we have, for all $\theta > 0$, $\sigma > 0$, $u, v \in [a_1, \varsigma]$, with $v \leq u$, and $a_1 < \varsigma \leq a_2$, the following

$$((u-a_1)^\sigma - (v-a_1)^\sigma)(\omega^\theta(u) - \omega^\theta(v)) \geq 0. \tag{12.28}$$

So, from (12.28) we get

$$(u-a_1)^\sigma \omega^\theta(u) + (v-a_1)^\sigma \omega^\theta(v) \geq (u-a_1)^\sigma \omega^\theta(v) + (v-a_1)^\sigma \omega^\theta(u). \tag{12.29}$$

Multiplying first both members of (12.29) by $\frac{G(\mathbb{T}_+(u,\chi),\lambda)T(u)h(u)}{\lambda^\gamma k \Gamma_k(\gamma)(\mathbb{T}_+(u,\chi))^{1-\frac{\gamma}{k}}}$, and integrating the resulting inequality over (a_1, ς) with respect to u, then multiplying the above by $\frac{G(\mathbb{T}_+(v,\chi),\lambda)T(v)h(v)}{\lambda^\gamma k \Gamma_k(\gamma)(\mathbb{T}_+(v,\chi))^{1-\frac{\gamma}{k}}}$, and integrating the resulting inequality over (a_1, ς) with respect to v, we obtain the inequality (12.27). □

Remark 12.2.23. *If we take the kernel* $T(\varsigma) = \varsigma^s$, *with* $k > 0$, $s \in \mathbb{R}$, $s \neq -1$ *and* $\lambda = 1$, *the previous theorem becomes the Theorem 6 of [25].*

A variant of this theorem, considering a family of functions $\omega_i(t)$, $i = 1, 2, ...n$ is given in the following result.

Theorem 12.2.24. *Let us consider a family of positive, continuous and decreasing functions* $\omega_i(t)$, $i = 1, 2, ...n$ *defined on* $[a_1, a_2]$. *So the following inequality is true*

$$J_{T,a_1+}^{\frac{\gamma}{k},\lambda}\left[\prod_{i \neq k}^n f_i^{\theta_i}(\varsigma)f_p^\sigma(\varsigma)\right]J_{T,a_1+}^{\frac{\gamma}{k},\lambda}\left[(\varsigma-a_1)^\delta\prod_{i=1}^n f_i^{\theta_i}(\varsigma)\right] \geq$$
$$J_{T,a_1+}^{\frac{\gamma}{k},\lambda}\left[(\varsigma-a_1)^\delta\prod_{i \neq k}^n f_i^{\theta_i}(\varsigma)f_p^\sigma(\varsigma)\right]J_{F,a_1+}^{\frac{\gamma}{k},\lambda}\left[\prod_{i=1}^n f_i^{\theta_i}(\varsigma)\right] \quad (12.30)$$

for $a_1 < \varsigma \leq a_2$, $\gamma > 0$, $\delta > 0$, $\sigma \geq \theta_p$ *with* p *is some integer fixed in* $i = 1, 2, ..., n$.

Proof. The proof of this theorem follows the previous one, where instead of the inequality (12.28) the following

$$((u-a_1)^\sigma - (v-a_1)^\sigma)(f_p^{\sigma-\theta_p}(u) - f_p^{\sigma-\theta_p}(v)) \geq 0$$

multiply by $\frac{G(\mathbb{T}_+(u,\chi),\lambda)T(u)h(u)}{\lambda^\gamma k\Gamma_k(\gamma)(\mathbb{F}_+(u,\chi))^{1-\frac{\gamma}{k}}}$ and integrate with respect to u to finish the proof. □

Remark 12.2.25. *If in this result we make* $k = 1$, $T = 1$ *and* $\lambda = 1$ *we obtain Theorem 3.1 of [14].*

Using two functional parameters, we can generalize the Theorem 12.2.22.

Theorem 12.2.26. *Assuming* $a_1 < \varsigma \leq a_2$, $\theta > 0, \sigma > 0, \gamma > 0$, *and* $\beta > 0$, *and that* ω *is a continuous, positive, non-decreasing function, defined on* $[a_1, a_2]$. *Then for* $h : [a_1, a_2] \to \mathbb{R}^+$ *a positive continuous function, the following inequality is satisfied*

$$J_{T,a_1+}^{\frac{\gamma}{k},\lambda}[(\varsigma-a_1)^\sigma h(\varsigma)\omega^\theta(\varsigma)]J_{T,a_1+}^{\frac{\beta}{k},\lambda}h(\varsigma)+$$
$$J_{T,a_1+}^{\frac{\beta}{k},\lambda}[(\varsigma-a_1)^\sigma h(\varsigma)\omega^\theta(\varsigma)]J_{T,a_1+}^{\frac{\gamma}{k},\lambda}h(\varsigma) \geq$$
$$J_{T,a_1+}^{\frac{\beta}{k},\lambda}[h(\varsigma)\omega^\theta(\varsigma)]J_{T,a_1+}^{\frac{\gamma}{k},\lambda}[(\varsigma-a_1)^\sigma h(\varsigma)]$$
$$+ J_{T,a_1+}^{\frac{\beta}{k},\lambda}[(\varsigma-a_1)^\sigma h(\varsigma)]J_{T,a_1+}^{\frac{\gamma}{k},\lambda}[h(\varsigma)\omega^\theta(\varsigma)]. \quad (12.31)$$

Proof. After multiplying both sides by $\frac{G(\mathbb{T}_+(v,\chi),\lambda)T(v)h(v)}{\lambda^\gamma k\Gamma_k(\gamma)(\mathbb{T}_+(v,\chi))^{1-\frac{\gamma}{k}}}$, and integrate the result with respect to v over (a_1, ς), we arrive at (12.31). □

Remark 12.2.27. *If in this result we do $\gamma = \beta$, we achieved Theorem 12.2.26.*

Remark 12.2.28. *The Theorem 7 of [25] is obtained from previous result, under the same conditions as the Remark 12.2.23.*

We can generalize the Theorem 12.2.22, considering a certain weight function, as follows.

Theorem 12.2.29. *Considering $a_1 < \varsigma \leq a_2$, $\theta > 0, \sigma > 0, \alpha > 0$, be ω and ψ two positive continuous functions defined on $[a_1, a_2]$ such that ω is non-decreasing, ψ is non-increasing. Then if $h : [a_1, a_2] \to \mathbb{R}^+$ is a positive continuous function, we obtain the following inequality*

$$J_{T,a_1+}^{\frac{\gamma}{k},\lambda}[h(\varsigma)\psi^\sigma(\varsigma)]J_{T,a_1+}^{\frac{\gamma}{k},\lambda}[h(\varsigma)\omega^\theta(\varsigma)] \geq J_{T,a_1+}^{\frac{\gamma}{k},\lambda}h(\varsigma)J_{T,a_1+}^{\frac{\gamma}{k},\lambda}[h(\varsigma)\omega^\theta(\varsigma)\psi^\sigma(\varsigma)]. \tag{12.32}$$

Proof. Taking into account the properties of the functions ω and ψ on $[a_1, a_2]$, the following inequality is valid for all $\theta > 0$, $\sigma > 0$, $u, v \in [a_1, \varsigma]$, with $v \leq u$, and $a_1 < \varsigma \leq a_2$

$$\omega^\theta(u)\psi^\sigma(v) + \omega^\theta(v)\psi^\sigma(u) \geq \omega^\theta(u)\psi^\sigma(u) + \omega^\theta(v)\psi^\sigma(v). \tag{12.33}$$

If we now multiply the inequality (12.33) by $\frac{G(\mathbb{T}_+(u,\chi),\lambda)T(u)h(u)}{\lambda^\gamma k \Gamma_k(\gamma)(\mathbb{T}_+(u,\chi))^{1-\frac{\gamma}{k}}}$, and we integrate with respect to u over (a_1,ς). Now, we multiple the above by $\frac{G(\mathbb{T}_+(v,\chi),\lambda)T(v)h(v)}{\lambda^\gamma k \Gamma_k(\gamma)(\mathbb{T}_+(v,\chi))^{1-\frac{\gamma}{k}}}$, then we integrate the resulting inequality with respect to v over (a_1,ς), we get the required inequality (12.32). □

Remark 12.2.30. *Considering $T(\varsigma, s) = \varsigma^s$, with $k > 0$ and $\lambda = 1$, this Theorem becomes the Theorem 9 of [25].*

Theorem 12.2.31. *Under the same assumptions as in the Theorem 12.2.29, about the functions ω, ψ, and h. Then the following inequality is valid for $a_1 < \varsigma \leq a_2, \theta > 0, \sigma > 0, \beta > 0, \gamma > 0$*

$$J_{T,a_1+}^{\frac{\gamma}{k},\lambda}[h(\varsigma)\psi^\sigma(\varsigma)]J_{T,a_1+}^{\frac{\gamma}{k},\lambda}[h(\varsigma)\omega^\theta(\varsigma)] + J_{T,a_1+}^{\frac{\gamma}{k},\lambda}[h(\varsigma)\omega^\theta(\varsigma)]J_{T,a_1+}^{\frac{\gamma}{k},\lambda}[h(\varsigma)\psi^\sigma(\varsigma)]$$
$$\geq J_{T,a_1+}^{\frac{\gamma}{k},\lambda}h(\varsigma)J_{T,a_1+}^{\frac{\gamma}{k},\lambda}[h(\varsigma)\omega^\theta(\varsigma)\psi^\sigma(\varsigma)] + J_{T,a_1+}^{\frac{\gamma}{k},\lambda}[h(\varsigma)\omega^\theta(\varsigma)\psi^\sigma(\varsigma)]J_{T,a_1+}^{\frac{\gamma}{k},\lambda}h(\varsigma). \tag{12.34}$$

Proof. The proof follows the same scheme as the previous ones, it multiplies by $\frac{G(\mathbb{T}_+(v,\chi),\lambda)T(v)h(v)}{\lambda^\gamma k \Gamma_k(\gamma)(\mathbb{T}_+(v,\chi))^{1-\frac{\gamma}{k}}}$, and the result obtained is integrated with respect to v over (a_1,ς), we arrived (12.34). □

Remark 12.2.32. *Putting $\alpha = \beta$ in the Theorem 12.2.31, this is reduced to Theorem 12.2.29.*

Remark 12.2.33. *We can get the Theorem 10 of [25] making $T(\varsigma) = \varsigma^s$, with $k > 0$ and $s \in \mathbb{R}$ in the Theorem 12.2.31.*

Generalized Fractional Operators 217

Theorem 12.2.34. *For $a_1 < \varsigma \leq a_2$, $\delta \geq \theta > 0, \sigma > 0, \gamma > 0$, the following inequality is satisfied by taking ω as a positive decreasing continuous function on $[a_1, a_2]$ and $h: [a_1, a_2] \to \mathbb{R}^+$ a positive continuous function*

$$J_{T,a_1+}^{\frac{\gamma}{k},\lambda}[(\varsigma-a_1)^\sigma h(\varsigma)\omega^\theta(\varsigma)] J_{T,a_1+}^{\frac{\gamma}{k},\lambda}[h(\varsigma)\omega^\delta(\varsigma)]$$
$$\geq J_{T,a_1+}^{\frac{\gamma}{k},\lambda}[(\varsigma-a_1)^\sigma h(\varsigma)\omega^\delta(\varsigma)] J_{T,a_1+}^{\frac{\gamma}{k},\lambda}[h(\varsigma)\omega^\theta(\varsigma)]. \quad (12.35)$$

Proof. Using the positivity of the function ω then

$$(u-a_1)^\sigma \omega^{\delta-\theta}(v) + (v-a_1)^\sigma \omega^{\delta-\theta}(u) \geq (u-a_1)^\sigma \omega^{\delta-\theta}(u) + (v-a_1)^\sigma \omega^{\delta-\theta}(v), \quad (12.36)$$

for all $\delta - \theta \geq 0$, $\sigma > 0$, $u, v \in [a_1, \varsigma]$, with $v \leq u$, and $a_1 < \varsigma \leq a_2$. Now we multiply (12.36) by $\frac{G(\mathbb{T}_+(u,\chi),\lambda)T(u)h(u)}{\lambda^\gamma k \Gamma_k(\gamma)(\mathbb{T}_+(u,\chi))^{1-\frac{\gamma}{k}}}$, and we integrate the result with respect to u over (a_1, ς). Then we multiply the above by $\frac{G(\mathbb{T}_+(v,\chi),\lambda)T(v)h(v)}{\lambda^\gamma k \Gamma_k(\gamma)(\mathbb{T}_+(v,\chi))^{1-\frac{\gamma}{k}}}$, and the result obtained is integrated with respect to v over (a_1, ς), which leads us to get the inequality (12.35). □

Remark 12.2.35. *Taking into account the same considerations of the previous Remarks, the Theorem 12 of [25] is obtained as a particular case.*

The results presented below contain, like the Theorem 12.2.24, the product of a family of functions.

Theorem 12.2.36. *Let us consider a family of positive, continuous and decreasing functions $\omega_i(t)$, $i = 1, 2, ...n$ defined on $[a_1, a_2]$ and $h: [a_1, a_2] \to \mathbb{R}^+$ a positive and continuous function. So, we have the following inequality*

$$J_{T,a_1+}^{\frac{\gamma}{k},\lambda}\left[h(\varsigma)\omega_r^\delta(\varsigma)\prod_{j\neq r}^n \omega_j^{\theta_j}(\varsigma)\right] J_{T,a_1+}^{\frac{\gamma}{k},\lambda}\left[(\varsigma-a_1)^\sigma h(\varsigma)\prod_{j=1}^n \omega_j^{\theta_j}(\varsigma)\right] \geq$$
$$J_{T,a_1+}^{\frac{\gamma}{k},\lambda}\left[h(\varsigma)\prod_{j=1}^n \omega_j^{\theta_j}(\varsigma)\right] J_{T,a_1+}^{\frac{\gamma}{k},\lambda}\left[(\varsigma-a_1)^\sigma h(\varsigma)\omega_r^\delta(\varsigma)\prod_{j\neq r}^n \omega_j^{\theta_j}(\varsigma)\right], \quad (12.37)$$

for $a_1 < \varsigma \leq a_2$, $\gamma > 0$, $\sigma > 0$ and $\delta \geq \theta_r > 0$ with $r \in \{1, 2, \ldots, n\}$.

Proof. From the properties of functions ω_r on $[a_1, a_2]$

$$((u-a_1)^\sigma - (v-a_1)^\sigma)(\omega_r^{\delta-\theta_r}(v) - \omega_r^{\delta-\theta_r}(u)) \geq 0. \quad (12.38)$$

is valid for all $\sigma > 0$, $a_1 < \varsigma \leq a_2$, $u, v \in [a_1, \varsigma]$, if $v \leq u$, $\delta \geq \theta_r > 0$, and for any fixed $r \in \{1, 2, \ldots, n\}$. From inequality (12.38), we deduce that

$$(u-a_1)^\sigma \omega_r^{\delta-\theta_r}(v) + (v-a_1)^\sigma \omega_r^{\delta-\theta_r}(u) \geq (u-a_1)^\sigma \omega_r^{\delta-\theta_r}(u) + (v-a_1)^\sigma \omega_r^{\delta-\theta_r}(v). \quad (12.39)$$

As before, multiplying (12.39) by $\frac{G(\mathbb{T}_+(u,\chi),\lambda)T(u)h(u)\prod_{j=1}^n \omega_j^{\theta_j}(u)}{\lambda^\gamma k \Gamma_k(\gamma)(\mathbb{T}_+(\chi,u))^{1-\frac{\gamma}{k}}}$, and integrating the result obtained with respect to u over (a_1,ς). Now, the previous result is multiplied by $\frac{G(\mathbb{T}_+(v,\chi),\lambda)T(v)h(v)\prod_{j=1}^n \omega_j^{\theta_j}(v)}{\lambda^\gamma k \Gamma_k(\gamma)(\mathbb{T}_+(\chi,v))^{1-\frac{\gamma}{k}}}$, and the result is integrated with respect to v over (a_1,ς). Thus, we arrive at inequality (12.37). □

Theorem 12.2.37. *Let us consider a family of positive, continuous and decreasing functions $\omega_i(t)$, $i = 1, 2, \ldots n$ defined on $[a_1, a_2]$ and $h : [a_1, a_2] \to \mathbb{R}^+$ a positive and continuous function. So, we have the following inequality*

$$J_{T,a_1+}^{\frac{\gamma}{k},\lambda}\left[(\varsigma-a_1)^\sigma h(\varsigma)\prod_{j=1}^n \omega_j^{\theta_j}(\varsigma)\right] J_{T,a_1+}^{\frac{\gamma}{k},\lambda}\left[h(\varsigma)\omega_r^\delta(\varsigma)\prod_{j\neq r}^n \omega_j^{\theta_j}(\varsigma)\right]$$

$$+ J_{T,a_1+}^{\frac{\gamma}{k},\lambda}\left[h(\varsigma)\omega_r^\delta(\varsigma)\prod_{j\neq r}^n \omega_j^{\theta_j}(\varsigma)\right] J_{T,a_1+}^{\frac{\gamma}{k},\lambda}\left[(\varsigma-a_1)^\sigma h(\varsigma)\prod_{j=1}^n \omega_j^{\theta_j}(\varsigma)\right] \geq$$

$$J_{T,a_1+}^{\frac{\gamma}{k},\lambda}\left[(\varsigma-a_1)^\sigma h(\varsigma)\omega_r^\delta(\varsigma)\prod_{j\neq r}^n \omega_j^{\theta_j}(\varsigma)\right] J_{T,a_1+}^{\frac{\gamma}{k},\lambda}\left[h(\varsigma)\prod_{j=1}^n \omega_j^{\theta_j}(\varsigma)\right]$$

$$+ J_{T,a_1+}^{\frac{\gamma}{k},\lambda}\left[h(\varsigma)\prod_{j=1}^n \omega_j^{\theta_j}(\varsigma)\right] J_{T,a_1+}^{\frac{\gamma}{k},\lambda}\left[(\varsigma-a_1)^\sigma h(\varsigma)\omega_r^\delta(\varsigma)\prod_{j\neq r}^n \omega_j^{\theta_j}(\varsigma)\right],$$

(12.40)

for $a_1 < \varsigma \leq a_2$, $\alpha > 0$, $\beta > 0$, $\sigma > 0$ and $\delta \geq \theta_r > 0$ with $r \in \{1, 2, \ldots, n\}$.

Proof. The inequality (12.40) is obtained, after multiplying (12.39) by $\frac{G(\mathbb{T}_+(v,\chi),\lambda)T(v)h(v)\prod_{j=1}^n \omega_j^{\theta_j}(v)}{\lambda^\gamma k \Gamma_k(\gamma)(\mathbb{T}_+(\chi,v))^{1-\frac{\gamma}{k}}}$. □

Remark 12.2.38. *Putting $\alpha = \beta$ in the Theorem 12.2.37, we obtain the Theorem 12.2.36.*

Theorem 12.2.39. *If we consider a family of positive, continuous, and decreasing functions $\omega_i(t)$, $i = 1, 2, \ldots n$ defined on $[a_1, a_2]$ and $h : [a_1, a_2] \to \mathbb{R}^+$ a positive and continuous function. Then the following inequality is valid*

$$J_{T,a_1+}^{\frac{\gamma}{k},\lambda}\left[h(\varsigma)\omega_r^\delta(\varsigma)\prod_{j\neq r}^n \omega_j^{\theta_j}(\varsigma)\right] J_{T,a_1+}^{\frac{\gamma}{k},\lambda}\left[\psi^\sigma(\varsigma)h(\varsigma)\prod_{j=1}^n \omega_j^{\theta_j}(\varsigma)\right] \geq$$

$$J_{T,a_1+}^{\frac{\gamma}{k},\lambda}\left[h(\varsigma)\prod_{j=1}^n \omega_j^{\theta_j}(\varsigma)\right] J_{T,a_1+}^{\frac{\gamma}{k},\lambda}\left[\psi^\sigma(\varsigma)h(\varsigma)\omega_r^\delta(\varsigma)\prod_{j\neq r}^n \omega_j^{\theta_j}(\varsigma)\right], \quad (12.41)$$

with $a_1 < \varsigma \leq a_2$, $\gamma > 0$, $\sigma > 0$ and $\delta \geq \theta_r > 0$ with $r \in \{1, 2, \ldots, n\}$.

Proof. Taking into account that the functions ω_r and ψ are positive and continuous on $[a_1, a_2]$ with ω_r decreasing and ψ increasing; then

$$\psi^\sigma(u)\omega_r^{\delta-\theta_r}(v) + \psi^\sigma(v)\omega_r^{\delta-\theta_r}(u) \geq \psi^\sigma(u)\omega_r^{\delta-\theta_r}(u) + \psi^\sigma(v)\omega_r^{\delta-\theta_r}(v). \quad (12.42)$$

for all $\sigma > 0$, $a_1 < \varsigma \leq a_2$, $u, v \in [a_1, \varsigma]$, with $v \leq u$, and $\delta \geq \theta_r > 0$, for any fixed $r \in \{1, 2, \ldots, n\}$. If we multiply the inequality (12.42) by $\frac{G(\mathbb{T}_+(u,\chi),\lambda)T(u)h(u)\prod_{j=1}^n \omega_j^{\theta_j}(u)}{\lambda^\gamma k\Gamma_k(\gamma)(\mathbb{T}_+(\chi,u))^{1-\frac{\gamma}{k}}}$, and we integrate the result obtained with respect to u over (a_1, ς). Then, to obtain the inequality (12.41), we multiply the above by $\frac{G(\mathbb{T}_+(v,\chi),\lambda)T(v)h(v)\prod_{j=1}^n \omega_j^{\theta_j}(v)}{\lambda^\gamma k\Gamma_k(\gamma)(\mathbb{T}_+(\chi,v))^{1-\frac{\gamma}{k}}}$, and the result obtained is integrated with respect to v over (a_1, ς). □

Theorem 12.2.40. *The following inequality is valid*

$$J_{T,a_1+}^{\frac{\gamma}{k},\lambda}\left[h(\varsigma)\omega_r^\delta(\varsigma)\prod_{\substack{j=1\\j\neq r}}^n \omega_j^{\theta_j}(\varsigma)\right] J_{T,a_1+}^{\frac{\gamma}{k},\lambda}\left[\psi^\sigma(\varsigma)h(\varsigma)\prod_{j=1}^n \omega_j^{\theta_j}(\varsigma)\right]$$

$$+ J_{T,a_1+}^{\frac{\gamma}{k},\lambda}\left[h(\varsigma)\omega_r^\delta(\varsigma)\prod_{\substack{j=1\\j\neq r}}^n \omega_j^{\theta_j}(\varsigma)\right] J_{T,a_1+}^{\frac{\gamma}{k},\lambda}\left[\psi^\sigma(\varsigma)h(\varsigma)\prod_{j=1}^n \omega_j^{\theta_j}(\varsigma)\right] \geq$$

$$J_{T,a_1+}^{\frac{\gamma}{k},\lambda}\left[h(\varsigma)\prod_{j=1}^n \omega_j^{\theta_j}(\varsigma)\right] J_{T,a_1+}^{\frac{\gamma}{k},\lambda}\left[\psi^\sigma(\varsigma)h(\varsigma)\omega_r^\delta(\varsigma)\prod_{\substack{j=1\\j\neq r}}^n \omega_j^{\theta_j}(\varsigma)\right]$$

$$+ J_{T,a_1+}^{\frac{\gamma}{k},\lambda}\left[h(\varsigma)\prod_{j=1}^n \omega_j^{\theta_j}(\varsigma)\right] J_{T,a_1+}^{\frac{\gamma}{k},\lambda}\left[\psi^\sigma(\varsigma)h(\varsigma)\omega_r^\delta(\varsigma)\prod_{\substack{j=1\\j\neq r}}^n \omega_j^{\theta_j}(\varsigma)\right], \quad (12.43)$$

for ω_j, $j = 1, 2, \ldots, n$ and ψ positive and continuous functions on $[a_1, a_2]$, such that ψ is increasing, ω_j, $j = 1, 2, \ldots, n$ are decreasing on $[a_1, a_2]$ and $h : [a_1, a_2] \to \mathbb{R}^+$ is a positive continuous function, with $a_1 < \varsigma \leq a_2$, $\beta > 0$, $\gamma > 0$, $s \neq -1$, $\sigma > 0$ and $\delta \geq \theta_r > 0$ with $r \in \{1, 2, \ldots, n\}$.

Proof. To obtain the inequality (12.43), it is enough to multiply by $\frac{G(\mathbb{T}_+(v,\chi),\lambda)T(v)h(v)\prod_{j=1}^n \omega_j^{\theta_j}(v)}{\lambda^\gamma k\Gamma_k(\gamma)(\mathbb{T}_+(\chi,v))^{1-\frac{\gamma}{k}}}$ and integrate the result obtained with respect to v over (a_1, ς). □

Remark 12.2.41. *Theorem 12.2.39 is obtained from Theorem 12.2.40 by putting $\alpha = \beta$.*

Remark 12.2.42. *Theorems 18, 19, 21, and 22 of [25] are particular cases of Theorems 12.2.36, 12.2.37, 12.2.39, and 12.2.40, respectively.*

Remark 12.2.43. *Theorem 12.2.40, contains as a particular case Theorem 3.6 of [14].*

12.2.3 Weighted integral operators

A variant of the previous approaches is focused on considering a certain "weight" that, under very general conditions, can contain several results known from the literature. Next we present a "variant" of these operators.

Definition 12.2.44. (see [37]) Let $\phi \in L_1[a_1, a_2]$ and let w be a continuous and positive function, $w : I \to \mathbb{R}$, with first and second order derivatives piecewise continuous on I and $w(0) = w(1) = 0$. Then the weighted fractional integrals are defined by (right and left respectively):

$$^w I_{a_1+}\phi(x) = \frac{1}{\Gamma(\alpha)} \int_{a_1}^{x} w''\left(\frac{x-t}{a_2-a_1}\right)\phi(t)\,dt, \quad x > a_1 \qquad (12.44)$$

$$^w I_{a_2-}\phi(x) = \frac{1}{\Gamma(\alpha)} \int_{x}^{a_2} w''\left(\frac{t-x}{a_2-a_1}\right)\phi(t)\,dt, \quad x < a_2. \qquad (12.45)$$

Remark 12.2.45. It is easy to notice that if in the previous definition, we put $w(t) = \frac{(a_2-a_1)^{\alpha-1} t^{\alpha+1}}{\alpha(\alpha+1)\Gamma(\alpha)}$, then we obtain the Riemann-Liouville fractional integral operators. Other known fractional integral operators can be derived, without difficulty, from the previous definition.

Theorem 12.2.46. Let ϕ a real function defined on some interval I, $I \subset \mathbb{R}$, twice differentiable on $\mathring{I}$, $a, b \in \mathring{I}$, $a < b$. If $\phi'' \in L[a,b]$, and $|\phi''|$ is quasi-convex on $[a,b]$, then the following inequality is valid

$$\left|(w'(0)\phi(a_2) - w'(1)\phi(a_1)) + \frac{1}{a_2-a_1}{}^w I_{a_1+}\phi(a_2)\right|$$
$$\leq B(a_2-a_1)^2 \max\{|\phi''(a_1)|, |\phi''(a_2)|\} \qquad (12.46)$$

with $B = \int_0^1 w(t)\,dt$.

Lemma 12.2.47. Let ϕ a real function defined on some interval I, $I \subset \mathbb{R}$, twice differentiable on I°, $a_1, a_2 \in I^\circ$, $a < b$. If $\phi'' \in L[a_1, a_2]$ and $w(0) = w(1) = 0$, then we have the following equality:

$$(w'(0)\phi(a_2) - w'(1)\phi(a_1)) + \frac{1}{\Delta}\left[{}^w_\Delta I_{a_1^+}\phi(a_2)\right]$$
$$= \Delta^2 \int_0^1 w(t)\phi''(ta_1 + (1-t)a_2)\,dt. \qquad (12.47)$$

Proof. Integrating by parts, we obtain and putting $z = t\,a_1 + (1-t)\,a_2$, so $dz = (a_1 - a_2)dt$, with this change of variables and rearranging the terms, we obtain the inequality sought. □

Similarly, it can be proved that the following lemma is true:

Lemma 12.2.48. *Let ϕ a real function defined on some interval I, $I \subset \mathbb{R}$, twice differentiable on $I°$, $a_1, a_2 \in I°$, $a < b$. If $\phi'' \in L[a_1, a_2]$ and $w(0) = w(1) = 0$, then we have the following equality:*

$$w'(0)\phi(a_1) - w'(1)\phi(a_2) + \frac{1}{\Delta}\left[{}^w_\Delta I_{a_2^-}\phi(a_1)\right] = \Delta^2 \int_0^1 w(t)\phi''((1-t)a_1 + ta_2)dt. \tag{12.48}$$

Remark 12.2.49. *If we take $w(t) = 1 - (1-t)^{\alpha+1} - t^{\alpha+1}$, then from (12.47) we get result in [32] (see Lemma 2.2).*

Remark 12.2.50. *The above result contains as a particular case Lemma 1 of [2] and Lemma 4 of [17], with $w(t) = t(1-t)$.*

For simplicity let us denote

$$w'(0)\phi(a_2) - w'(1)\phi(a_1) + \frac{1}{\Delta}\left[{}^w_\Delta I_{a_1^+}\phi(a_2)\right] = L^+(HH) \tag{12.49}$$

$$w'(0)\phi(a_1) - w'(1)\phi(a_2) + \frac{1}{\Delta}\left[{}^w_\Delta I_{a_2^-}\phi(a_1)\right] = L^-(HH)$$

Theorem 12.2.51. *If in addition to the conditions of Lemma 12.2.7, $|\phi''|$ is quasi-convex on $[a_1, a_2]$, then the following inequality is valid*

$$\left|L^+(HH)\right| \leq \Delta^2 \cdot B \cdot \max\{|\phi''(a_1)|, |\phi''(a_2)|\}. \tag{12.50}$$

with $B = \int_0^1 w(t)dt$, $\Delta = a_2 - a_1$.

Proof. From the quasi-convexity of $|\phi''|$ and Lemma 12.2.7, we get the required inequality. This completes the proof. □

Remark 12.2.52. *If we consider that $w(t) = t(1-t)$, then this result becomes Theorem 3 of [2].*

We can improve the previous result, if we impose additional conditions to the quasi-convexity of $|\phi''|$.

Theorem 12.2.53. *Under assumptions Lemma 12.2.7 if $|\phi''|^q$ is quasi-convex on $[a_1, a_2]$, for $q > 1$, the following inequality holds*

$$\left|L^+(HH)\right| \leq B_p \cdot \Delta^2 \cdot \left(\max\{|\phi''(a_1)|^q, |\phi''(a_2)|^q\}\right)^{\frac{1}{q}}. \tag{12.51}$$

with $B_p = \left(\int_0^1 w^p(t)\right)^{\frac{1}{p}}$ and $\Delta = a_2 - a_1$.

Proof. From Hölder's Inequality, in its integral version, and Lemma 12.2.7 we have the conclusion sought. □

Remark 12.2.54. *Theorem 4 of [2] is easily obtained from the previous result if we put $w(t) = t(1-t)$.*

A variant more general of the previous Theorem, is given in the following result.

Theorem 12.2.55. *Under assumptions Lemma 12.2.7 if $|\phi''|^q$ is quasi-convex on $[a_1, a_2]$, for $q \geq 1$, the following inequality holds*

$$|L^+(HH)| \leq B \cdot \Delta^2 \cdot \left(\max\left\{|\phi''(a_1)|^q, |\phi''(a_2)|^q\right\}\right)^{\frac{1}{q}}, \qquad (12.52)$$

with B and Δ it's like in Theorem 12.2.51.

Proof. Taking into account the Lemma 12.2.7 and the power mean inequality for $q \geq 1$, we have the required inequality (12.52). □

Remark 12.2.56. *We can verify, without much difficulty that Theorem 5 of [2] is a particular case of the previous result, if we make $w(t) = t(1-t)$.*

The following theorem is obvious.

Theorem 12.2.57. *If in addition to the conditions of Lemma 12.2.7, $|\phi''|$ is convex on $[a_1, a_2]$, then the following inequality is valid*

$$|L^+(HH) + L^-(HH)| \leq \Delta^2 \cdot B \cdot (|\phi''(a_1)| + |\phi''(a_2)|)$$

with $B = \int_0^1 w(t)dt$ and $\Delta = a_2 - a_1$.

Proof. The proof follows from equalities (12.47), (12.48), (12.49) and from the condition that the function $|\phi''|$ is convex. □

Remark 12.2.58. *We can verify, without much difficulty, that Theorem 3.1 (for $s = 1$ and $m = 1$) of [6] is a special case of the Theorem 12.2.57, if we make $w(t) = t^\alpha(1-t)$.*

12.3 A general formulation of the notion of convex function

Note that in the inequality (12.1) (obviously the reciprocal for concave functions) the expression $tx + (1-t)y$ is the weighted average of the numbers of the arithmetic mean x e y with the weights t and $1-t$ and there is such a mean value of the weighted mean power [26, 27]:

$$F^{(\lambda, 1-\lambda)}_{a_1, a_2}(x) = \begin{cases} (\lambda a_1^x + (1-\lambda)a_2^x)^{\frac{1}{x}}, & x \neq 0 \\ a_1^\lambda a_2^{1-\lambda}, & x = 0 \end{cases} \qquad (12.53)$$

of positive values a_1 and a_2 for $x = 1$; for such x, the function given in (12.53) is defined for any value a_1 and a_2. Consequently, in terms of weighted average power (12.53) the inequality (12.7) can be represented as

$$\phi\left(F_{a_1,a_2}^{(\lambda,1-\lambda)}(1)\right) \leq F_{\phi(a_1),\phi(a_2)}^{(\lambda,1-\lambda)}(1) \tag{12.54}$$

The last circumstance inspires the introduction of the concept of function (α, β)-convex which generalizes the concept of convex function in the usual sense.

Definition 12.3.1. *A continuous and positive function $\phi : I \subseteq \mathbb{R} \to \mathbb{R}$, on the interval $I \subset (0, +\infty)$, is called (α, β)-convex if for any interval $[a_1, a_2] \subset I$ and any number $\lambda \in (0, 1)$ the following inequality holds:*

$$\phi\left(F_{a,b}^{(\lambda,1-\lambda)}(\alpha)\right) \leq F_{\phi(a),\phi(b)}^{(\lambda,1-\lambda)}(\beta). \tag{12.55}$$

If the inequality is not strict, then we will call it (α, β)-convex non-strict.

Remark 12.3.2. *It is obvious that under the above definition, the convex functions are $(1, 1)$-convex. Similarly, note that the $(1, 0)$-convexity of a function in the interval means its logarithmic convexity in this interval, the $(0, 1)$-convexity is a GA-convexity, and the $(-1, 1)$-convexity is the harmonic convexity.*

Definition 12.3.3. *A continuous and positive function $\phi : I \subseteq \mathbb{R} \to \mathbb{R}$, over the interval $I \subset (0, +\infty)$, is called (p, q)-type of function (A, B)-convex, if for any interval $[a_1, a_2] \subset I$ and any number $\lambda \in (0, 1)$ the following inequality holds:*

$$\phi\left(F_{a_1,a_2}^{(p_1(\lambda),p_2(1-\lambda))}(A)\right) \leq F_{\phi(a_1),\phi(a_2)}^{(q_1(\lambda),q_2(1-\lambda))}(B). \tag{12.56}$$

12.4 Integral inequalities and fractional derivatives

As we know, by manipulating simple algebraic identities, we can follow the idea of fractional differential operators of Riemann-Liouville or Caputo type. So we consider $\alpha = 1 + \alpha - 1$ or $\alpha = \alpha - 1 + 1$, respectively. In 1967, Michele Caputo gave a new definition of fractional derivative in [10] with an interesting property: when solving differential equations using Caputo's definition, it is not necessary to define the fractional order initial conditions. This derivative has been used in the study of many applied problems (see [10]).

The *Caputo derivative* of a differentiable function f of order $0 < \alpha < 1$ is defined as

$$^C D^\alpha \phi(x) = \frac{1}{\Gamma(1-\alpha)} \int_0^x \frac{\phi'(t)}{(x-t)^\alpha} dt. \tag{12.57}$$

An extension of $^{C}D^{\alpha}$ is the so-called *Caputo-Fabricio derivative* (see [7, 12]), given by:

$$^{CF}D_a^{\alpha}\phi(x) = \frac{M(\alpha)}{1-\alpha} \int_a^x \phi'(t) e^{-\frac{\alpha(x-t)}{1-\alpha}} dt, \qquad (12.58)$$

where $M(\alpha)$ is a normalization function such that $M(0) = M(1) = 1$.

A more recent extension is the *Atangana-Baleanu derivative*, defined in [5] by

$$^{ABC}D_a^{\alpha}\phi(x) = \frac{M(\alpha)}{1-\alpha} \int_a^x \phi'(t) E_{\alpha}\left(-\frac{\alpha(x-t)^{\alpha}}{1-\alpha}\right) dt, \qquad (12.59)$$

where

$$E_{\alpha}(z) = \sum_{k=0}^{\infty} \frac{z^k}{\Gamma(\alpha k + 1)}$$

is the *Mittag-Leffler function*.

And the derivative of order n is (see [18]).

Definition 12.4.1. *Let $\alpha > 0$ non integer, $k \geq 1$ and $\alpha \neq 1, 2, 3, ...$, $n = [\alpha] + 1$, $\phi \in C^{(n)}[a_1, a_2]$. The right-sided and left-sided Caputo k-fractional derivatives of order as follows:*

$$\left(^{C}D_{a_1+}^{\alpha,k}\phi\right)(x) = \frac{1}{k\Gamma_k(n-\frac{\alpha}{k})} \int_{a_1}^x \frac{\phi^{(n)}(t)}{(x-t)^{\frac{\alpha}{k}-n+1}} dt, x > a_1$$

and

$$\left(^{C}D_{a_2-}^{\alpha,k}\phi\right)(x) = \frac{1}{k\Gamma_k(n-\frac{\alpha}{k})} \int_x^{a_2} \frac{\phi^{(n)}(t)}{(t-x)^{\frac{\alpha}{k}-n+1}} dt, a_2 > x.$$

In [8] the following results are obtained.

Lemma 12.4.2. *Let $\phi : [a_1, a_2] \to \mathbb{R}$, be such that $\phi \in C^n[a_1, a_2]$. If $\phi^{(n+1)} \in L[a_1, a_2]$, then the following equality for fractional derivatives holds:*

$$\frac{\phi^{(n)}(a_1) + \phi^{(n)}(a_2)}{2}$$

$$- \frac{\Gamma_k\left(n - \frac{\alpha}{k} + k\right)}{2(a_2 - a_1)^{n-\frac{\alpha}{k}}} \left[\left(^{C}D_{a_1+}^{\alpha,k}\phi\right)(a_2) + (-1)^n \left(^{C}D_{a_2-}^{\alpha,k}\phi\right)(a_1)\right]$$

$$= \frac{a_2 - a_1}{2} \int_0^1 \left[(1-t)^{n-\frac{\alpha}{k}} - t^{n-\frac{\alpha}{k}}\right] \phi^{(n+1)}(ta_1 + (1-t)a_2) dt$$

Theorem 12.4.3. *Let $\phi : I \subset \mathbb{R} \to \mathbb{R}$, $\phi \in C^{n+1}[I]$ be such that If $\phi^{(n+1)} \in L[a_1, a_2]$, where $a_1, a_2 \in I$ with $a_1 \leq x < y \leq a_2$. If $\phi^{(n+1)}$ is η-quasi convex on $[x, y]$ for $t \in [0, 1]$. Then for all $\alpha > 0$, $k \geq 1$*

$$\frac{1}{y-x} \phi^{(n)}(y) - \frac{(-1)^n \Gamma_k\left(n - \frac{\alpha}{k} + k\right)}{(y-x)^{n-\frac{\alpha}{k}+k}} \left(^{C}D_{a+}^{\alpha,k}\phi\right)(x)$$

$$\leq \frac{1}{(n - \frac{\alpha}{k} + k)} \max\left\{\phi^{(n+1)}(y), \phi^{(n+1)}(y) + \eta\left(\phi^{(n+1)}(x), \phi^{(n+1)}(y)\right)\right\}$$

where $k\Gamma_k(\alpha) = \Gamma_k(\alpha + k)$.

Bibliography

[1] Ahmad, B., Alsaedi, A., Kirane, M. and Toberek, B.T. Hermite-Hadamard, Hermite-Hadamard-Fejer, Dragomir-Agarwal and Pachpatte type inequalities for convex functions via new fractional integrals, *ArXiv: 1701.00092*

[2] Alomari, M., Darus, M. Dragomir, S.S. 2009. New inequalities of Hermite-Hadamard type for functions whose second derivatives absolute values are quasi-convex, *RGMIA Res. Rep. Coll.* 12:17.

[3] Akkurt, A., Yildirim, M.E. and Yildirim, H. 2016. On some integral inequalities for (k, h)-Riemann-Liouville fractional integral, *New Trends Math. Sci.* 4:1: 138–146.

[4] Ali, M.A., Nápoles Valdés, J.E., Kashuri, A. and Zhang, Z. 2021. Fractional non conformable Hermite-Hadamard inequalities for generalized-convex functions, *Fasciculi Mathematici*, 64:1: 5–16. DOI:10.21008/j.0044-4413.2020.0007.

[5] Atangana, A. and Baleanu, D.2016. New fractional derivatives with non-local and non-singular kernel. Theory and application to heat transfer model. *Therm Sci.* 20:2: 763–769.

[6] Bayraktar, B. 2020. Some Integral Inequalities Of Hermite Hadamard Type For Differentiable (s, m)-convex Functions via Fractional Integrals, *TWMS J. Appl. Eng. Math.* 10:3: 625–637.

[7] Baleanu, D., Diethelm, K., Scalas, E. and Trujillo, J.J. 2016. *Fractional Calculus: Models and Numerical Methods*, Vol.5. World Scientific Publishing Co. Pte. Ltd., Series on Complexity, Nonlinearity and Chaos, Second Ed.

[8] Butt, S.I., Özdemir, M.E., Umar, M. and Çelik, B. 2021. Several new integral inequalities via Caputo k-fractional derivative operators, *Asian-European J. Math.* 14:9: 2150150.

[9] Bohner, M., Kashuri, A., Mohammed, P.O. and Nápoles Valdés, J.E. 2022. Hermite-Hadamard-type Inequalities for Integrals arising in Conformable Fractional Calculus, *Hacettepe J. Math. Stat.* 51:3: 775–786.

[10] Caputo, M. 1967. Linear model of dissipation whose Q is almost frequency independent II. *Geophysical J. Int.* 13:5: 529–539.

[11] Caputo, M. 1969. ElasticitÄ e Dissipazione, Bologna: Zanichelli.

[12] Caputo, M. and Fabrizio, M. 2015. A new definition of fractional derivative without singular kernel. Natural Sciences Publishing Cor., Progress in Fractional Differentiation and Applications, 1: 73–85.

[13] Chebyshev, P.L. 1882. Sur les expressions approximatives des integrales definies par les autres prises entre les mêmes limites, *Proc. Math. Soc. Charkov.* 2: 93–98.

[14] Dahmani, Z. 2014. New classes of integral inequalities of fractional order, *Le Matematiche* 69:1: 227–235.

[15] Dragomir, S.S. and Agarwal, P.A. 1998. Two Inequalities for Differentiable Mappings and Applications to Special Means of Real Numbers and to Trapezoidal Formula, *Appl. Math. Letters* 11:5: 91–95.

[16] Dragomir, S.S. and Fitzpatrick, S. 1999. The Hadamard's inequality for s-convex functions in the second sense, *Demonstration Math.* 32:4: 687–696.

[17] Dragomir, S.S. and Pearce, C.E.M. 2000. Selected Topics on Hermite-Hadamard Inequalities, RGMIA Monographs, Victoria University.

[18] Farid, G., Javed, A. and Rehman, A. 2017. On Hadamard Inequalities for n-times differentiable functions which are relative convex via Caputo k-fractional derivatives, *Nonlinear Anal. Forum* 22:2: 17–28.

[19] Fleitas, A., Méndez, J.A., Nápoles Valdés, J.E. and Sigarreta, J.M. 2019. On fractional Liénard-type systems, *Revista Mexicana de FÃsica* 65:6: 618–625.

[20] Gordji, M.E., Delavar, M.R., Sen, M.D.L. 2016. On η-Convex Functions, *J. Math. Inequa.* 10:1: 173–183.

[21] Guzmán, P.M., Lugo, L.M., Nápoles Valdes, J.E. and Vivas-Cortez, M. 2020. On a new generalized integral operator and certain operating properties, *Axioms* 9:69: doi:10.3390/axioms9020069

[22] Guzmán, P.M., Kórus, P. and Nápoles Valdes, J.E. 2020. Generalized Integral Inequalities of Chebyshev Type, *Fractal Fract.* 4:10: doi:10.3390/fractalfract4020010.

[23] Guzmán, P.M., Nápoles Valdes, J.E. and Gasimov, Y. 2021. Integral inequalities within the framework of generalized fractional integrals, *Fractional Differential Calculus* 11:1: 69–84.

[24] Helms, L.L. 1969. *Introduction To Potential Theory*, Wiley-Interscience, New York.

[25] Houas, M., Dahmani, Z. and Sarikaya, M.Z. 2018. Some integral inequalities for (k, s)-Riemann-Liouville fractional operators, *J. Interdisciplinary Math.* 21:7-8: 1–11.

[26] Kalinin, S.I. 2017. (0,0)-Convex functions and their properties, *Results of Science and Technology. Ser. Modern Mat. Appl. Topic. Review* 142: 81–87 (Russian).

[27] Kalinin, S.I. 2019. (0,0)-convex functions and their properties, *J. Math. Sci.* 241:6: 727–734.

[28] Katugampola, U.N. 2011. New Approach Generalized Fractional Integral, *Appl. Math. Comp.* 218: 860–865.

[29] Katugampola, U.N. 2014. A new approach to generalized fractional derivatives. *Bull. Math. Anal. Appl.* 6: 1–15.

[30] Kilbas, A.A., Marichev, O.I. Samko, S.G. 1993. *Fractional Integrals and Derivatives. Theory and Applications.* Gordon & Breach, Switzerland.

[31] Kilbas, A.A., Srivastava, H.M. and Trujillo, J.J. 2006. *Theory and Applications of Fractional Differential Equations*, Amsterdam, Netherlands, Elsevier.

[32] Boukerrioua, K. Chiheb, T. and Meftah, B. 2016. Fractional Hermite-Hadamard Type Inequalities For Functions Whose Second Derivative Are (s,r)−Convex in The Second Sense, *Kragujevac J. Math.* 40:2: 172–191.

[33] Mubeen, S. and Habibullah, G.M. 2012. k-fractional integrals and applications. *Int. J. Contemp. Math. Sci.* 7: 89–94.

[34] Nápoles Valdes, J.E. A Generalized k-Proportional Fractional Integral Operators with General Kernel, submitted.

[35] Nápoles Valdes, J.E., and Bayraktar, B. 2021. On The Generalized Inequalities Of The Hermite-Hadamard Type, *Filomat* 35:14: 4917–4924.

[36] Nápoles Valdes, J.E., Rabossi, F. and Samaniego, A.D. 2020. Convex Functions: Ariadne's Tread or Charlotte's Spider Web?, *Advanced Math. Models & Appl.* 5:2: 176–191.

[37] Nápoles Valdés, J.E., Bayraktar, B. and Butt, S.I. 2021. New integral inequalities of Hermite-Hadamard type in a generalized context, *Punjab University J. Math.* 53:11: 765–777. DOI:10.52280/Pujm.2021.531101.

[38] Nápoles Valdés, J.E., Rodríguez, J.M. and Sigarreta, J.M. 2019. On Hermite-Hadamard type inequalities for non-conformable integral operators. *Symmetry* 11: 1108.

[39] C. P. Niculescu, L. E. Persson, Convex Functions and their Applications, A Contemporary Aproach, CMS Books in Mathematics, Vol. 23, Springer-Verlag, New York, 2006.

[40] Rahman, G., Abdeljawad, T., Jarad, F., Khan, A. and Nisar, K.S. 2019. Certain inequalities via generalized proportional Hadamard fractional integral operators, *Adv. Diff. Equ.* 2019: 454.

[41] Rashid, S., Jarad, F., Noor, M.A., Kalsoom, H. and Chu, Y.M. 2020. Inequalities by Means of Generalized Proportional Fractional Integral Operators with Respect to Another Function, *Mathematics* 7:1225: doi:10.3390/math7121225

[42] Sarikaya, M.Z., Saglam, A. and Yildirim, H. 2008. On some Hadamard-type inequalities for h-convex functions, *J. Math. Inequa.* 2:3: 335–341.

[43] Sarikaya, M.Z., Set, E., Yaldiz, H. and Basak, N. 2013. Hermite–Hadamard's inequalities for fractional integrals and related fractional inequalities, *Math. Comput. Model.* 57:9-10: 2403–2407.

[44] Seto, K., Kuroiwa, D. and Popovici, N. 2018. A systematization of convexity and quasi-convexity concepts for set-valued maps, defined by l-type and u-type preorder relations. *Optimization* 67:7: 1077–1094.

[45] Samko, S.G., Kilbas, A.A. and Marichev, O.I. 1993. *Fractional Integrals and Derivatives*, Gordon & Breach Science, Yverdon.

[46] Tunc, M. 2013. On new inequalities for h-convex functions via Riemann-Liouville fractional integration, *Filomat*, 27:4: 559–565.

[47] Varosneac, S. 2007. On h-convexity, *J. Math. Anal. Appl.* 326: 303–311.

[48] Xiang, R. 2015. Refinements of Hermite-Hadamard type inequalities for convex functions via fractional integrals, *J. Appl. Math. & Informatics* 33:1-2: 119–125.

[49] Yang, G.S. and Tseng, K.L. 1999. On certain integral inequalities related to Hermite-Hadamard inequalities, *J. Math. Anal. Appl.* 239: 180–187.

Chapter 13

Exponentially biconvex functions and bivariational inequalities

Muhammad Aslam Noor

Khalida Inayat Noor

13.1	Introduction	229
13.2	Preliminary results	231
13.3	Properties of exponentially biconvex functions	234
13.4	Bivariational inequalities	244
	Bibliography	248

13.1 Introduction

Convexity theory describes a broad spectrum of very interesting developments involving a link among various fields of mathematics, physics, economics, and engineering sciences. Some of these developments have made mutually enriching contacts with other fields. Ideas explaining these concepts led to the developments of various fields of pure and applied sciences. This theory provides us several new, powerful and novel techniques to solve a wide class of linear and nonlinear problems. It reveals the fundamental facts on the qualitative behavior of solutions (regarding its existence, uniqueness, and regularity) to important classes of problems. Convexity also enabled us to develop highly efficient and powerful new numerical methods to solve nonlinear problems, see [1, 9, 10, 12, 14, 16, 17, 26, 31].

In recent years, various extensions and generalizations of convex functions and convex sets have been considered and studied using innovative ideas and techniques. It is known that more accurate and inequalities can be obtained using the algorithmically convex functions than the convex functions. Closely related to the log-convex functions, we have the concept of exponentially convex(concave) functions, the origin of exponentially convex functions can be traced back to Bernstein [8]. Avriel [4, 5] introduced and studied the concept of r-convex functions, whereas the (r,p)-convex functions were studied by Antczak [3]. For further properties of the r-convex functions, see [8, 12, 30, 32] and the references therein. Exponentially convex functions have important applications in information theory, big data analysis, machine learning and statistic. See [2–4, 12, 30, 32] and the references therein. Noor et al. [19–25, 27, 28] considered the concept of exponentially convex functions and discussed the basic properties. It is worth mentioning that these exponentially convex functions introduced in [19–23] are distinctly different from the exponentially convex functions considered and studied by Bernstein [8] and Awan et al. [6,7]. For example, $F(x) = \log x$ is exponentially convex in Noor's sense but not in Bernstein's sense, since it is not convex.

It is worth mentioning that variational inequalities represent the optimality conditions for the differentiable convex functions on the convex sets in normed spaces, which were introduced and considered in early 1960s by Stampacchia [33]. In recent years, considerable interest has been shown in developing various generalizations of variational inequalities and generalized convexity, both for their own sake and their applications. We use the auxiliary principle technique to suggest some iterative methods. This technique deals with finding the auxiliary variational inequality and proving that the solution of the auxiliary problem is the solution of the original problem by using the fixed-point approach. Glowinski et al. [11] used this technique to study the existence of a solution of mixed variational inequalities. Noor [18] and Noor et al. [26–29] have used this technique to suggest some predictor-corrector methods for solving various classes of variational inequalities. It is well known that a substantial number of numerical methods can be obtained as special cases from this technique.

Inspired by the research work going in this field, we introduce and consider some new classes of nonconvex functions with respect to an arbitrary bifunction, which are called the exponentially biconvex functions. Several new concepts of monotonicity are introduced. We establish the relationship between these classes and derive some new results under some mild conditions. We have shown that the exponentially biconvex functions enjoy the same interesting properties which exponentially convex functions have. Optimality conditions for differentiable exponentially biconvex functions are characterized by a class of variational inequalities, which is called exponential variational inequalities. We use the auxiliary principle technique to suggest some new iterative methods for solving exponential bivariational inequalities. As special cases, one can obtain various new and refined versions of known results. It is expected that the ideas and techniques of this paper may stimulate further research in this field.

13.2 Preliminary results

Let K_β be a nonempty closed set in a real Hilbert space H. We denote by $\langle \cdot, \cdot \rangle$ and $\|\cdot\|$ be the inner product and norm, respectively. Let $F : K_\beta \to R$ be a continuous function and $\beta(.,.) : K_\beta \times K_\beta \to R$ be an arbitrary continuous bifunction.

For the sake of completeness and the convenience of the readers, we recall the following known results, which are mainly due to Noor and Noor [19–24].

Definition 13.2.1. *The set K in H is said to be convex set, if*

$$u + \lambda(v - u) \in K, \quad \forall u, v \in K, \lambda \in [0, 1].$$

Definition 13.2.2. *A function F is said to be convex, if*

$$F((1-\lambda)u + tv) \leq (1-\lambda)F(u) + \lambda F(v), \quad \forall u, v \in K, \quad \lambda \in [0, 1]. \quad (13.1)$$

Definition 13.2.3. *[19] A function F is said to be exponentially convex function, if*

$$e^{F((1-\lambda)u + \lambda v)} \leq (1-\lambda)e^{F(u)} + \lambda e^{F(v)}, \quad \forall u, v \in K, \quad \lambda \in [0, 1].$$

We remark that Definition 13.2.3 can be rewritten in the following equivalent way, which is due to Antczak [3].

Definition 13.2.4. *A function F is said to be exponentially convex function, if*

$$F((1-\lambda)u + \lambda v) \leq \log[(1-\lambda)e^{F(a)} + \lambda e^{F(b)}], \quad \forall a, b \in K, \lambda \in [0, 1]. \quad (13.2)$$

A function is called the exponentially concave function f, if $-f$ is exponentially convex function.

It is obvious that two concepts are equivalent. This equivalent have been used to discuss various aspects of the exponentially convex functions. It is worth mentioning that one can also deduce the concept of exponentially convex functions from r-convex functions, which were considered by Avriel [4,5] and Bernstein [8]. For the applications of the exponentially convex functions in the mathematical programming and information theory, see Antczak [3], Alirezaei and Mathar [2], and Pal et al. [30].

Definition 13.2.5. *[19] A function F is said to be exponentially affine convex function, if*

$$e^{F((1-\lambda)u + \lambda v)} = (1-\lambda)e^{F(u)} + \lambda e^{F(v)}, \quad \forall u, v \in K, \quad \lambda \in [0, 1].$$

Definition 13.2.6. *The function F on the convex set K is said to be exponentially quasi-convex, if*

$$e^{F(u+\lambda(v-u))} \leq \max\{e^{F(u)}, e^{F(v)}\}, \quad \forall u, v \in K, \lambda \in [0, 1].$$

Definition 13.2.7. *The function F on the convex set K is said to be exponentially log-convex, if*

$$e^{F(u+\lambda(v-u))} \le e^{(F(u))^{1-\lambda}} e^{(F(v))^{\lambda}}, \quad \forall u,v \in K, \lambda \in [0,1],$$

where $F(\cdot) > 0$.

From the above definitions, we have

$$\begin{aligned}e^{F(u+\lambda(v-u))} &\le e^{(F(u))^{1-\lambda}} e^{(F(v))^{\lambda}} \le (1-\lambda)e^{F(u)} + \lambda e^{F(v)})\\&\le \max\{e^{F(u)}, e^{F(v)}\}, \quad \forall u,v \in K, \lambda \in [0,1].\end{aligned}$$

This shows that every exponentially log convex function is a exponentially convex function and every convex function is a exponentially quasi-convex function. However, the converse is not true. For the properties of the exponentially convex functions in variational inequalities and equilibrium problems, see Noor [15].

We now introduce the concepts of biconvex set, exponentially biconvex functions and their variant forms.

Definition 13.2.8. [15] *The set K_β in H is said to be biconvex set with respect to an arbitrary bifunction $\beta(\cdot,\cdot)$, if*

$$u + \lambda\beta(v-u) \in K_\beta, \quad \forall u,v \in K_\beta, \lambda \in [0,1].$$

The biconvex set K_β is also called β-connected set. If $\eta(v,u) = v - u$, then the biconvex set K_β is a convex set, but the converse is not true.

Remark 13.2.9. *We would like to emphasize that, if $u + \beta(v-u) = v$, $\forall u,v \in K_\beta$, then $\beta(v-u) = v - u$. Consequently, the β-biconvex set reduces to the convex set K. Thus, $K_\beta \subset K$. This implies that every convex set is a biconvex set.*

Definition 13.2.10. *The function F on the biconvex set K_β is said to be a exponentially biconvex function with respect to the bifunction $\beta(\cdot - \cdot)$, if*

$$e^{F(u+\lambda\beta(v-u))} \le (1-\lambda)e^{F(u)} + \lambda e^{F(v)}, \forall u,v \in K_\beta, \lambda \in [0,1]. \qquad (13.3)$$

Note that every exponentially convex function is a exponentially biconvex, but the converse is not true.

The function F is said to be exponentially biconcave, if and only if, $-F$ is a biconvex function. Consequently, we have a new concept.

Definition 13.2.11. *A function F is said to be exponentially affine biconvex involving an arbitrary bifunction $\beta(. - .)$, if*

$$e^{F(u+\lambda\beta(v-u))} = (1-\lambda)e^{F(u)} + \lambda e^{F(v)}, \forall u,v \in K_\beta, \lambda \in [0,1]. \qquad (13.4)$$

Exponentially biconvex functions 233

Definition 13.2.12. *The function F on the biconvex set K_β is said to be exponentially quasi biconvex with respect to the bifunction $\beta(\cdot - \cdot)$, if*

$$e^{F(u+\lambda\beta(v-u))} \leq \max\{e^{F(u)}, e^{F(v)}\}, \quad \forall u, v \in K_\beta, \lambda \in [0,1].$$

Definition 13.2.13. *The function F on the biconvex set K_β is said to be exponentially log-biconvex with respect to the bifunction $\beta(\cdot - \cdot)$, if*

$$e^{F(u+\lambda\beta(v-u))} \leq (e^{F(u)})^{1-\lambda}(e^{F(v)})^\lambda, \quad \forall u, v \in K_\beta, \lambda \in [0,1].$$

where $F(\cdot) > 0$, or equivalency.

Definition 13.2.14. *The function F on the biconvex set K_β is said to be exponentially log-biconvex with respect to the bifunction $\beta(\cdot - \cdot)$, if*

$$\log F(u + \lambda\beta(v - u)) \leq (1 - \lambda)\log F(u) + \lambda \log F(v), \quad \forall u, v \in K_\beta, \lambda \in [0,1].$$

where $F(\cdot) > 0$,

This equivalent definition can be used to discus the properties of the differentiable exponentially log-biconvex functions.
From the above definitions, we have

$$\begin{aligned} e^{F(u+\lambda\beta(v-u))} &\leq (e^{F(u)})^{1-\lambda}(e^{F(v)})^\lambda \leq (1-\lambda)e^{F(u)} + \lambda e^{F(v)} \\ &\leq \max\{e^{F(u)}, e^{F(v)}\}, \quad \forall u, v \in K_\beta, \lambda \in [0,1]. \end{aligned}$$

This shows that every exponentially log-biconvex function is exponentially biconvex function and every exponentially biconvex function is a exponentially quasi-biconvex function. However, the converse is not true.
For $\lambda = 1$, Definition 13.2.10 and 13.2.13 reduce to the following condition.
Condition A. The bifunction $\beta(. - .)$ satisfies

$$e^{F(u+\beta(v-u))} \leq e^{F(v)}, \quad \forall u, v \in K_\beta,$$

which is known as the condition A and plays an important role in the derivation of the results.
We now define the exponentially biconvex functions on the interval $K_\beta = I_\beta = [a, a + \beta(b - a)]$.

Definition 13.2.15. *Let $I = [a, a + \beta(b - a)]$. Then F is an exponentially biconvex function, if and only if,*

$$\begin{vmatrix} 1 & 1 & 1 \\ a & x & a+\beta(b-a) \\ e^{F(a)} & e^{F(x)} & e^{F(b)} \end{vmatrix} \geq 0; \quad a \leq x \leq a + \beta(b-a).$$

13.3 Properties of exponentially biconvex functions

In this section, we consider some basic properties of exponentially biconvex functions and their variant forms.

Theorem 13.3.1. *Let F be a strictly exponentially biconvex function. Then any local minimum of F is a global minimum.*

Proof. Let the exponentially biconvex function F have a local minimum at $u \in K_\beta$. Assume the contrary, that is, $F(v) < F(u)$ for some $v \in K_\beta$. Since F is a exponentially biconvex function, so

$$e^{F(u+\lambda\beta(v-u))} < \lambda e^{F(v)} + (1-\lambda)e^{F(u)}, \text{ for } 0 < \lambda < 1.$$

Thus

$$e^{F(u+\lambda\beta(v-u))} - e^{F(u)} < \lambda[e^{F(v)} - e^{F(u)}] < 0,$$

from which it follows that

$$e^{F(u+\lambda\beta(v-u))} < e^{F(u)},$$

for arbitrary small $\lambda > 0$, contradicting the local minimum. □

Theorem 13.3.2. *If the function F on the biconvex set K_β is exponentially biconvex, then the level set*

$$L_\alpha = \{u \in K : e^{F(u)} \leq \alpha, \alpha \in R\}$$

is a biconvex set.

Proof. Let $u, v \in L_\alpha$. Then $e^{F(u)} \leq \alpha$ and $e^{F(v)} \leq \alpha$. Now, $\forall \lambda \in (0,1), w = u + \lambda\beta(v-u) \in K_\beta$, since K_β is a biconvex set. Thus, by the biconvexity of F, we have

$$e^{F(u+\lambda\beta(v-u))} \leq (1-\lambda)e^{F(u)} + \lambda e^{F(b)} \leq (1-t)\alpha + t\alpha = \alpha,$$

from which it follows that $u + t\beta(v-u) \in L_\alpha$ Hence, L_α is a biconvex set. □

Theorem 13.3.3. *A positive function F is a exponentially biconvex, if and only if,*

$$epi(F) = \{(u, \alpha) : u \in K_\beta : e^{F(u)} \leq \alpha, \alpha \in R\}$$

is a biconvex set.

Proof. Assume that F is a exponentially biconvex function. Let (u, α), $(v, \beta) \in epi(F)$. Then it follows that $e^{F(u)} \leq \alpha$ and $e^{F(v)} \leq \beta$. Thus, $\forall \lambda \in [0, 1]$, $u, v \in K_\beta$, we have

$$e^{F(u+\lambda\beta(v-u))} \leq (1-\lambda)e^{F(u)} + \lambda e^{F(v)} \leq (1-t)\alpha + t\beta,$$

which implies that
$$(u + \lambda\beta(v - u), (1 - \lambda)\alpha + \lambda\beta) \in epi(F).$$

Thus $epi(F)$ is a biconvex set. Conversely, let $epi(F)$ be a biconvex set. Let $u, v \in K_\beta$. Then $(u, e^{F(u)}) \in epi(F)$ and $(v, e^{F(v)}) \in epi(F)$. Since $epi(F)$ is a biconvex set, we must have
$$(u + \lambda\beta(v - u), (1 - \lambda)e^{F(u)} + \lambda e^{F(v)}) \in epi(F),$$
which implies that
$$e^{F(u+\lambda\beta(v-u))} \leq (1 - \lambda)e^{F(u)} + \lambda e^{F(v)}.$$

This shows that F is a exponentially biconvex function. $\square$

Theorem 13.3.4. *A positive function F is exponentially quasi biconvex, if and only if, the level set*
$$L_\alpha = \{u \in K_\beta, \alpha \in R : e^{F(u)} \leq \alpha\}$$
is a biconvex set.

Proof. Let $u, v \in L_\alpha$. Then $u, v \in K_\beta$ and $\max(eF(u), e^{F(v)}) \leq \alpha$. Now for $\lambda \in (0, 1)$, $w = u + \lambda\beta(v - u) \in K_\beta$, We have to prove that $u + \lambda\beta(v - u) \in L_\alpha$. By the exponentially quasi biconvexity of F, we have
$$e^{F(u+\lambda\beta(v-u))} \leq \max\left(e^{F(u)}, e^{F(v)}\right) \leq \alpha,$$
which implies that $u + \lambda\beta(v - u) \in L_\alpha$, showing that the level set L_α is indeed a biconvex set.
Conversely, assume that L_α is a biconvex set. Then $\forall u, v \in L_\alpha, \lambda \in [0, 1]$,
$$u + \lambda\beta(v - u) \in L_\alpha.$$
Let $u, v \in L_\alpha$ for
$$\alpha = \max\left(e^{F(u)}, e^{F(v)}\right) \text{ and } e^{F(v)} \leq e^{F(u)}.$$
From the definition of the level set L_α, it follows that
$$e^{F(u+\lambda\beta(v-u))} \leq \max\left(e^{F(u)}, e^{F(v)}\right) \leq \alpha.$$

Thus F is a exponentially quasi biconvex function. This completes the proof. $\square$

Theorem 13.3.5. *Let F be a exponentially biconvex function. Let $\mu = \inf_{u \in K_\beta} F(u)$. Then the set*
$$E = \{u \in K_\beta : e^{F(u)} = \mu\}$$
is a biconvex set of K_β. If F is strictly exponentially biconvex, then E is a singleton.

Proof. Let $u, v \in E$. For $0 < \lambda < 1$, let $w = u + \lambda \beta(v - u)$. Since F is a exponentially biconvex function, so

$$F(w) = e^{F(u+\lambda\beta(v-u))} \leq (1-\lambda)e^{F(u)} + \lambda e^{F(v)} = \lambda\mu + (1-\lambda)\mu = \mu,$$

which implies that $w \in E$ and hence E is a biconvex set. For the second part, assume to the contrary that $e^{F(u)} = e^{F(v)} = \mu$. Since K_β is a biconvex set, then for $0 \leq \lambda \leq 1$, $u + \lambda\beta(v-u) \in K_\beta$. Further, since F is strictly exponentially biconvex,

$$e^{F(u+\lambda\beta(v-u))} < (1-\lambda)e^{F(u)} + \lambda e^{F(v)} = (1-t)\mu + t\mu = \mu.$$

This contradicts the fact that $\mu = \inf\limits_{u \in K_\beta} e^{F(u)}$ and hence the result follows. □

Theorem 13.3.6. *If F is a biconvex function such that $e^{F(v)} < e^{F(u)}, \forall u, v \in K_\beta$, then F is a strictly exponentially quasi biconvex function.*

Proof. By the biconvexity of the function F, $\forall u, v \in K_\beta, \lambda \in [0, 1]$, we have

$$e^{F(u+\lambda\beta(v-u))} \leq (1-\lambda)e^{F(u)} + \lambda e^{F(v)} < e^{F(u)},$$

since $e^{F(v)} < e^{F(u)}$, which shows that the function F is strictly exponentially quasi biconvex. □

We now derive some properties of the differentiable log-biconvex functions. To derive the main results, we need the following assumption regarding the bifunction $\beta(\cdot - \cdot)$.

Condition M. The bifunction $\beta(, -,)$ is said to satisfy the following assumptions:

(i) $\beta(\gamma\beta(v-u)) = \gamma\beta(v-u), \forall u, v \in K_\beta, \gamma \in R^n$.

(ii) $\beta(v - u - \gamma\beta(v-u)) = (1-\gamma)\beta(v-u), \forall u, v \in K_\beta$.

Remark 13.3.7. Let $\beta(\cdot - \cdot) : K_\beta \times K_\beta \to H$ satisfy the assumption

$$\beta(v - u) = \beta(v - z) + \beta(z - u), \forall u, v, z \in K_\beta. \quad (13.5)$$

One can easily show that $\beta(v - u) = 0 \Leftrightarrow u = v$, $\forall u, v \in K_\beta$. Consequently $\beta(0) = 0$, for $v = u \in K_\beta$. Also $\beta(v-u) + \beta(u-v) = 0$, $\forall u, v, z \in K_\beta$. This implies that the bifunction $\beta(. - .)$ is skew symmetric.

Theorem 13.3.8. *Let F be a differentiable function on the biconvex set K_β and let the condition M hold. Then the function F is essentially log-biconvex function, if and only if,*

$$\log F(v) - \log F(u) \geq \langle \frac{F'(u)}{F(u)}, \beta(v-u) \rangle, \quad \forall v, u \in K_\beta. \quad (13.6)$$

Proof. Let F be a log-biconvex function. Then

$$\log F(u + \lambda\beta(v-u)) \leq (1-\lambda)\log F(u) + \lambda \log F(v), \quad \forall u, v \in K_\beta,$$

which can be written as

$$\log F(v) - \log F(u) \geq \{\frac{\log F(u + \lambda\beta(v-u)) - \log F(u)}{\lambda}\}.$$

Taking the limit in the above inequality as $\lambda \to 0$, we have

$$\log F(v) - \log F(u) \geq \langle \frac{F'(u)}{F(u)}, \beta(v-u)\rangle,$$

which is (13.6), the required result.
Conversely, let (13.6) hold. Then $\forall u, v \in K_\beta, \lambda \in [0,1]$, $v_\lambda = u + \lambda\beta(v-u) \in K_\beta$ and using the condition M, we have

$$\log F(v) - \log F(v_\lambda) \geq \langle \frac{F'(v_\lambda)}{F(v_\lambda)}, \beta(v-v_\lambda))\rangle = (1-\lambda)\langle \frac{F'(v_\lambda)}{F(v_\lambda)}, \beta(v-u)\rangle. \quad (13.7)$$

In a similar way, we have

$$\log F(u) - \log F(v_\lambda) \geq \langle \frac{F'(v_\lambda)}{F(v_\lambda)}, \beta(u-v_\lambda)\rangle = -\lambda\langle \frac{F'(v_\lambda)}{F(v_\lambda)}, \beta(v-u)\rangle. \quad (13.8)$$

Multiplying (13.7) by λ and (13.8) by $(1-\lambda)$ and adding the resultant, we have

$$\log F(u + \lambda\beta(v-u)) \leq (1-\lambda)\log F(u) + \lambda \log F(v),$$

showing that F is a log-biconvex function. □

Remark 13.3.9. *From (13.6), we have*

$$F(v) \geq F(u) exp\{\langle \frac{F'(u)}{F(u)}, \beta(v-u)\rangle\}, \quad u, v \in K_\beta.$$

Changing the role of u and v in the above inequality, we also have

$$F(u) \geq F(v) exp\{\langle \frac{F'(v)}{F(v)}, \beta(u-v)\rangle\}, u, v \in K_\beta.$$

Thus, we can obtain the following inequality

$$F(u) + F(v) \geq F(v)\exp\{\langle \frac{F'(v)}{F(v)}, \beta(u-v)\rangle\},$$
$$+ F(u)\exp\{\langle \frac{F'(u)}{F(u)}, \beta(v-u)\rangle\} u, v \in K_\beta.$$

Definition 13.3.10. *The differentiable function F on the biconvex set K_β is said to be biconvex function with respect to the bifunction $\beta(\cdot - \cdot)$, if*

$$\log F(v) - \log F(u) \geq \langle \frac{F'(u)}{F(u)}, \beta(v-u) \rangle, \forall u, v \in K_\beta, p \geq 1,$$

where $F'(u))$ is the differential of F at u.

Theorem 13.3.11. *Let F be a differentiable function on the biconvex set K_β and Condition M hold. Then the function F is log-biconvex function, if and only if,*

$$\langle \frac{F'(u)}{F(u)}, \beta(v-u) \rangle + \langle \frac{F'(v)}{F(v)}, \beta(u-v) \rangle \leq 0, \quad \forall v, u \in K_\beta. \tag{13.9}$$

Proof. Let F be a differentiable function on the biconvex set K_β. Then from Theorem 13.3.8, it follows that

$$\log F(v) - \log F(u) \geq \langle \frac{F'(u)}{F(u)}, \beta(v-u) \rangle, \quad \forall v, u \in K_\beta. \tag{13.10}$$

Changing the role of u and v in (13.10), we have

$$\log F(u) - \log F(v) \geq \langle \frac{F'(v)}{F(v)}, \beta(v-u) \rangle, \forall v, u \in K_\beta. \tag{13.11}$$

Adding (13.10) and (13.11), we have

$$\langle \frac{F'(u)}{F(u)}, \beta(v-u) \rangle + \langle \frac{F'(v)}{F(v)}, \beta(u-v) \rangle \leq 0, \forall v, u \in K_\beta,$$

which is the required (13.9).
Since K_β is a biconvex set, so, $\forall u, v \in K_\beta$, $\lambda \in [0,1]$, $v_\lambda = u + \lambda \beta(v-u) \in K_\beta$. Conversely, from (13.9), we have

$$\langle \frac{F'(v_\lambda)}{F(v_\lambda)}, \beta(u-v_\lambda) \rangle \leq \langle \frac{F'(u)}{F(u)}, \beta(u-v_\lambda) \rangle$$
$$= -\lambda \langle \frac{F'(u)}{F(u)}, \beta(v-u) \rangle, \tag{13.12}$$

which implies that

$$\langle \frac{F'(v_\lambda)}{F(v_\lambda)}, \beta(v-u) \rangle \geq \langle \frac{F'(u)}{F(u)}, \beta(v-u) \rangle. \tag{13.13}$$

Consider the auxiliary function

$$\xi(\lambda) = \log F(u + \lambda(v-u)),$$

Exponentially biconvex functions

from which, we have

$$\xi(1) = \log F(v), \quad \xi(0) = \log F(u).$$

Then, from (13.13), we have

$$\xi'(\lambda) = \langle \frac{F'(v_\lambda)}{F(v_\lambda)}, \beta(v-u) \rangle \geq \langle \frac{F'(u)}{F(u)}, \beta(v-u) \rangle. \tag{13.14}$$

Integrating (13.14) between 0 and 1, we have

$$\xi(1) - \xi(0) = \int_0^1 \xi'(t)dt \geq \langle \frac{F'(u)}{F(u)}, \beta(v-u) \rangle.$$

Thus it follows that

$$\log F(v) - \log F(u) \geq \langle \frac{F'(u)}{F(u)}, \beta(v-u) \rangle,$$

which is the required (13.6). □

Definition 13.3.12. *An operator* $T: K_\beta \to H$ *is said to be*

1. *exponentially β-monotone, iff,*

$$\langle e^{Tu}, \beta(v-u) \rangle + \langle e^{Tv}, \beta(u-v) \rangle \leq 0, \quad \forall u, v \in K_\beta.$$

2. *exponentially β-pseudomonotone, iff,*

$$\langle e^{Tu}, \beta(v-u) \rangle \geq 0 \Rightarrow -\langle e^{Tv}, \beta(u-v) \rangle \geq 0, \forall u, v \in K_\beta.$$

3. *exponentially relaxed β-pseudomonotone, iff,*

$$\langle e^{Tu}, \beta(v-u) \rangle \geq 0 \Rightarrow -\langle e^{Tv}, \beta(u-v) \rangle \geq 0, \forall u, v \in K_\beta.$$

4. *exponentially strictly β-monotone, iff,*

$$\langle e^{Tu}, \beta(v-u) \rangle + \langle e^{Tv}, \beta(u-v) \rangle < 0, \forall u, v \in K_\beta.$$

5. *exponentially β-pseudomonotone, iff,*

$$\langle e^{Tu}, \beta(v-u) \rangle \geq 0 \Rightarrow \langle e^{Tv}, \eta(u,v) \rangle \leq 0, \forall u, v \in K_\beta.$$

6. *exponentially quasi β-monotone, iff,*

$$\langle e^{Tu}, \beta(v-u) \rangle > 0 \Rightarrow \langle e^{Tv}, \beta(u-v) \rangle \leq 0, \forall u, v \in K_\beta.$$

7. *exponentially strictly β-pseudomonotone, iff,*

$$\langle e^{Tu}, \beta(v-u) \rangle \geq 0 \Rightarrow \langle e^{Tv}, \beta(u-v) \rangle < 0, \forall u, v \in K_\beta.$$

Definition 13.3.13. *A differentiable function F on the biconvex set K_β is said to be exponentially pseudo β-biconvex function, iff,*

$$\langle e^{F(u)} F'(u), \beta(v-u) \rangle \geq 0 \Rightarrow e^{F(v)} - e^{F(u)} \geq 0, \forall u, v \in K_\beta.$$

Definition 13.3.14. *A differentiable function F on K_β is said to be exponentially quasi-biconvex function, iff,*

$$e^{F(v)} \leq e^{F(u)} \Rightarrow \langle e^{F(u)} F'(u), \beta(v-u) \rangle \leq 0, \forall u, v \in K_\beta.$$

Definition 13.3.15. *The function F on the set K_β is said to be exponentially pseudo-biconvex, iff,*

$$\langle e^{F(u)} F'(u), \beta(v-u) \rangle \geq 0 \Rightarrow e^{F(v)} \geq e^{F(u)}, \forall u, v \in K_\beta.$$

Definition 13.3.16. *The differentiable function F on the K_β is said to be exponentially quasi-biconvex function, iff,*

$$e^{F(v)} \leq e^{F(u)} \Rightarrow \langle e^{F(u)} F'(u), \beta(v-u) \rangle \leq 0, \forall u, v \in K_\beta.$$

We remark that the concepts introduced in this paper represent significant improvement of the previously known ones. All these new concepts may play important and fundamental part in the mathematical programming and optimization.

Theorem 13.3.17. *Let F be a differentiable function on the biconvex set K_β in H and let the condition M hold. Then the function F is a exponentially biconvex function, if and only if, the inequality*

$$e^{F(v)} - e^{F(u)} \geq \langle e^{F(u)} F'(u), \beta(v-u)) \rangle, \forall u, v \in K_\beta. \tag{13.15}$$

holds.

Proof. Let F be a biconvex function on the biconvex set K_β. Then

$$e^{F(u+\lambda\beta(v-u))} \leq (1-\lambda) e^{F(u)} + \lambda e^{F(v)} \quad \forall u, v \in K_\beta, \lambda \in [0,1].$$

which can be written as

$$e^{F(v)} - e^{F(u)} \geq \{ \frac{e^{F(u+\lambda\beta(v-u))} - e^{F(u)}}{\lambda} \}.$$

Taking the limit in the above inequality as $\lambda \to 0$, we have

$$e^{F(v)} - e^{F(u)} \geq \langle e^{F(u)} F'(u), \beta(v-u)) \rangle,$$

which is the required (13.15).

Conversely, let F be a exponentially biconvex function on the biconvex set K_β. Then, $\forall u, v \in K_\beta, \lambda \in [0,1]$, $v_t = u + \lambda\beta(v-u) \in K_\beta$ and using the condition M, we have

$$e^{F(v)} - e^{F(u+\lambda\beta(v-u))}$$
$$\geq \langle e^{F(u+\lambda\beta(v-u))} F'(u+\lambda\beta(v-u)), \beta(v-u+\lambda\beta(v-u))\rangle$$
$$= (1-\lambda)F'(u+\lambda\beta(v-u)), \beta(v-u)\rangle. \quad (13.16)$$

In a similar way, we have

$$e^{F(u)} - e^{F(u+\lambda\beta(v-u))}$$
$$\geq \langle e^{F(u+\lambda\beta(v-u))} F'(u+\lambda\beta(v-u)), \beta(u, u+\lambda\beta(v-u))\rangle$$
$$= -\lambda e^{F(u+\lambda\beta(v-u))} F'(u+\lambda\beta(v-u)), \beta(v-u)\rangle. \quad (13.17)$$

Multiplying (13.16) by λ and (13.17) by $(1-\lambda)$ and adding the resultant, we have

$$e^{F(u+\lambda\beta(v-u))} \leq (1-\lambda)e^{F(u)} + \lambda e^{F(v)}, \quad \forall u, v, z \in K_\beta$$

showing that F is a exponentially biconvex function. □

Theorem 13.3.18. *Let F be a differentiable exponentially biconvex function on the biconvex set K_β. If F is a exponentially biconvex function, then*

$$\langle e^{F(u)} F'(u), \beta(v-u))\rangle + \langle e^{F(v)} F'(v), \beta(u-v)\rangle \leq 0, \forall u, v \in K_\beta. \quad (13.18)$$

Proof. Let F be a exponentially biconvex function on the biconvex set K_β. Then

$$e^{F(v)} - e^{F(u)} \geq \langle e^{F(u)} F'(u), \beta(v-u))\rangle, \quad \forall u, v \in K_\beta. \quad (13.19)$$

Changing the role of u and v in (13.19), we have

$$e^{F(u)} - e^{F(v)} \geq \langle e^{F(v)} F'(v), \beta(u-v)\rangle, \quad \forall u, v \in K_\beta. \quad (13.20)$$

Adding (13.19) and (13.20), we have

$$\langle e^{F(u)} F'(u), \beta(v-u))\rangle + \langle e^{F(v)} F'(v), \beta(u-v)\rangle \leq 0, \forall u, v \in K_\beta, \quad (13.21)$$

which shows that $F'(.)$ is a exponentially β-monotone operator. □

Theorem 13.3.19. *If the differential $F'(.)$ is a exponentially β-monotone, then*

$$e^{F(v)} - e^{F(u)} \geq \langle e^{F(u)} F'(u), \beta(v-u)\rangle, \quad \forall u, v \in K_\beta.$$

Proof. Let $F'(.)$ be a exponentially β-monotone. From (13.21), we have

$$\langle e^{F(v)} F'(v), \beta(u-v)\rangle \leq \langle e^{F(u)} F'(u), \beta(v-u)\rangle\}. \tag{13.22}$$

Since K_β is an biconvex set, $\forall u, v \in K_\beta$, $\lambda \in [0,1]$ $v_t = u + \lambda\beta(v-u) \in K_\beta$. Taking $v = v_\lambda$ in (13.22) and using Condition M, we have

$$\begin{aligned}\langle e^{F(v_\lambda)} F'(v_\lambda), \beta(u - u - \lambda\beta(v-u))\rangle &\leq \langle e^{F(u)} F'(u), \eta(u+\lambda\beta(v-u)-u))\rangle \\ &\quad + \|\beta(u - u - \lambda\beta(v-u))\|^p\} \\ &= -\lambda \langle e^{F(u)} F'(u), \beta(v-u)\rangle,\end{aligned}$$

which implies that

$$\langle e^{F(v_\lambda)} F'(v_\lambda), \beta(v-u)\rangle \geq \langle e^{F(u)} F'(u), \beta(v-u)\rangle. \tag{13.23}$$

Let $\xi(\lambda) = e^{F(u+\lambda\beta(v-u))}$. Then, from (13.23), we have

$$\begin{aligned}\xi'(\lambda) &= \langle e^{F(u+\lambda\beta(v-u))} F'(u+\lambda\beta(v-u)), \beta(v-u)\rangle \\ &\geq \langle e^{F(u+\lambda\beta(v-u))} F'(u), \beta(v-u)\rangle.\end{aligned} \tag{13.24}$$

Integrating (13.24) between 0 and 1, we have

$$\xi(1) - \xi(0) \geq \langle e^{F(u)} F'(u), \beta(v-u)\rangle,$$

that is,

$$e^{F(u+\beta(v-u))} - e^{F(u)} \geq \langle e^{F(u)} F'(u), \beta(v-u)\rangle.$$

By using Condition A, we have

$$e^{F(v)} - e^{F(u)} \geq \langle e^{F(u)} F'(u), \beta(v-u)\rangle, \quad \forall u, v \in K_\beta.$$

the required result. $\square$

We now give a necessary condition for exponentially β-pseudo-biconvex function.

Theorem 13.3.20. *Let $F'(.)$ be a relaxed exponentially β-pseudomonotone operator and Conditions A and M hold. Then F is a exponentially β-pseudo-biconvex function.*

Proof. Let F' be a relaxed exponentially β-pseudomonotone. Then, $\forall u, v \in K_\beta$,

$$\langle e^{F(u)} F'(u), \beta(v-u)\rangle \geq 0,$$

implies that

$$-\langle e^{F(v)} F'(v), \beta(u-v)\rangle \geq 0. \tag{13.25}$$

Exponentially biconvex functions 243

Since K is a biconvex set, $\forall u, v \in K_\eta$, $\lambda \in [0,1]$, $v_\lambda = u + \lambda\beta(v-u) \in K_\beta$. Taking $v = v_\lambda$ in (13.25) and using condition Condition M, we have

$$-\langle e^{F(u+\lambda\beta(v-u))} F'(u+\lambda\beta(v-u)), \beta(u-v)\rangle \geq 0. \tag{13.26}$$

Let

$$\xi(\lambda) = F(u+\lambda\beta(v-u)), \quad \forall u,v \in K_\beta, \lambda \in [0,1].$$

Then, using (13.26), we have

$$\xi'(\lambda) = \langle E^{F(u+\lambda\beta(v-u))} F'(u+\lambda\beta(v-u)), \beta(u-v)\rangle \geq 0.$$

Integrating the above relation between 0 to 1, we have

$$\xi(1) - \xi(0) \geq 0,$$

that is,

$$e^{F(u+\lambda\beta(v-u))} - e^{F(u)} \geq 0,$$

which implies, using Condition A,

$$e^{F(v)} - e^{F(u)} \geq 0, \quad \forall u, v, z \in K_\beta$$

showing that F is a exponentially β-pseudo-biconvex function. $\square$

Definition 13.3.21. *The function F is said to be sharply exponentially pseudo biconvex, if*

$$\langle e^{F(u)} F'(u), \beta(v-u)\rangle \geq 0 \quad \Rightarrow \quad e^{F(v)} \geq e^{F(v+\lambda\beta(v-u))}, \forall u,v \in K_\beta, \lambda \in [0,1].$$

Theorem 13.3.22. *Let F be a sharply exponentially pseudo biconvex function on K_β. Then*

$$-\langle e^{F(v)} F'(v), \beta(v-u)\rangle \geq 0, \quad \forall u, v \in K_\beta.$$

Proof. Let F be a sharply exponentially pseudo biconvex function on K_β. Then

$$e^{F(v)} \geq e^{F(v+\lambda\beta(v-u))}, \quad \forall u,v \in K_\beta, \lambda \in [0,1],$$

from which we have

$$\frac{e^{F(v+\lambda\beta(v-u))} - e^{F(v)}}{\lambda} \leq 0.$$

Taking limit in the above-mentioned inequality, as $\lambda \to 0$, we have

$$-\langle e^{F(v)} F'(v), \beta(v-u)\rangle \geq 0,$$

the required result. $\square$

13.4 Bivariational inequalities

In this section, we consider the exponential bivariational inequalities and suggest some iterative methods by using the auxiliary principle techniques involving the Bregman distance functions and without Bregman distance functions. For the applications of the Bregman distance functions, see [9,28,29,34]. For strongly convex function F, the Bregman distance function is defined as

$$B(v,u) = F(v) - F(u) - \langle F'(u), v - u \rangle \geq \alpha \|v - u\|^2, \forall u, v \in K. \quad (13.27)$$

It is important to emphasize that various types of function F gives different Bregman distance functions. It is a challenging problem to explore the applications of Bregman distance function for other types of nonconvex functions as biconvex, k-convex functions, preinvex, and harmonic functions.

We now discuss the optimality conditions for the differentiable exponentially biconvex functions.

Theorem 13.4.1. *Let F be a differentiable biconvex function with modulus $\mu > 0$. If $u \in K_\beta$ is the minimum of the function F, if and only if, $u \in K_\beta$ satisfies the*

$$\langle e^{F(u)} F'(u), \beta(v - u) \rangle \geq 0, \quad \forall u, v \in K_\beta. \quad (13.28)$$

Remark 13.4.2. *We would like to mention that, if $u \in K_\beta$ satisfies the inequality*

$$\langle e^{F(u)} F'(u), \beta(v, u) \rangle = \langle \left(e^{F(u)}\right)', \beta(v, u) \rangle \geq 0, \quad \forall u, v \in K_\beta, \quad (13.29)$$

then $u \in K_\beta$ is the minimum of the differentiable exponentially biconvex function F. The inequality of the type (13.29) is called the exponentially bivariational inequality and appears to new one.

It is worth mentioning that inequalities of the type (13.29) may not arise as the minimization of the biconvex functions. This motivated us to consider a more general exponentially bivariational inequality of which (13.29) is a special case.

For a given operator T, bifunction $\beta(.-.)$, consider the problem of finding $u \in K_\beta$, such that

$$\langle e^{Tu}, \beta(v - u) \rangle \geq 0, \forall v \in K_\beta, \quad (13.30)$$

which is called the exponential bivariational inequality. It is worth mentioning that for suitable and appropriate choice of the operators, bifunction, biconvex sets, and spaces, one can obtain a wide class of variational inequalities and optimization problems. This shows that the exponential bivariational inequalities are quite flexible and unified ones.

Due to the structure of the exponentially bivariational inequalities, the projection method and its variant form can not be used to suggest the iterative methods for solving these bivariational inequalities. To overcome these drawback, one may use the auxiliary principle technique of Glowinski et al. [11] as developed by Noor [18] and Noor et al. [27–29] to suggest and analyze some iterative methods for solving the bivariational inequalities (13.30). This technique does not involve the concept of the projection, which is the main advantage of this technique. We use the auxiliary principle technique to suggest some iterative methods for solving bivariational inequalities.

For a given $u \in K_\beta$ satisfying the bivariational inequality (13.30), we consider the auxiliary problem of finding a $w \in K$ such that

$$\langle \rho T w, \beta(v-w) \rangle + \langle e^{E(w)} E'(w) - e^{E(u)} E'(u), \beta(v-w) \rangle$$
$$\geq 0, \quad \forall v \in K_\beta, \qquad (13.31)$$

where $\rho > 0$ is a constant and $E'(u)$ is the differential of a strongly biconvex function $E(u)$ at $u \in K_\beta$.

Remark 13.4.3. *The function*

$$B(w,u) = e^{E(w)} - e^{E(u)} - \langle e^{E(u)} E'(u), \beta(w,u) \rangle$$
$$\geq \nu \|\beta(w-u)\|^2, \quad \forall u, w \in K_\beta.$$

associated with the biconvex function $E(u)$ is called the exponentially Bregman distance function. By the strongly biconvexity of the function $E(u)$, the Bregman function $B(.,.)$ is nonnegative and $B(w,u) = 0$, if and only if, $u = w, \forall u, w \in K_\beta$. For the applications of the Bregman function in solving variational inequalities and complementarity problems, see [27–29, 34].

Algorithm 13.4.4. *For a given $u_0 \in H$, compute the approximate solution u_{n+1} by the iterative scheme*

$$\langle \rho T u_{n+1}, \beta(v-u_{n+1}) \rangle + \langle e^{E(u_{n+1})} E'(u_{n+1}) - e^{E(u_n)} E'(u_n), \beta(v-u_{n+1}) \rangle$$
$$\geq 0, \forall v \in K_\beta, \qquad (13.32)$$

where $\rho > 0$ is a constant. Algorithm 13.4.4 is called the proximal method for solving exponential bivariational inequalities (13.30). In passing we remark that the proximal point method was suggested in the context of convex programming problems as a regularization technique.

It is well-known that to implement the proximal point methods, one has to find the approximate solution implicitly, which is itself a difficult problem. To overcome this drawback, we now consider another method for solving the exponential bivariational inequality(13.30) using the auxiliary principle technique.

For a given $u \in K_\eta$, find $w \in K_\eta$ such that

$$\langle \rho e^{T(u)}, \beta(v-w)\rangle + \langle e^{E(w)}E'(w) - e^{E(u)}E'(u), \beta(v-w)\rangle$$
$$\geq 0, \quad \forall v \in K_\beta, \qquad (13.33)$$

where $E'(u)$ is the differential of a exponentially biconvex function $E(u)$ at $u \in K_\beta$. Problem (13.30) has a unique solution, since E is strongly exponentially biconvex function. Note that problems (13.33) and (13.32) are quite different problems.

It is clear that for $w = u$, w is a solution of (13.30). This fact allows us to suggest and analyze another iterative method for solving the exponential bivariational inequality (13.30).

Algorithm 13.4.5. *For a given $u_0 \in H$, compute the approximate solution u_{n+1} by the iterative scheme*

$$\langle \rho e^{Tu_n}, \beta(v - u_{n+1})\rangle + \langle e^{E(u_{n+1})}E'(u_{n+1}) - e^{E(u_n)}E'(u_n), \beta(v - u_{n+1})\rangle$$
$$\geq 0, \quad \forall v \in K_\beta, \qquad (13.34)$$

for solving the exponential bivariational inequality (13.30).

If $\beta(v, u)) = v - y$, Algorithm 13.4.5 collapses to:

Algorithm 13.4.6. *For a given $u_0 \in H$, compute the approximate solution u_{n+1} by the iterative schemes*

$$\rho\langle Tu_n, \beta(v - u_{n+1})\rangle + \langle E'(u_{n+1}) - E'(u_n), \beta(v - u_{n+1})\rangle \geq 0, \quad \forall v \in K_\beta,$$

for solving the exponential variational inequalities and appears to be a new one.

We again use the auxiliary principle technique to suggest some iterative method, which do not involve Bregman distance function.

For given $u \in K_\beta$ satisfying (13.30), consider the problem of finding $w \in K_\beta$, such that

$$\langle \rho e^{Tw}, \beta(v-w)\rangle + \langle w - u + \alpha(u-u), v - w\rangle \geq 0, \quad \forall v \in K_\beta, \quad (13.35)$$

where $\rho > 0$, α are parameters. The problem (13.35) is called the auxiliary exponential bivariational inequality. It is clear that the relation (13.35) defines a mapping connecting the problems (13.30) and (13.35). We not that, if $w(u) = u$, then w is a solution of problem (13.30). This simple observation enables to suggest an iterative method for solving (13.30).

Algorithm 13.4.7. *For given $u_0 \in K_\beta$, find the approximate solution u_{n+1} by the scheme*

$$\langle \rho e^{Tu_{n+1}}, \beta(v - u_{n+1})\rangle + \langle u_{n+1} - u_n + \alpha(u_n - u_{n-1}), v - u_{n+1}\rangle$$
$$\geq 0. \quad \forall v \in K_\beta. \qquad (13.36)$$

The Algorithm 13.4.7 is known as the implicit method. Such type of methods have been studied extensively for various classes of variational inequalities. See [13, 15] and the reference therein. If $\nu = 0$, then Algorithm 13.4.7 reduces to:

Algorithm 13.4.8. *For given $u_0 \in K$, find the approximate solution u_{n+1} by the scheme*

$$\langle \rho e^{Tu_{n+1}}, \beta(v - u_{n+1}) \rangle + \langle u_{n+1} - u_n + \alpha(u_n - u_{n-1}), v - u_{n+1} \rangle$$
$$\geq 0, \quad \forall v \in K_\beta, \tag{13.37}$$

which appears to be new ones even for solving the exponential variational inequalities.

In order to implement the implicit Algorithm 13.4.7, one uses the predictor-corrector technique. Consequently, Algorithm 13.4.7 for the case $\alpha = 0$, is equivalent to the following two-step iterative method for solving the exponentially bivariational inequality (13.30).

Algorithm 13.4.9. *For a given $u_0 \in K_\beta$, find the approximate solution u_{n+1} by the schemes*

$$\langle \rho e^{Tu_n}, \beta(v - y_n) \rangle + \langle y_n - u_n, v - y_n \rangle \geq 0, \forall v \in K_\beta$$
$$\langle \rho e^{Ty_n}, \beta(u_n - y_n) \rangle + \langle u_n - y_n, v - y_n \rangle \geq 0, \forall v \in K_\beta.$$

Algorithm 13.4.9 is called the predictor-corrector iterative method and appears to be a new one.

Using the auxiliary principle technique, we now suggest an other iterative method for solving the exponential bivariational inequalities and related optimization problems.

For a given $u \in K_\beta$ satisfying (13.30), consider the problem of finding $w \in H : h(w) \in K$, such that

$$\langle \rho e^{Tu}, \beta(v - w) \rangle + \langle w - u, v - w \rangle \geq 0, \forall v \in K_\beta, \tag{13.38}$$

where $\rho > 0$ is a parameter. The problem (13.38) is called the auxiliary exponential bivariational inequality.

It is clear that the relation (13.38) defines a mapping connecting the problems (13.30) and (13.38). We note that, if $w(u) = u$, then w is a solution of problem (13.30). This simple observation enables to suggest an iterative method for solving (13.30).

Algorithm 13.4.10. *For given $u_0 \in K_\beta$, find the approximate solution u_{n+1} by the scheme*

$$\langle \rho e^{Tu_n}, \beta(v - u_{n+1}) \rangle + \langle u_{n+1} - u_n, v - u_{n+1} \rangle \geq 0, \quad \forall v \in K_\beta, \tag{13.39}$$

which is an explicit algorithm.

It is worth mentioning that the auxiliary principle technique can be used efficiently to suggest a wide class of iterative methods for solving strongly exponential bivariational inequalities. We have only given some glimpses of the exponential bivariational inequalities. It is an interesting problem to explore the applications of such type exponential bivariational inequalities in various fields of pure and applied sciences.

Bibliography

[1] Akhiezer, N.I. 1965. *The Classical Moment Problem and Some Related Questions in Analysis*, Oliver and Boyd, Edinburgh, U.K.

[2] Alirezaei, G. and Mazhar, R. 2018. On exponentially concave functions and their impact in information theory, *J. Inform. Theory Appl.* 9:5: 265–274.

[3] Antczak, T. 2001. On (p,r)-invex sets and functions, *J. Math. Anal. Appl.* 263: 355–379.

[4] Avriel, M. 1972. r-Convex functions. *Math. Program.* 2: 309–323.

[5] Avriel, M. 1973. Solution of certain nonlinear programs involving r-convex functions *J. Optimization Theory Appl.* 11: 159–174.

[6] Awan, M.U., Noor, M.A. and Noor, K.I. 2018. Hermite-Hadamard inequalities for exponentially convex functions, *Appl. Math. Inform. Sci.* 12:2: 405–409.

[7] Awan, M.U., Noor, M.A., Set, E. and Mihai, M.V. 2019. On strongly (p,h)-convex functions, *TWMS J. Pure Appl. Math.* 10:2: 145–153.

[8] Bernstein, S.N. 1929. Sur les fonctions absolument monotones, *Acta Math.* 52: 1–66.

[9] Bregman, L.M. 1967. The relaxation method for finding common points of convex sets and its application to the solution of problems in convex programming. *USSR Comput. Math. Math. Phys.* 7: 200–217.

[10] Cristescu, G. and Lupsa, L. 2002. *Non-Connected Convexities and Applications*, Kluwer Academic Publisher, Dordrechet.

[11] Glowinski, R., Lions, J. L. and Tremileres, R., 1981. *Numerical Analysis of Variational Inequalities*, NortHolland, Amsterdam, Holland.

[12] Huang, C.-H., Huang, H.-L. and Chen, J.-S. 2017. Examples of rconvex functions and characterizations of r-convex functions associated with second-order cone, *Linear Nonlinear Anal.* 3:3: 367–384.

[13] Mohsen, B.B., Noor, M.A., Noor, K.I. and Postolache, M. 2019. Strongly convex functions of higher order involving bifunction, *Mathematics*, 7:11:1028: doi:10.3390/math7111028

[14] Niculescu, C.P. and Persson, L.E. 2018. *Convex Functions and Their Applications*, Springer-Verlag, New York.

[15] Noor, M.A. *Advanced Convex Analysis, Lecture Notes*, COMSATS University Islamabad, Islamabad, Pakistan, (2008-2021).

[16] Noor, M.A. 1988. General variational inequalities, *Appl. Math. Letters* 1: 119–121.

[17] Noor, M.A. 2000. New approximation schemes for general variational inequalities, *J. Math. Anal. Appl.* 251: 217–229.

[18] Noor, M.A. 2004. Some developments in general variational inequalities, *Appl. Math. Comput.* 152: 199–277.

[19] Noor, M.A. and Noor, K.I. 2019. Exponentially convex functions, *J. Orisa Math. Soc.* 38:1-2: 33–51.

[20] Noor, M.A. and Noor, K.I. 2019. Strongly exponentially convex functions, *It U.P.B. Bull Sci. Appl. Math. Series A.* 81:4: 75–84.

[21] Noor, M.A. and Noor, K.I. 2019. Strongly exponentially convex functions and their properties, *J. Advanc. Math. Studies* 9:2: 180–188.

[22] Noor, M.A. and Noor, K.I. 2019. On generalized strongly convex functions involving bifunction, *Appl. Math. Inform. Sci.* 13:3: 411–416.

[23] Noor, M.A. and Noor, K.I. 2019. Some properties of exponentially preinvex functions, *FACTA Universitat (NIS) Ser. Math. Inform.* 34:5: 939–953.

[24] Noor, M.A. and Noor, K.I. 2019. New classes of strongly exponentially preinvex functions, *AIMS Math.* 4:6: 1554–1565.

[25] Noor, M.A. and Noor, K.I. 2020. Higher-order strongly-generalized convex functions, *Appl. Math. Inf. Sci.* 14:1: 133–139.

[26] Noor, M.A. and Noor, K.I. 2020. Higher order strongly general convex functions and variational inequalities, *AIMS Math.* 5:4: 3646–3663.

[27] Noor, M.A. and Noor, K.I. 2021. Higher order general convex functions and general variational inequalities, *Canad. J. Appl. Math.* 3:1: 1–17.

[28] Noor, M.A., Noor, K.I. and Rassias, M.Th. 2020. New trends in general variational inequalities, *Acta Appl. Mathematica* 170:1: 981–1064.

[29] Noor, M.A., Noor, K.I. and Rassias, M.Th. 1993. Some aspects of variational inequalities, *J. Appl. Math. Comput.* 47: 485–512.

[30] Pal, S. and Wong, T.K. 2018. On exponentially concave functions and a new information geometry, *Annals. Prob.* 46:2: 1070–1113.

[31] Pecaric, J.E., Proschan, F. and Tong, Y.L. 1992. *Convex Functions and Statistical Applications*, Academic Press, New York.

[32] Qu, G. and Li, N. 2019. On the exponentially stability of primal-dual gradient dynamics, *IEEE Control Syst. Letters* 3:1: 43–48.

[33] Stampacchia, G. 1964. Formes bilieaires coercives sur les ensembles convexes, Comput. *Rend. Acad. Sciences, Paris* 258: 4413–4416.

[34] Zu, D.L. and Marcotte, P. 1996. Co-coercivity and its role in the convergence of iterative schemes for solving variational inequalities. *SIAM J. Optim.* 6:3: 714–726.

Chapter 14

On a certain subclass of analytic functions defined by Bessel functions

B. Venkateswarlu

P. Thirupathi Reddy

Shashikala A

14.1	Introduction	251
14.2	Coefficient bounds	254
14.3	Neighborhood property	256
14.4	Partial sums	258
	Bibliography	262

14.1 Introduction

Let A be the class of mappings ϑ normalized by

$$\vartheta(\varsigma) = \varsigma + \sum_{\imath=2}^{\infty} a_\imath \varsigma^\imath \qquad (14.1)$$

and T indicate the class of mappings in the type of

$$\vartheta(\varsigma) = \varsigma - \sum_{\imath=2}^{\infty} \varrho_\imath \varsigma^\imath, (\varrho_\imath \geq 0), \qquad (14.2)$$

which are regular in the open unit disk $U = \{\varsigma : \varsigma \in \mathbb{C} \text{ and } |\varsigma| < 1\}$. This subclass was given in [14,15]. Let $T^*(\nu)$ and $C(\nu)$ be indicate star-shaped and convex mappings of order $\nu, (0 \leq \nu < 1)$, respectively.

The classes $UCV(\nu, \sigma)$ and and $SP(\nu, \sigma)$ consists of uniform σ−convex and parabolic σ− star-shaped mappings of order $\nu - 1 < \nu \leq 1, \sigma \geq 0$, generalizes

DOI: 10.1201/9781003330868-14

the class UCV and SP, respectively, were given in [10] such that

$$UCV(\nu,\sigma) = \left\{ \vartheta \in A : \Re\left\{1 + \frac{\varsigma\vartheta''(\varsigma)}{\vartheta'(\varsigma)} - \nu\right\} > \sigma\left\{\frac{\varsigma\vartheta''(\varsigma)}{\vartheta'(\varsigma)}\right\}, \varsigma \in U \right\} \quad (14.3)$$

and

$$SP(\nu,\sigma) = \left\{ f \in A : \Re\left\{\frac{\varsigma\vartheta'(\varsigma)}{\vartheta(\varsigma)} - \nu\right\} > \sigma\left\{\frac{\varsigma\vartheta'(\varsigma)}{\vartheta(\varsigma)} - 1\right\}, \varsigma \in U \right\}. \quad (14.4)$$

It is obvious from (14.3) and (14.4) that $\vartheta \in UCV(\nu,\sigma)$ if and only if $\varsigma\vartheta'(\varsigma) \in SP(\nu,\sigma)$.

A few attractive occurrences of the category of starlike and convex of order ν connected with Bessel mappings (as hypergeometric mapping), locating requirements on the triple p, b and ℓ such that the mapping $u_{p,b,\ell}$ is star-shaped and convex of order ν, and finding situations on the parameters for which Gaussian hypergeometric mappings pertain to the specific categories of mapping have been deliberated in the references [4, 5, 7, 16, 17].

Let us take into consideration second order linear homogenous differential equation (see [6])

$$\varsigma^2 w''(\varsigma) + b\varsigma w'(\varsigma) + [\ell\varsigma^2 - p^2 + (1-b)p]w(\varsigma) = 0, \quad (p,b,\ell \in \mathbb{C}). \quad (14.5)$$

As a particular solution of (14.5) generalized Bessel mapping of the first kind of order p, is indicated in [6] as following:

$$w(\varsigma) = w_{p,b,\ell}(\varsigma) = \sum_{\imath=0}^{\infty} \frac{(-1)^{\imath}\ell^{\imath}}{\imath!\hbar\left(p + \imath + \frac{b+1}{2}\right)} \left(\frac{\varsigma}{2}\right)^{2\imath+p}, \quad \varsigma \in \mathbb{C}, \quad (14.6)$$

where $\hbar$ stands for the Euler gamma mapping and $\tau = p + \frac{b+1}{2} \notin \varsigma_0 = \{0, -1, -2, \cdots\}$.

Though the series given in (14.6) is convergent everywhere, the mapping $w_{p,b,\ell}$ is not univalent in U.

Specially, choosing $b = \ell = 1$ in (14.6), we get Bessel mapping of the first kind of order p given in [6] as

$$J_p(\varsigma) = \sum_{\imath=0}^{\infty} \frac{(-1)^{\imath}}{\imath!\hbar(p + \imath + 1)} \left(\frac{\varsigma}{2}\right)^{2\imath+p}, \varsigma \in \mathbb{C}. \quad (14.7)$$

Selecting $b = 1$ and $\ell = -1$ in (14.6), we get the modified Bessel mapping of the first kind order of p given in [6] as

$$I_p(\varsigma) = \sum_{\imath=0}^{\infty} \frac{1}{\imath!\hbar(p + \imath + 1)} \left(\frac{\varsigma}{2}\right)^{2\imath+p}, \varsigma \in \mathbb{C}. \quad (14.8)$$

Further choosing $b = 2$ and $\ell = 1$ in (14.6), the mappings $w_{p,b,\ell}$ reduces to $\sqrt{\frac{2}{\pi}}j_p$, where j_p is the spherical Bessel mapping of the first kind of order p, given in [6] as

$$j_p(\varsigma) = \sqrt{\frac{\pi}{2}} \sum_{\imath=0}^{\infty} \frac{(-1)^{\imath}}{\imath!\hbar(p+\imath+\frac{3}{2})} \left(\frac{\varsigma}{2}\right)^{2\imath+p}, \varsigma \in \mathbb{C}. \qquad (14.9)$$

The mapping $\vartheta_{p,b,\ell}$ is indicated in [8] as

$$\vartheta_{p,b,\ell}(\varsigma) = 2^p \hbar\left(p + \frac{b+1}{2}\right) \varsigma^{1-\frac{p}{2}} w_{p,b,\ell}(\sqrt{\varsigma}) \qquad (14.10)$$

in terms of generalized Bessel mapping $w_{p,b,\ell}$.
With the assistance of Pochhammer symbol, Gamma mapping as specified by
$(\lambda)_\mu = \frac{\hbar(\lambda+\imath)}{\hbar(\lambda)} = \begin{cases} 1, & \text{if } \mu = 0, \lambda \in \mathbb{C} \setminus \{0\}; \\ \lambda(\lambda+1)\cdots(\lambda+\imath-1), & \text{if } \mu = \imath \in \mathbb{N}, \lambda \in \mathbb{C}; \end{cases}$
$(\lambda)_0 = 1$ and we get $\vartheta_{p,b,\ell}$ given in (14.12) as

$$\vartheta_{p,b,\ell}(\varsigma) = \varsigma + \sum_{\imath=1}^{\infty} \frac{(-\ell)^{\imath}}{4^{\imath}(\tau)_{\imath} \imath!}, \qquad (14.11)$$

where $\tau = p + \frac{b+1}{2} \notin \varsigma$ and $\mathbb{N} = \{1, 2, 3, \cdots\}$.
We will write $\vartheta_{\tau,\ell}(\varsigma) = \vartheta_{p,b,\ell}(\varsigma)$ for convenience. Now, we think about S^{ℓ}_{τ} operator given as

$$S^{\ell}_{\tau} \vartheta(\varsigma) = \vartheta_{\tau,\ell}(\varsigma) * \vartheta(\varsigma) = \varsigma + \sum_{\imath=1}^{\infty} \frac{(-\ell)^{\imath} \varrho_{\imath+1}}{4^{\imath}(\tau)_{\imath} \imath!} \varsigma^{\imath+1}$$

$$= \varsigma + \sum_{\imath=2}^{\infty} \frac{(-\ell)^{\imath-1} \varrho_{\imath}}{4^{\imath-1}(\tau)_{\imath-1}(\imath-1)!} \varsigma^{\imath} = \varsigma + \sum_{\imath=2}^{\infty} M(\ell, \tau, \imath) \varrho_{\imath} \varsigma^{\imath},$$

where

$$M(\ell, \tau, \imath) = \frac{(-\ell)^{\imath-1}}{4^{\imath-1}(\tau)_{\imath-1}(\imath-1)!},$$

$$\tau = \left(p + \frac{b+1}{2}\right) \neq 0, -1, -2, \cdots.$$

In [13], Sakaguchi assigned the category S_s of star-shaped mappings with respect to symmetric points as describes.

Let $\vartheta \in A$. Then ϑ is said to be starlike with respect to symmetric points in $U \Leftrightarrow$

$$\Re\left\{\frac{2\varsigma \vartheta'(\varsigma)}{\vartheta(\varsigma) - \vartheta(-\varsigma)}\right\} > 0, \ (\varsigma \in U).$$

Recently, Owa et al. [11] defined the class $S_s(\nu, x)$ as describes:

$$\Re\left\{\frac{(1-x)\varsigma \vartheta'(\varsigma)}{\vartheta(\varsigma) - \vartheta(x\varsigma)}\right\} > \nu, \ (\varsigma \in U),$$

where $0 \leq \nu < 1, |x| \leq 1, x \neq 1$. Note that $S_s(0,-1) = S_s$ and $S_s(\nu,-1) = S_s(\nu)$ is called Sakaguchi mapping of order ν.

Now, we specified a new subclass of mappings pertaining to the class A.

Definition 14.1.1. *A mapping $\vartheta \in A$ is said to be in the class $\wp - US_s(\tau,\hbar,x)$ if for all $\varsigma \in U$*

$$\Re\left\{\frac{(1-x)\varsigma\left(S_\tau^\ell \vartheta(\varsigma)\right)'}{S_\tau^\ell \vartheta(\varsigma) - S_\tau^\ell \vartheta(x\varsigma)}\right\} \geq \wp\left|\frac{(1-x)\varsigma\left(S_\tau^\ell \vartheta(\varsigma)\right)'}{S_\tau^\ell \vartheta(\varsigma) - S_\tau^\ell \vartheta(t\varsigma)} - 1\right| + \hbar,$$

for $\wp \geq 0, |x| \leq 1, x \neq 1, 0 \leq \hbar < 1$.

Additionally, we say that a mapping $\vartheta \in \wp - US_s(\tau,\hbar,x)$ is in the subclass $\wp - \widetilde{U}S_s(\tau,\hbar,x)$ if ϑ is of the following type

$$\vartheta(\varsigma) = \varsigma - \sum_{\imath=2}^{\infty} \varrho_\imath \varsigma^\imath, \quad \varrho_\imath \geq 0, \imath \in \mathbb{N}, \varsigma \in U. \qquad (14.12)$$

The focus of this section is to explore the class's coefficient bounds, partial sums, and such neighborhoods results $\wp - \widetilde{U}S_s(\tau,\hbar,x)$.

To begin, consider the following lemmas [3].

Lemma 14.1.2. *Let $\aleph = u + iv$. Then*

$$\Re\left(\aleph\right) \geq \nu \Leftrightarrow |\aleph - (1+\nu)| \leq |\aleph + (1-\nu)|.$$

Lemma 14.1.3. *Let $\aleph = u + iv$ and $\nu, \hbar$ be real numbers. Then*

$$\Re\left(\aleph\right) > \nu|\aleph - 1| + \hbar \Leftrightarrow \Re\{\aleph(1+\nu e^{i\theta}) - \nu e^{i\theta}\} > \hbar.$$

14.2 Coefficient bounds

Theorem 14.2.1. *The mapping ϑ defined by (14.12) is in $\wp - \widetilde{U}S_s(\tau,\hbar,x) \Leftrightarrow$*

$$\sum_{\imath=2}^{\infty} M(\ell,\tau,\imath)|\imath(\wp+1) - u_\imath(\wp+\hbar)|a_\imath \leq 1 - \hbar, \qquad (14.13)$$

where $\wp \geq 0, |x| \leq 1, x \neq 1, 0 \leq \hbar < 1$ and $u_\imath = 1 + x + \cdots + x^{\imath-1}$. The result is sharp for the mapping $\vartheta(\varsigma)$ supplied by

$$\vartheta(\varsigma) = \varsigma - \frac{1-\hbar}{M(\ell,\tau,\imath)|\imath(\wp+1) - u_\imath(\wp+\hbar)|}\varsigma^\imath.$$

Proof. By Definition 14.1.1, we get

$$\Re\left\{\frac{(1-x)\varsigma\left(S_\tau^\ell\vartheta(\varsigma)\right)'}{S_\tau^\ell\vartheta(\varsigma)-S_\tau^\ell\vartheta(x\varsigma)}\right\} \geq \wp\left|\frac{(1-x)\varsigma\left(S_\tau^\ell\vartheta(\varsigma)\right)'}{S_\tau^\ell\vartheta(\varsigma)-S_\tau^\ell\vartheta(x\varsigma)}-1\right|+\hbar.$$

Then by Lemma 14.1.3, we have

$$\Re\left\{\frac{(1-x)\varsigma\left(S_\tau^\ell\vartheta(\varsigma)\right)'}{S_\tau^\ell\vartheta(\varsigma)-S_\tau^\ell\vartheta(x\varsigma)}(1+\wp e^{i\theta})-\wp e^{i\theta}\right\} \geq \hbar, \quad -\pi<\theta\leq\pi$$

or equivalently

$$\Re\left\{\frac{(1-x)\varsigma\left(S_\tau^\ell\vartheta(\varsigma)\right)'(1+\wp e^{i\theta})}{S_\tau^\ell\vartheta(\varsigma)-S_\tau^\ell\vartheta(x\varsigma)}-\frac{\wp e^{i\theta}\left[S_\tau^\ell\vartheta(\varsigma)-S_\tau^\ell\vartheta(t\varsigma)\right]}{S_\tau^\ell\vartheta(\varsigma)-S_\tau^\ell\vartheta(x\varsigma)}\right\} \geq \hbar. \quad (14.14)$$

Let $F(\varsigma) = (1-x)\varsigma\left(S_\tau^\ell\vartheta(\varsigma)\right)'(1+\wp e^{i\theta}) - \wp e^{i\theta}\left[S_\tau^\ell\vartheta(\varsigma) - S_\tau^\ell\vartheta(x\varsigma)\right]$
and $M(\varsigma) = S_\tau^\ell\vartheta(\varsigma) - S_\tau^\ell\vartheta(x\varsigma)$.
By Lemma 14.1.2, (14.14) is equivalent to

$$|F(\varsigma)+(1-\hbar)M(\varsigma)| \geq |F(\varsigma)-(1+\hbar)M(\varsigma)|, \text{ for } 0 \leq \hbar < 1.$$

But

$$|F(\varsigma)+(1-\hbar)M(\varsigma)| = \left|(1-x)\left\{(2-\hbar)\varsigma - \sum_{\imath=2}^{\infty}M(\ell,\tau,\imath)(\imath+u_\imath(1-\hbar))\varrho_\imath\varsigma^\imath\right.\right.$$
$$\left.\left. - \wp e^{i\theta}\sum_{\imath=2}^{\infty}M(\ell,\tau,\imath)(\imath-u_\imath)\varrho_\imath\varsigma^\imath\right\}\right|$$
$$\geq |1-x|\left\{(2-\hbar)|\varsigma| - \sum_{\imath=2}^{\infty}M(\ell,\tau,\imath)|\imath+u_\imath(1-\hbar)||\varrho_\imath||\varsigma^\imath|\right.$$
$$\left. - \wp\sum_{\imath=2}^{\infty}M(\ell,\tau,\imath)|\imath-u_\imath||\varrho_\imath||\varsigma^\imath|\right\}.$$

Also

$$|F(\varsigma)-(1+\hbar)M(\varsigma)| = \left|(1-x)\left\{-\hbar\varsigma - \sum_{\imath=2}^{\infty}M(\ell,\tau,\imath)(\imath-u_\imath(1+\hbar))\varrho_\imath\varsigma^\imath\right.\right.$$
$$\left.\left. - \wp e^{i\theta}\sum_{\imath=2}^{\infty}M(\ell,\tau,\imath)(\imath-u_\imath)\varrho_\imath\varsigma^\imath\right\}\right|$$
$$\leq |1-x|\left\{\hbar|\varsigma| + \sum_{\imath=2}^{\infty}M(\ell,\tau,\imath)|\imath-u_\imath(1+\hbar)||\varrho_\imath||\varsigma^\imath|\right.$$
$$\left. + \wp\sum_{\imath=2}^{\infty}M(\ell,\tau,\imath)|\imath-u_\imath||\varrho_\imath||\varsigma^\imath|\right\}.$$

So

$$|F(\varsigma) + (1-\hbar)M(\varsigma)| - |F(\varsigma) - (1+\hbar)M(\varsigma)|$$
$$\geq |1-x|\{2(1-\hbar)|\varsigma| -$$
$$\sum_{\imath=2}^{\infty} M(\ell,\tau,\imath)\big[|\imath + u_\imath(1-\hbar)| + |\imath - u_\imath(1+\hbar)| + 2\wp|\imath - u_\imath|\big]\varrho_\imath|\varsigma^\imath|\}$$
$$\geq 2(1-\hbar)|\varsigma| - \sum_{\imath=2}^{\infty} 2M(\ell,\tau,\imath)|\imath(\wp+1) - u_\imath(\wp+\hbar)|\varrho_\imath|\varsigma^\imath| \geq 0$$

or

$$\sum_{\imath=2}^{\infty} M(\ell,\tau,\imath)|\imath(\wp+1) - u_\imath(\wp+\hbar)|\varrho_\imath \leq 1-\hbar.$$

Conversely, suppose that (14.13) holds. So we should explain

$$\Re\left\{\frac{(1-x)\varsigma \left(S^\ell_\tau\vartheta(\varsigma)\right)'(1+\wp e^{i\theta}) - \wp e^{i\theta}\left[S^\ell_\tau\vartheta(\varsigma) - S^\ell_\tau\vartheta(t\varsigma)\right]}{S^\ell_\tau\vartheta(\varsigma) - S^\ell_\tau\vartheta(x\varsigma)}\right\} \geq \hbar.$$

When selecting on the values of ς on the positive real axis where $0 \leq |\varsigma| = r < 1$, the above inequality decreases to

$$\Re\left\{\frac{(1-\hbar) - \sum_{\imath=2}^{\infty} M(\ell,\tau,\imath)[\imath(1+\wp e^{i\theta}) - u_\imath(\hbar + \wp e^{i\theta})]\varrho_\imath\varsigma^{\imath-1}}{1 - \sum_{\imath=2}^{\infty} M(\ell,\tau,\imath)u_\imath\varrho_\imath\varsigma^{\imath-1}}\right\} \geq 0.$$

Since $\Re(-e^{i\theta}) \geq -|e^{i\theta}| = -1$, the above inequality reduces to

$$\Re\left\{\frac{(1-\hbar) - \sum_{\imath=2}^{\infty} M(\ell,\tau,\imath)[\imath(1+\wp) - u_\imath(\hbar + \wp)]\varrho_\imath r^{\imath-1}}{1 - \sum_{\imath=2}^{\infty} M(\ell,\tau,\imath)u_\imath\varrho_\imath r^{\imath-1}}\right\} \geq 0.$$

Letting $r \to 1^-$, we have desired conclusion. □

14.3 Neighborhood property

Continuing Goodman's [9], Ruscheweyh [12] and Altinta et al. [1,2] previous studies (which were based on the well-known principle of analytic mapping neighbourhoods) we defined as:

Definition 14.3.1. Let $\wp \geq 0, |x| \leq 1, x \neq 1, 0 \leq \hbar < 1, \nu \geq 0$ and $u_\imath = 1 + x + \cdots + x^{\imath-1}$. We define the ν-neighborhood of a mapping $\vartheta \in A$ and indicate by $N_\nu(\vartheta)$ consisting of all mappings $g(\varsigma) = \varsigma - \sum\limits_{\imath=2}^{\infty} b_\imath \varsigma^\imath \in S(b_\imath \geq 0, \imath \in \mathbb{N})$ satisfying

$$\sum_{\imath=2}^{\infty} \frac{M(\ell,\tau,\imath)|\imath(\wp+1) - u_\imath(\wp+\hbar)|}{1-\hbar}|\varrho_\imath - b_\imath| \leq 1 - \nu.$$

Theorem 14.3.2. Let $\vartheta(\varsigma) \in \wp - \widetilde{U}S_s(\tau,\hbar,x)$ and for all real θ, we have $\hbar(e^{i\theta} - 1) - 2e^{i\theta} \neq 0$. For any complex number ϵ with $|\epsilon| < \nu (\nu \geq 0)$, if f fulfills the following requirement

$$\frac{\vartheta(\varsigma) + \epsilon\varsigma}{1+\epsilon} \in \wp - \widetilde{U}S_s(\tau,\hbar,x)$$

then $N_\nu(\vartheta) \subset \wp - \widetilde{U}S_s(\tau,\hbar,x)$.

Proof. It is obvious that

$$\vartheta \in \wp - \widetilde{U}S_s(\tau,\hbar,x)$$
$$\Leftrightarrow \left|\frac{(1-x)\varsigma\left(S_\tau^\ell\vartheta(\varsigma)\right)'(1+\wp e^{i\theta}) - (\wp e^{i\theta} + 1 + \hbar)\left(S_\tau^\ell\vartheta(\varsigma) - S_\tau^\ell\vartheta(t\varsigma)\right)}{(1-x)\varsigma\left(S_\tau^\ell\vartheta(\varsigma)\right)'(1+\wp e^{i\theta}) + (1 - \wp e^{i\theta} - \hbar)\left(S_\tau^\ell\vartheta(\varsigma) - S_\tau^\ell\vartheta(x\varsigma)\right)}\right|$$
$$< 1, (-\pi \leq \theta \leq \pi)$$

for any complex number s with $|s| = 1$, we have

$$\frac{(1-x)\varsigma\left(S_\tau^\ell\vartheta(\varsigma)\right)'(1+\wp e^{i\theta}) - (\wp e^{i\theta} + 1 + \hbar)\left(S_\tau^\ell\vartheta(\varsigma) - S_\tau^\ell\vartheta(x\varsigma)\right)}{(1-x)\varsigma\left(S_\tau^\ell\vartheta(\varsigma)\right)'(1+\wp e^{i\theta}) + (1 - \wp e^{i\theta} - \hbar)\left(S_\tau^\ell\vartheta(\varsigma) - S_\tau^\ell\vartheta(t\varsigma)\right)} \neq s.$$

To put it another way, we must have

$$(1-s)(1-x)\varsigma\left(S_\tau^\ell\vartheta(\varsigma)\right)'(1+\wp e^{i\theta})$$
$$- (\wp e^{i\theta} + 1 + \hbar + s(-1 + \wp e^{i\theta} + \hbar))\left(S_\tau^\ell\vartheta(\varsigma) - S_\tau^\ell\vartheta(x\varsigma)\right) \neq 0,$$

it is the same as

$$\varsigma - \sum_{\imath=2}^{\infty} \frac{M(\ell,\tau,\imath)\left((\imath - u_\imath)(1 + \wp e^{i\theta} - s\wp e^{i\theta}) - s(\imath + u_\imath) - u_\imath \hbar(1-s)\right)}{\hbar(s-1) - 2s}\varsigma^\imath \neq 0.$$

However, $\vartheta \in \wp - \widetilde{U}S_s(\tau,\hbar,x) \Leftrightarrow \frac{(\vartheta * h)}{\varsigma} \neq 0, \varsigma \in U - \{0\}$, where

$$h(\varsigma) = \varsigma - \sum_{\imath=2}^{\infty} \ell_\imath \varsigma^\imath$$

and
$$\ell_\imath = \frac{M(\ell,\tau,\imath)\left((\imath-u_\imath)(1+\wp e^{i\theta}-s\wp e^{i\theta})-s(\imath+u_\imath)-u_\imath\hbar(1-s)\right)}{\hbar(s-1)-2s}$$

we note that
$$|\ell_\imath| \leq \frac{M(\ell,\tau,\imath)|\imath(1+\wp)-u_\imath(\wp+\hbar)|}{1-\hbar}$$

since $\frac{\vartheta(\varsigma)+\epsilon\varsigma}{1+\epsilon} \in \wp - \tilde{U}S_s(\tau,\hbar,x)$, therefore $\varsigma^{-1}\left(\frac{\vartheta(\varsigma)+\epsilon\varsigma}{1+\epsilon} * h(\varsigma)\right) \neq 0$, which is equivalent to

$$\frac{(\vartheta * h)(\varsigma)}{(1+\epsilon)\varsigma} + \frac{\epsilon}{1+\epsilon} \neq 0. \tag{14.15}$$

Now suppose that $\left|\frac{(\vartheta*h)(\varsigma)}{\varsigma}\right| < \nu$. Then by (14.15), we must have

$$\left|\frac{(\vartheta * h)(\varsigma)}{(1+\epsilon)\varsigma} + \frac{\epsilon}{1+\epsilon}\right| \geq \frac{|\epsilon|}{|1+\epsilon|} - \frac{1}{|1+\epsilon|}\left|\frac{(\vartheta * h)(\varsigma)}{\varsigma}\right|$$
$$> \frac{|\epsilon|-\nu}{|1+\epsilon|} \geq 0,$$

this is a contradiction by $|\epsilon| < \nu$ and however, we have $\left|\frac{(\vartheta*h)(\varsigma)}{\varsigma}\right| \geq \nu$. If $g(\varsigma) = \varsigma - \sum_{\imath=2}^{\infty} b_\imath \varsigma^\imath \in N_\nu(\vartheta)$, then

$$\nu - \left|\frac{(g*h)(\varsigma)}{\varsigma}\right| \leq \left|\frac{((\vartheta-g)*h)(\varsigma)}{\varsigma}\right| \leq \sum_{\imath=2}^{\infty}|a_\imath - b_\imath|\|\ell_\imath\||\varsigma^\imath|$$
$$< \sum_{\imath=2}^{\infty}\frac{M(\ell,\tau,\imath)|\imath(1+\wp)-u_\imath(\wp+\hbar)|}{1-\hbar}|\varrho_\imath - b_\imath| \leq \nu.$$

□

14.4 Partial sums

In this part, we discuss partial sums.

Theorem 14.4.1. *If ϑ of the form (14.1) fulfills the requirement (14.13), then*

$$\Re\left\{\frac{\vartheta(\varsigma)}{\vartheta_\psi(\varsigma)}\right\} \geq 1 - \frac{1}{\varpi_{\psi+1}} \tag{14.16}$$

and
$$\varpi_\imath = \begin{cases} 1, & \text{if } \imath = 2, 3, \cdots, \psi; \\ \varpi_{\psi+1}, & \text{if } \imath = \psi+1, \psi+2, \cdots, \end{cases} \quad (14.17)$$

where
$$\varpi_\imath = \frac{M(\ell, \tau, \imath)|\imath(1+\wp) - u_\imath(\wp + \hbar)|}{1 - \hbar}. \quad (14.18)$$

The result in (14.16) is sharp for every ψ, with the extremal mapping
$$\vartheta(\varsigma) = \varsigma + \frac{\varsigma^{\psi+1}}{\varpi_{\psi+1}}. \quad (14.19)$$

Proof. Define the mapping $\aleph$, as
$$\frac{1+\aleph(\varsigma)}{1-\aleph(\varsigma)} = \varpi_{\psi+1}\left\{\frac{\vartheta(\varsigma)}{\vartheta_\psi(\varsigma)} - \left(1 - \frac{1}{\varpi_{\psi+1}}\right)\right\} \quad (14.20)$$
$$= \left\{\frac{1 + \sum_{\imath=2}^{\psi} \varrho_\imath \varsigma^{\imath-1} + \varpi_{\psi+1} \sum_{\imath=\psi+1}^{\infty} \varrho_\imath \varsigma^{\imath-1}}{1 + \sum_{\imath=2}^{\psi} a_\imath \varsigma^{\imath-1}}\right\}.$$

Then, from (14.20), we acquire
$$\aleph(\varsigma) = \frac{\varpi_{\psi+1} \sum_{\imath=\psi+1}^{\infty} a_\imath \varsigma^{\imath-1}}{2 + 2\sum_{\imath=2}^{\psi} \varrho_\imath \varsigma^{\imath-1} + \varpi_{\psi+1} \sum_{\imath=\psi+1}^{\infty} \varrho_\imath \varsigma^{\imath-1}}$$

and
$$|\aleph(\varsigma)| \leq \frac{\varpi_{\psi+1} \sum_{\imath=\psi+1}^{\infty} \varrho_\imath}{2 - 2\sum_{\imath=2}^{\psi} \varrho_\imath - \varpi_{\psi+1} \sum_{\imath=\psi+1}^{\infty} \varrho_\imath}.$$

Now $|\aleph(\varsigma)| \leq 1$ if
$$2\varpi_{\psi+1} \sum_{\imath=\psi+1}^{\infty} \varrho_\imath \leq 2 - 2\sum_{\imath=2}^{\psi} \varrho_\imath,$$

which is equivalent to
$$\sum_{\imath=2}^{\psi} \varrho_\imath + \varpi_{\psi+1} \sum_{\imath=\psi+1}^{\infty} \varrho_\imath \leq 1. \quad (14.21)$$

It is suffices to show that the left hand side of (14.21) is bounded above by $\sum_{\imath=2}^{\infty} \varpi_\imath \varrho_\imath$, which is equivalent to

$$\sum_{\imath=2}^{\psi}(\varpi_\imath - 1)\varrho_\imath + \sum_{\imath=\psi+1}^{\infty}(\varpi_\imath - \varpi_{\psi+1})\varrho_\imath \geq 0.$$

To see that the mapping given by (14.19) gives the sharp result, we notice that for $\varsigma = re^{i\pi/n}$,

$$\frac{\vartheta(\varsigma)}{\vartheta_\psi(\varsigma)} = 1 + \frac{\varsigma^\psi}{\varpi_{\psi+1}}. \tag{14.22}$$

Taking $\varsigma \to 1^-$, we have $\frac{\vartheta(\varsigma)}{\vartheta_\psi(\varsigma)} = 1 - \frac{1}{\varpi_{\psi+1}}$. This concludes the proof. □

We next determine bounds for $\frac{\vartheta_\psi(\varsigma)}{\vartheta(\varsigma)}$.

Theorem 14.4.2. *If ϑ of the form (14.1) fulfills the requirement (14.13) then*

$$\Re\left\{\frac{\vartheta_\psi(\varsigma)}{\vartheta(\varsigma)}\right\} \geq \frac{\varpi_{\psi+1}}{1 + \varpi_{\psi+1}}. \tag{14.23}$$

The result is sharp with the mapping indicated by (14.19).

Proof. We may write

$$\frac{1 + \aleph(\varsigma)}{1 - \aleph(\varsigma)} = (1 + \varpi_{\psi+1})\left\{\frac{\vartheta_\psi(\varsigma)}{\vartheta(\varsigma)} - \frac{\varpi_{\psi+1}}{1 + \varpi_{\psi+1}}\right\}$$

$$= \left\{\frac{1 + \sum_{\imath=2}^{\psi}\varrho_\imath \varsigma^{\imath-1} - \varpi_{\psi+1}\sum_{\imath=\psi+1}^{\infty}\varrho_\imath\varsigma^{\imath-1}}{1 + \sum_{\imath=2}^{\infty}\varrho_\imath\varsigma^{\imath-1}}\right\}, \tag{14.24}$$

where

$$\aleph(\varsigma) = \frac{(1 + \varpi_{\psi+1})\sum_{\imath=\psi+1}^{\infty}\varrho_\imath\varsigma^{\imath-1}}{-\left(2 + 2\sum_{\imath=2}^{\psi}\varrho_\imath\varsigma^{\imath-1} - (1 - \varpi_{\psi+1})\sum_{\imath=\psi+1}^{\infty}\varrho_\imath\varsigma^{\imath-1}\right)}$$

and

$$|\aleph(\varsigma)| \leq \frac{(1 + \varpi_{\psi+1})\sum_{\imath=\psi+1}^{\infty}\varrho_\imath}{2 - 2\sum_{\imath=2}^{\psi}\varrho_\imath + (1 - \varpi_{\psi+1})\sum_{\imath=\psi+1}^{\infty}\varrho_\imath} \leq 1. \tag{14.25}$$

This last inequality is equivalent to

$$\sum_{\imath=2}^{\psi} \varrho_\imath + \varpi_{\psi+1} \sum_{\imath=\psi+1}^{\infty} \varrho_\imath \leq 1. \qquad (14.26)$$

It is suffices to show that the left hand side of (14.26) is bounded above by $\sum_{\imath=2}^{\infty} \varpi_\imath \varrho_\imath$, which is the same as

$$\sum_{\imath=2}^{\psi} (\varpi_\imath - 1)\varrho_\imath + \sum_{\imath=\psi+1}^{\infty} (\varpi_\imath - \varpi_{\psi+1})\varrho_\imath \geq 0.$$

This completes the proof. □

We next turn to ratios involving derivatives.

Theorem 14.4.3. *If ϑ of the form (14.1) satisfies the condition (14.13), then*

$$\Re \left\{ \frac{\vartheta'(\varsigma)}{\vartheta'_\psi(\varsigma)} \right\} \geq 1 - \frac{\psi+1}{\varpi_{\psi+1}}, \qquad (14.27)$$

$$\Re \left\{ \frac{\vartheta'_\psi(\varsigma)}{\vartheta'(\varsigma)} \right\} \geq \frac{\varpi_{\psi+1}}{1+\psi+\varpi_{\psi+1}}, \qquad (14.28)$$

where

$$\varpi_\imath \geq \begin{cases} 1, & \text{if } \imath = 1, 2, 3, \cdots, \psi; \\ \imath \frac{\varpi_{\psi+1}}{\psi+1}, & \text{if } \imath = \psi+1, \psi+2, \cdots \end{cases}$$

and ϖ_n is defined by (14.18). The estimates in (14.26) and (14.27) are sharp with the extremal mapping given by (14.19).

Proof. Firstly, we give proof of (14.27). We write

$$\frac{1+\aleph(\varsigma)}{1-\aleph(\varsigma)} = \varpi_{\psi+1} \left\{ \frac{\vartheta'(\varsigma)}{\vartheta'_\psi(\varsigma)} - \left(1 - \frac{\psi+1}{\varpi_{\psi+1}}\right) \right\}$$

$$= \left\{ \frac{1 + \sum_{\imath=2}^{\psi} \imath a_\imath \varsigma^{\imath-1} + \frac{\varpi_{\psi+1}}{\psi+1} \sum_{\imath=\psi+1}^{\infty} \imath \varrho_\imath \varsigma^{\imath-1}}{1 + \sum_{\imath=2}^{\psi} \varrho_\imath \varsigma^{\imath-1}} \right\},$$

where

$$\aleph(\varsigma) = \frac{\frac{\varpi_{\psi+1}}{\psi+1} \sum_{\imath=\psi+1}^{\infty} \imath \varrho_\imath \varsigma^{\imath-1}}{2 + 2\sum_{\imath=2}^{\psi} \imath \varrho_\imath \varsigma^{\imath-1} + \frac{\varpi_{\psi+1}}{\psi+1} \sum_{\imath=\psi+1}^{\infty} \imath \varrho_\imath \varsigma^{\imath-1}}$$

and

$$|\aleph(\varsigma)| \leq \frac{\frac{\varpi_{\psi+1}}{\psi+1}\sum_{\imath=\psi+1}^{\infty}\imath\varrho_\imath}{2-2\sum_{\imath=2}^{\psi}\imath\varrho_\imath + \frac{\varpi_{\psi+1}}{\psi+1}\sum_{\imath=\psi+1}^{\infty}\imath\varrho_\imath}.$$

Now $|\aleph(\varsigma)| \leq 1$ if and only if

$$\sum_{\imath=2}^{\psi}\imath\varrho_\imath + \frac{\varpi_{m+1}}{m+1}\sum_{\imath=m+1}^{\infty}\imath\varrho_\imath \leq 1, \qquad (14.29)$$

since the left hand side of (14.29) is bounded above by $\sum_{\imath=2}^{\infty}\varpi_\imath\varrho_\imath$. The proof of (14.29) continues the trend stated in Theorem (14.17). This concludes the proof. □

Theorem 14.4.4. *If $\vartheta(\varsigma) \in \wp - \widetilde{U}S_s(\tau, \hbar, x)$ then*

$$\varrho_\imath \leq \frac{1-\hbar}{M(\ell,\tau,\imath)|\imath(\wp+1) - u_\imath(\wp+\hbar)|},$$

where $\wp \geq 0, |x| \leq 1, x \neq 1, 0 \leq \hbar < 1$ and $u_\imath = 1 + x + \cdots + x^{\imath-1}$.

Bibliography

[1] Altinta, S.O. and Owa, S. 1996. Neighborhoods of certain analytic mappings with negative coefficients, *Int. J. Math. and Math. Sci.* 19: 797–800.

[2] Altinta, S.O., Ozkan, E. and Srivastava, H.M. 2000. Neighborhoods of a class of analytic mappings with negative coefficients, *Appl. Math. Letters* 13: 63–67.

[3] Aqlan, E., Jhangiri, J.M. and Kulkarni, S.R. 2004. Class of k-uniformly convex and starlike mappings, *Tamkang J. Math.* 35: 1–7.

[4] Baricz, A. 2008. Geometric properties of generalized Bessel mapping, *Publ. Math. Debrecan* 73: 155–178.

[5] Baricz, A., Generalized Bessel mappings of the first kind, Ph.D thesis, Babes-Bolyai University, Cluj-Napoca, 2008.

[6] Baricz, A., Generalized Bessel mappings of the first kind, Lecture Notes in Math., 1994, Springer, Berlin, 2010.

[7] Baricz, A. and Frasin, B. A. 2010. Univalence of integral operators involving Bessel mappings, *Appl. Math. Letters* 23: 371–376.

[8] Deniz, E., Orhan, H. and Srivastava, H.M. 2011. Some sufficient conditions for univalence of certain families of integral operators involving generalized Bessel mappings, *Taiwan. J. Math.* 15:2: 883–917.

[9] Goodman, A. W. 1957. Univalent mappings and non-analytic curves, *Proc. Amer. Math. Soc.* 8: 598–601.

[10] Kanas, S. and Wisniowska, A. 1999. Conic regions and k-uniforn convexity, *Comput. Appl. Math.* 105: 327–336.

[11] Owa, S., Sekine, T. and Yamakawa. R. 2007. On Sakaguchi type mappings, *Appl. Math. Comput.* 187: 356–361;

[12] Ruscheweyh, S. 1981. Neighborhoods of univalent mappings, *Proc. Amer. Math. Soc.* 81:4: 521–527.

[13] Sakaguchi, K. 1959. On a certain univalent mapping, *J. Math. Soc. Japan* 11: 72–75.

[14] Silverman, H. 1975. Univalent mappings with negative coefficients, *Proc. Amer. Math. Soc.* 51: 109–116.

[15] Silverman, H. 1997. Partial sums of starlike and convex mappings, *J. Anal. Appl.* 209: 221–227.

[16] Thirupathi Reddy, P. and Venkateswarlu, B. 2018. On a certain subclass of uniformly convex mappings defined by Bessel mappings, *Transylvanian J. Math.* 10:1: 43–49.

[17] Thirupathi Reddy, P. and Venkateswarlu, B. 2019. A certain subclass of uniformly convex functions defined by Bessel mappings, *Proyecciones* 38:4: 719–731.

Chapter 15

A note on meromorphic functions with positive coefficients defined by differential operator

B. Venkateswarlu

P. Thirupathi Reddy

Sujatha

15.1	Introduction	265
15.2	Coefficient inequality	268
15.3	Distortion theorem	269
15.4	Integral operators	271
15.5	Convex linear combinations and convolution properties	272
15.6	Neighborhood property	275
	Bibliography	276

15.1 Introduction

Complex analysis (complex function theory) was begun in the eighteenth century and has since become one of the major topics of mathematics. Euler, Riemann, and Cauchy are well-known complicated analyzers. Because of its effective applicability to a wide range of concepts and issues, this domain has had a significant impact on a wide range of research areas, including engineering, physics, and mathematics. Researchers have discovered some unexpected links between ostensibly disparate study fields. The study of the unusual and intriguing combination of geometry and complex analysis is known as Geometric Analytic Function Theory. It concerns the structure of analytic functions in the complex domain, with geometries such as starlike, close-to-starlike, convex, close-to-convex, spiral, and so on.

This study analyses and proposes a novel differential operator of morphometric functions with positive coefficients using regular techniques. Further, significant criteria were defined and analyzed for specific formulations of this new operator to yield some fundamental Geometric function theory.

The theory of exceptional functions in one variable has a long and entertaining history; the increasing popularity of exceptional mappings in multiple variables is relatively recent. There is now considerable progress, notably in the domains of special mappings with symmetries and harmonic analysis connected to root systems. This study is prompted by several generalizations of the theory of symmetric spaces, the mappings of which may be expressed as special mappings relying on specific sets of parameters.

Later, academics investigated these operators using various classes and formulas of analytic functions. Differential operators play an important role in functions theory and its information in a complex domain. They used it to explain geometric interpolation of regular mapping in a complex domain. Later, scholars study these operators using other classes and formulae of regular mappings.

Let A indicate the category of mapping of the type

$$\vartheta(o) = o + \sum_{\imath=2}^{\infty} a_\imath o^\imath, \qquad (15.1)$$

in the open unit disk $E = \{o : |o| < 1\}$ which are analytic and fulfill the following standard normalization condition $\vartheta(0) = \vartheta'(0) - 1 = 0$. The subcategory of A consisting of functions $\vartheta(o)$ that are all univalent in E is denoted by S. A function $\vartheta \in A$ is a star-shaped mapping by the order $\hbar, 0 \leq \hbar < 1$ if it satisfy

$$\Re\left\{\frac{o\vartheta'(o)}{\vartheta(o)}\right\} > \hbar \ (o \in E). \qquad (15.2)$$

We indicate this category with $S^*(\hbar)$.

A function $\vartheta \in A$ is a convex mapping by the order $\hbar, 0 \leq \hbar < 1$ if it fulfill

$$\Re\left\{1 + \frac{o\vartheta''(o)}{\vartheta'(o)}\right\} > \hbar \ (o \in E). \qquad (15.3)$$

We indicate this category $K(\hbar)$.

Let T indicate the category of mappings analytic in E that are of the type

$$\vartheta(o) = o - \sum_{\imath=2}^{\infty} \varrho_\imath o^\imath, (\varrho_\imath \leq 0, o \in E) \qquad (15.4)$$

and let $T^*(\hbar) = T \cap S^*(\hbar), C(\hbar) = T \cap K(\hbar)$. The category $T^*(\hbar)$ and allied categories possess Silverman [17] and others have meticulously researched certain remarkable aspects.

A function $\vartheta \in A$ is said to be in the category of uniformly convex mappings of order $\aleph$ and type $\wp$, indicated by $UCV(\wp, \aleph)$, if

$$\Re\left\{1 + \frac{o\vartheta''(o)}{\vartheta'(o)} - \aleph\right\} > \wp\left|\frac{o\vartheta''(o)}{\vartheta'(o)}\right|, \tag{15.5}$$

where $\wp \geq 0, \aleph \in [-1, 1)$ and $\wp + \aleph \geq 0$, and is said to be in category corresponding category indicated by $SP(\wp, \aleph)$ if

$$\Re\left\{\frac{o\vartheta'(o)}{\vartheta(o)} - \aleph\right\} > \wp\left|\frac{o\vartheta'(o)}{\vartheta(o)} - 1\right|, \tag{15.6}$$

where $\wp \geq 0, \aleph[-1, 1)$ and $\wp + \aleph \geq 0$. Indeed it follows from (15.5) and (15.6) that

$$\vartheta \in UCV(\aleph, \wp) \Leftrightarrow o\vartheta' \in SP(\aleph, \wp). \tag{15.7}$$

For $\wp = 0$ we get respectively, the categories $K(\aleph)$ and $S^*(\aleph)$. The mapping of the category $UCV(1, 0) \equiv UCV$ are known as uniformly convex mappings, and they were first proposed by Goodman [5, 6] with a geometric meaning. The category $SP(1, 0) \equiv SP$ is defined by Ronning [13, 14]. The categories $UCV(1, \aleph) \equiv UCV(\aleph)$ and $SP(1, \aleph) \equiv SP(\aleph)$ are interested by Ronning [12]. For $\aleph = 0$, the $UCV(\wp, 0) \equiv \wp - UCV$ and $SP(\wp, 0) \equiv \wp - SP$ are specified respectively, by Kanas and Wisniowska [8, 9].

Further, researchers, see [1, 2, 10] and others have gone into and explored the categories connected to $UCV(\wp, \aleph)$ and $SP(\wp, \aleph)$.

Let Σ indicate the category of mappings of the type

$$\vartheta(o) = \frac{1}{o} + \sum_{\imath=1}^{\infty} \varrho_\imath o^\imath \tag{15.8}$$

which are regular in $E = \{o : 0 < |o| < 1\}$, 1 in residue and a single pole at the origin.

Let $\Sigma_s, \Sigma^*(\hbar)$ and $\Sigma_k(\hbar)(0 \leq \hbar < 1)$ indicate the subcategoryes of Σ that are univalent, moromorphically star-shaped and convex of order $\hbar$, respectively. Analytically $\vartheta(o)$ of the type (15.8) is in

$$\Sigma^*(\hbar) \iff \Re\left\{-\frac{o\vartheta'(o)}{\vartheta(o)}\right\} > \hbar, (o \in E). \tag{15.9}$$

Similarly, $\vartheta \in \Sigma_k(\hbar) \iff \vartheta(o)$ is of the type (15.8) and satisfies

$$\Re\left\{-\left(1 + \frac{o\vartheta''(o)}{\vartheta'(o)}\right)\right\} > \hbar, (o \in E). \tag{15.10}$$

The categoryes $\Sigma^*(\hbar)$ and $\Sigma_k(\hbar)$ have been scrutinized by Pommerenke [11], Clunie [3], Royster [15], and others. Juneja and Reddy [7] originated the category Σ_p of mappings of the type

$$\vartheta(o) = \frac{1}{o} + \sum_{\imath=1}^{\infty} \varrho_\imath o^\imath, (\varrho_\imath \geq 0), \qquad (15.11)$$

$$\Sigma_p^*(\hbar) = \Sigma_p \cap \Sigma^*(\hbar).$$

For mapping $\vartheta(o)$ in the category Σ_p, we specify a linear operator D^τ by the following type

$$D^1\vartheta(o) = \frac{1}{o} + \sum_{\imath=1}^{\infty} (\imath\ell+2)^\tau \varrho_\imath o^{\imath\ell} = \frac{(o^2\vartheta(o))'}{o}$$

$$D^2\vartheta(o) = D(D'\vartheta(o)) \text{ and for } \tau = 1, 2, 3, \cdots,$$

$$D^\tau\vartheta(o) = D(D^{\tau-1}\vartheta(o)) = \frac{1}{o} + \sum_{\imath=1}^{\infty}(\imath\ell+2)^\tau \varrho_\imath o^\imath = \frac{(o^2 D^{\tau-1}\vartheta(o))'}{o}. \quad (15.12)$$

Now, we specify a new subcategory $\sigma_\xi^+(\hbar, \wp)$ of Σ_p. We establish a new subcategory motivated by Venkateswarlu et al. [18–20].

Definition 15.1.1. *For $-1 \leq \hbar < 1$, and $\wp \geq 1$, let $\sigma_\xi^+(\hbar, \wp)$ be the subcategory of Σ_p consisting of mapping of the type (15.11) and fulfilling the analytic criterion*

$$-\Re\left\{\frac{D^{\tau+1}\vartheta(o)}{D^\tau\vartheta(o)} + \hbar\right\} > \wp \left|\frac{D^{\tau+1}\vartheta(o)}{D^\tau\vartheta(o)} + 1\right|, \qquad (15.13)$$

where $D^\tau\vartheta(o)$ is given by (15.12).

The main goal of this chapter is to look at some of the most common geometric mapping theory aspects for the category, like coefficient limits, growth and distortion properties, radius of convexity, convex linear combination and convolution properties, integral operators, and ζ-neighborhoods.

15.2 Coefficient inequality

Theorem 15.2.1. *A mapping $\vartheta(o)$ of the type (15.11) is in $\sigma_\xi^+(\hbar, \wp)$ if*

$$\sum_{\imath=1}^{\infty}(\imath\ell+2)^\tau[(1+\wp)(\imath\ell+1)+1-\hbar]|\varrho_\imath| \leq (1-\hbar), -1 \leq \hbar < 1 \text{ and } \wp \geq 1. \quad (15.14)$$

Proof. It suffices to show that

$$\wp\left|\frac{D^{\tau+1}\vartheta(o)}{D^\tau\vartheta(o)} + 1\right| + \Re\left\{\frac{D^{\tau+1}\vartheta(o)}{D^\tau\vartheta(o)} + 1\right\} \leq 1 - \hbar.$$

We have

$$\wp\left|\frac{D^{\tau+1}\vartheta(o)}{D^\tau\vartheta(o)}+1\right|+\Re\left\{\frac{D^{\tau+1}\vartheta(o)}{D^\tau\vartheta(o)}+1\right\}$$

$$\leq (1+\wp)\left|\frac{D^{\tau+1}\vartheta(o)}{D^\tau\vartheta(o)}+1\right|$$

$$\leq \frac{(1+\wp)\sum_{i=1}^\infty (i\ell+2)^\tau(i\ell+1)|\varrho_i||o^i|}{\frac{1}{|o|}-\sum_{i=2}^\infty(i\ell+2)^\tau|\varrho_i|}.$$

Letting $o \to 1$ along the real axis, we obtain

$$\wp\left|\frac{D^{\tau+1}\vartheta(o)}{D^\tau\vartheta(o)}+1\right|+\Re\left\{\frac{D^{\tau+1}\vartheta(o)}{D^\tau\vartheta(o)}+1\right\}$$

$$\leq \frac{(1+\wp)\sum_{i=1}^\infty (i\ell+2)^\tau(i\ell+1)|\varrho_i|}{1-\sum_{i=2}^\infty(i\ell+2)^\tau|\varrho_i|}.$$

This last expression is bounded by $(1-\hbar)$ if

$$\sum_{i=1}^\infty (i\ell+2)^\tau[(1+\wp)(i\ell+1)+1-\hbar]|\varrho_i| \leq (1-\hbar).$$

The theorem is so completed. □

Corollary 15.2.2. Let the mapping $\vartheta(o)$ specified by (15.11) be in the category $\sigma_\xi^+(\hbar,\wp)$. Then

$$\varrho_i \leq \frac{(1-\hbar)}{\sum_{i=1}^\infty (i\ell+2)^\tau[(1+\wp)(i\ell+1)1-\hbar]}, (i \geq 1). \tag{15.15}$$

Equality holds for the mappings of the type

$$\vartheta_i(o) = \frac{1}{o}+\frac{(1-\hbar)}{(i\ell+2)^\tau[(1+\wp)(i\ell+1)+1-\hbar]}o^i. \tag{15.16}$$

15.3 Distortion theorem

Theorem 15.3.1. Let $\vartheta(o)$ is in $\sigma_\xi^+(\hbar,\wp)$. Then for $0<|o|=r<1$,

$$\frac{1}{o}-\frac{(1-\hbar)}{(\ell+2)^\tau[(1+\wp)(\ell+1)+1-\hbar]}r$$

$$\leq |\vartheta(o)| \leq \frac{1}{o}-\frac{(1-\hbar)}{(\ell+2)^\tau[(1+\wp)(\ell+1)+1-\hbar]}r \tag{15.17}$$

with equality for the mapping

$$\vartheta(o) = \frac{1}{o} + \frac{(1-\hbar)}{(\ell+2)^\tau[(1+\wp)(\ell+1)+1-\hbar]} o \text{ at } |o| = r. \qquad (15.18)$$

Proof. Suppose $\vartheta(o)$ is in $\sigma_\xi^+(\hbar,\wp)$. In view of Theorem 15.2.1, we have

$$(\ell+2)^\tau[(1+\wp)(\ell+1)+1-\hbar]\sum_{\imath=1}^\infty \varrho_\imath \leq \sum_{\imath=1}^\infty (\imath\ell+2)^\tau[(1+\wp)(\imath\ell+1)+1-\hbar] \leq (1-\hbar),$$

which evidently yields

$$\sum_{\imath=1}^\infty \varrho_\imath \leq \frac{(1-\hbar)}{(\ell+2)^\tau[(1+\wp)(\ell+1)+1-\hbar]}.$$

Consequently, we obtain

$$|\vartheta(o)| = \left|\frac{1}{o} + \sum_{\imath=1}^\infty \varrho_\imath o^\imath\right| \leq \left|\frac{1}{o}\right| + \sum_{\imath=1}^\infty \varrho_\imath |o|^\imath \leq \frac{1}{r} + r\sum_{\imath=1}^\infty \varrho_\imath$$
$$\leq \frac{1}{r} + \frac{1-\hbar}{(\ell+2)^\tau[(1+\wp)(\ell+1)+1-\hbar]} r$$

also,

$$|\vartheta(o)| = \left|\frac{1}{o} + \sum_{\imath=1}^\infty \varrho_\imath o^\imath\right| \geq \frac{1}{o} - \sum_{\imath=1}^\infty \varrho_\imath |o|^\imath \geq \frac{1}{r} - r\sum_{\imath=1}^\infty \varrho_\imath$$
$$\geq \frac{1}{r} - \frac{(1-\hbar)}{(\ell+2)^\tau[(1+\wp)(\ell+1)+1-\hbar]} r.$$

Hence the result (15.17) follow. $\square$

Theorem 15.3.2. *Let $\vartheta(o)$ is in $\sigma_\xi^+(\hbar,\wp)$. Then for $0 < |o| = r < 1$,*

$$\frac{1}{r^2} - \frac{(1-\hbar)}{(\ell+2)^\tau[(1+\wp)(\ell+1)+1-\hbar]} \leq |\vartheta'(o)| \leq \frac{(1-\hbar)}{(\ell+2)^\tau[(1+\wp)(\ell+1)+1-\hbar]}.$$

The result is sharp.

Proof. From Theorem 15.2.1, we have

$$(\ell+2)^\tau[(1+\wp)(1+\ell)+1-\hbar]\sum_{\imath=1}^\infty \imath\varrho_\imath \leq \sum_{\imath=1}^\infty (\imath+2)^2[(1+\wp)(\imath+1)+1-\hbar] \leq (1-\hbar)$$

which evidently yields

$$\sum_{\imath=1}^\infty \imath\varrho_\imath \leq \frac{(1-\hbar)}{(\ell+2)^\tau[(1+\wp)(\ell+1)+1-\hbar]}.$$

Consequently, we obtain

$$|\vartheta'(o)| \leq \frac{1}{r^2} + \sum_{i=1}^{\infty} i\varrho_i r^{i-1}$$

$$\leq \frac{1}{r^2} + \sum_{i=1}^{\infty} i\varrho_i$$

$$\leq \frac{1}{r^2} + \frac{(1-\hbar)}{(\ell+2)^\tau[(1+\wp)(\ell+1)+1-\hbar]}.$$

Also,

$$|\vartheta'(o)| \leq \frac{1}{r^2} - \sum_{i=1}^{\infty} i\varrho_i r^{i-1}$$

$$\leq \frac{1}{r^2} - \sum_{i=1}^{\infty} i\varrho_i$$

$$\leq \frac{1}{r^2} - \frac{(1-\hbar)}{(\ell+2)^\tau[(1+\wp)(\ell+1)+1-\hbar]}.$$

This completes the proof. □

15.4 Integral operators

In this part, we look at the type's (15.11) category-preserving integral operators.

Theorem 15.4.1. *Let $\vartheta(o)$ is in $\sigma_\xi^+(\hbar, \wp)$. Then*

$$F(o) = co^{-c-1}\int_0^o t^c\vartheta(t)dt = \frac{1}{o} + \sum_{i=1}^{\infty}\frac{c}{c+i+1}\varrho_i o^i, c > 0 \qquad (15.19)$$

is in $\sigma[\zeta(\hbar, \wp, n, c)]$, where

$$\zeta(\hbar, \wp, n, c) = \frac{3^\tau(3+2\wp-\hbar)(c+2)-(1-\hbar)c}{3^\tau(3+2\wp-\hbar)(c+2)+(1-\hbar)c}. \qquad (15.20)$$

The result sharp for

$$\vartheta(o) = \frac{1}{o} + \frac{(1-\hbar)}{(\ell+2)^\tau[(1+\wp)(\ell+1)+1-\hbar]}o.$$

Proof. Suppose $\vartheta(o) = \frac{1}{o} + \sum_{i=1}^{\infty} \varrho_i o^i$ is in $\sigma_\xi^+(\hbar, \wp)$. We have

$$F(o) = co^{-c-1} \int_0^o t^c \vartheta(t) dt = \frac{1}{o} + \sum_{i=1}^{\infty} \frac{c}{c+i+1} \varrho_i o^i, c > 0.$$

It is sufficient to affirm that

$$\sum_{i=1}^{\infty} \frac{i+\zeta}{1-\zeta} \cdot \frac{c\varrho_i}{i+c+1} \leq 1.. \qquad (15.21)$$

Since $\vartheta(o)$ is in $\sigma_\xi^+(\hbar, \wp)$, we have

$$\frac{\sum_{i=1}^{\infty} (i\ell+2)^\tau [(1+\wp)(i\ell+1) + 1 - \hbar]|\varrho_i|}{(1-\hbar)} \leq 1. \qquad (15.22)$$

Thus (15.21) will be satisfied if

$$\frac{(i+\zeta)c}{(1-\zeta)(i+c+1)} \leq \frac{(i\ell+2)^\tau [(1+\wp)(i\ell+1) + 1 - \hbar]}{(1-\hbar)}, \text{ for each } i$$

or

$$\zeta \leq \frac{(i\ell+2)^\tau [(1+\wp)(i\ell+1) + (1-\hbar)](c+i+1) - ic(1-\hbar)}{(i\ell+2)^\tau [(1+\wp)(i\ell+1)(1-\hbar)](c+i+1) + c(1-\hbar)} \qquad (15.23)$$

$$G(i) = \frac{(i\ell+2)^\tau [(1+\wp)(i\ell+1) + (1-\hbar)](c+i+1) - ic(1-\hbar)}{(i\ell+2)^\tau [(1+\wp)(i\ell+1)(1-\hbar)](c+i+1) + c(1-\hbar)}.$$

Then $G(i+1) - G(i) > 0$ for each i.
Hence $G(i)$ is an increasing mapping of i. Since

$$G(1) = \frac{3^\tau (3+2\wp-\hbar)(c+2) - c(1-\hbar)}{3^\tau (3+2\wp-\hbar)(c+2) + c(1-\hbar)}.$$

The result follows. $\square$

15.5 Convex linear combinations and convolution properties

Theorem 15.5.1. *If the mapping $\vartheta(o)$ is in $\sigma_\xi^+(\hbar, \wp)$, then $\vartheta(o)$ is meromorphically convex of order $\zeta (0 \leq \zeta < 1)$ in $|o| < \tilde{r} = r(\hbar, \wp, \zeta)$, where*

$$r(\hbar, \wp, \zeta) = \inf_{n \geq 1} \left\{ \frac{(1-\zeta)(i\ell+2)^\tau [(1+\wp)(1+i\ell) + 1 - \hbar]}{(1-\hbar)i(i+2-\zeta)} \right\}^{\frac{1}{i+1}}.$$

The result is sharp.

Proof. Let $\vartheta(o) \in \sigma_\xi^+(\hbar, \wp)$. Then by Theorem 15.2.1, we have

$$\sum_{i=1}^{\infty}(i\ell + 2)^\tau[(1+\wp)(i\ell+1)+1-\hbar]|\varrho_i| \leq (1-\hbar). \tag{15.24}$$

It is sufficient to demonstrate that

$$\left|2 + \frac{o\vartheta''(o)}{\vartheta'(o)}\right| \leq 1-\zeta.$$

For $|o| < r = r(\hbar, \wp, \zeta)$, where $r(\hbar, \wp, \zeta)$ is indicated in the theorem's assertion. Then

$$\left|2 + \frac{o\vartheta''(o)}{\vartheta'(o)}\right| = \left|\frac{\sum_{i=1}^{\infty} i(i+1)\varrho_i o^{i-1}}{\frac{-1}{o^2} + \sum_{i=1}^{\infty} i\varrho_i o^{i-1}}\right| \leq \frac{\sum_{i=1}^{\infty} i(i+1)\varrho_i |o|^{i+1}}{1 - \sum_{i=1}^{\infty} i\varrho_i |o|^{i+1}}.$$

This will be restricted by $(1-\zeta)$ if

$$\sum_{i=1}^{\infty} \frac{i(i+2-\zeta)}{(1-\zeta)} \varrho_i |o|^{i+1} \leq 1. \tag{15.25}$$

By (15.24), it follow that (15.25) is true if

$$\frac{i(i+2-\zeta)}{(1-\zeta)}|o|^{i+1} \leq \frac{(i\ell+2)^\tau[(1+\wp)(i\ell+1)+1-\hbar]}{(1-\hbar)}, i \geq 1$$

or

$$|o| \leq \left\{\frac{(1-\zeta)(i\ell+2)^\tau[(1+\wp)(1+i\ell)+1-\hbar]}{(1-\hbar)i(i+2-\zeta)}\right\}^{\frac{1}{i+1}}. \tag{15.26}$$

Setting $|o| = r(\hbar, \wp, \zeta)$ in (15.26), the conclusions are as stated. The outcome is sharp for the mapping

$$\vartheta_i(o) = \frac{1}{o} + \frac{(1-\hbar)}{(i\ell+2)^\tau[(1+\wp)(i\ell+1)+1-\hbar]} o^i, (i \geq 1).$$

$\square$

Theorem 15.5.2. Let $\vartheta_0(o) = \frac{1}{o}$ and

$$\vartheta_i(o) = \frac{1}{o} + \frac{(1-\hbar)}{(i\ell+2)^\tau[(1+\wp)(i\ell+1)+1-\hbar]} o^i, (i \geq 1).$$

Then

$$\vartheta(o) = \frac{1}{o} + \sum_{i=1}^{\infty} \varrho_i o^i$$

is in the category $\sigma_\xi^+(\hbar, \wp) \iff$ it can be expressed as

$$\vartheta(o) = \lambda_0 \vartheta_0(o) + \sum_{i=1}^{\infty} \lambda_i \vartheta_i(o),$$

where $\lambda_0 \geq 0, \lambda_i \geq 0 \ (i \geq 1)$ and $\lambda_0 + \sum_{i=1}^{\infty} \lambda_i = 1$.

Proof. Let $\vartheta(o) = \lambda_0 \vartheta_0(o) + \sum_{i=1}^{\infty} \lambda_i \vartheta_i(o)$ with $\lambda_0 \geq 0, \lambda_i \geq 0 \ (i \geq 1)$ and $\lambda_0 + \sum_{i=1}^{\infty} \lambda_i = 1$.
Then

$$\vartheta(o) = \lambda_0 \vartheta_0(o) + \sum_{i=1}^{\infty} \lambda_i \vartheta_i(o)$$

$$= \frac{1}{o} + \sum_{i=1}^{\infty} \lambda_i \frac{(1-\hbar)}{(i\ell+2)^\tau[(1+\wp)(i\ell+1)+1-\hbar]} o^i.$$

Since

$$\sum_{i=1}^{\infty} \frac{(i\ell+2)^\tau[(1+\wp)(i\ell+1)+1-\hbar]}{(1-\hbar)} \lambda_i \frac{(1-\hbar)}{(i\ell+2)^\tau(1+\wp)(i\ell+1)+1-\hbar}$$

$$= \sum_{i=1}^{\infty} \lambda_i = 1 - \lambda_0 \leq 1.$$

By Theorem 15.2.1 is in category $\sigma_\xi^+(\hbar, \wp)$. On the other hand $\vartheta(o)$ is in $\sigma_\xi^+(\hbar, \wp)$.
Since

$$\varrho_i \leq \frac{(1-\hbar)}{(i\ell+2)^\tau[(1+\wp)(i+1)+1-\hbar]}, (i \geq 1)$$

$$\lambda_i = \frac{(i+2)^\tau[(1+\wp)(i+1)+1-\hbar]}{(1-\hbar)} \varrho_i,$$

and $\lambda_0 = 1 - \sum_{i=1}^{\infty} \lambda_i$, as a result of $\vartheta(o) = \lambda_0 \vartheta_0(o) + \sum_{i=1}^{\infty} \lambda_i \vartheta_i(o)$. This accomplishes the theorem's explanation. □

For the mapping $\vartheta(o) = \frac{1}{o} + \sum_{i=1}^{\infty} \varrho_i o^i$ and $g(o) = \frac{1}{o} + \sum_{i=1}^{\infty} b_i o^i$ belong to Σ_p, we indicate by $(\vartheta * g)(o)$, the convolution of $\vartheta(o)$ and $g(o)$ specified as

$$(\vartheta * g)(o) = \frac{1}{o} + \sum_{i=1}^{\infty} \varrho_i b_i o^i.$$

Theorem 15.5.3. *If the mapping* $\vartheta(o) = \frac{1}{o} + \sum_{\imath=1}^{\infty} \varrho_\imath o^\imath$ *and* $g(o) = \frac{1}{o} + \sum_{\imath=1}^{\infty} b_\imath o^\imath$ *are in the category* $\sigma_\xi^+(\hbar, \wp)$, *then*

$$(\vartheta * g)(o) = \frac{1}{o} + \sum_{\imath=1}^{\infty} \varrho_\imath b_\imath o^\imath$$

is in the category $\sigma_\xi^+(\hbar, \wp)$.

Proof. Suppose $\vartheta(o)$ and $g(o)$ are in $\sigma_\xi^+(\hbar, \wp)$. By Theorem 15.2.1, we have

$$\sum_{\imath=1}^{\infty} \frac{(\imath\ell + 2)^\tau[(1+\wp)(\imath\ell+1) + 1 - \hbar]}{(1-\hbar)} \varrho_\imath \leq 1$$

and

$$\sum_{\imath=1}^{\infty} \frac{(\imath\ell + 2)^\tau[(1+\wp)(\imath\ell+1) + 1 - \hbar]}{(1-\hbar)} b_\imath \leq 1$$

since $\vartheta(o)$ and $g(o)$ are regular are in E, so is $(\vartheta * g)(o)$. Additionally

$$\sum_{\imath=1}^{\infty} \frac{(\imath\ell + 2)^\tau[(1+\wp)(\imath\ell+1) + 1 - \hbar]}{(1-\hbar)} \varrho_\imath b_\imath$$

$$\leq \sum_{\imath=1}^{\infty} \left\{ \frac{(\imath\ell + 2)^\tau[(1+\wp)(\imath\ell+1) + 1 - \hbar]}{(1-\hbar)} \right\}^2 \varrho_\imath b_\imath$$

$$\leq \left(\sum_{\imath=1}^{\infty} \frac{(\imath\ell + 2)^\tau[(1+\wp)(\imath\ell+1) + 1 - \hbar]}{(1-\hbar)} \varrho_\imath \right)$$

$$\times \left(\sum_{\imath=1}^{\infty} \frac{(\imath\ell + 2)^\tau[(1+\wp)(\imath\ell+1) + 1 - \hbar]}{(1-\hbar)} b_\imath \right)$$

$$\leq 1.$$

Hence by Theorem 15.2.1, $(\vartheta * g)(o)$ is in $\sigma_\xi^+(\hbar, \wp)$. $\square$

15.6 Neighborhood property

We construct the neighborhood for the category $\sigma_\xi^+(\hbar, \wp)$, as regards:

Definition 15.6.1. *A mapping* $\vartheta \in \Sigma_p$ *is in* $\sigma_\xi^+(\hbar, \wp, \aleph)$ *if* $\exists$ *a mapping* $g \in \sigma_p(\hbar, \wp)$ *as well as*

$$\left| \frac{\vartheta(o)}{g(o)} - a \right| < 1 - \aleph, (o \in E, 0 \leq \aleph < 1). \tag{15.27}$$

Continuing on from past efforts on neighborhoods of analytic mappings by Goodman [4] and Ruscheweyh [16], we specify the ζ-neighborhood of a mapping $\vartheta \in \Sigma_p$ by

$$N_\zeta(\vartheta) := \left\{ g \in \Sigma_p : g(o) = \frac{1}{o} + \sum_{\imath=1}^{\infty} b_\imath o^\imath : \sum_{\imath=1}^{\infty} \imath|\varrho_\imath - b_\imath| \leq \zeta \right\}. \quad (15.28)$$

Theorem 15.6.2. *If* $g \in \sigma_p(\hbar, \wp)$ *and*

$$\aleph = 1 - \frac{\zeta(3 + 2\wp - \hbar)}{2 + 2\wp} \quad (15.29)$$

then $N_\zeta(g) \subset \sigma_p(\hbar, \wp, \aleph)$.

Proof. Let $\vartheta \in N_\zeta(g)$. Then from (15.28) that

$$\sum_{\imath=1}^{\infty} \imath|\varrho_\imath - b_\imath| \leq \zeta$$

which signifies the coefficient inequality

$$\sum_{\imath=1}^{\infty} |\varrho_\imath - b_\imath| \leq \zeta, (\imath \in N)$$

since $g \in \sigma_p(\hbar, \wp)$, we have

$$\sum_{\imath=1}^{\infty} b_\imath < \frac{(1 - \hbar)}{3 + 2\wp - \hbar}.$$

So that

$$\left| \frac{\vartheta(o)}{g(o)} - 1 \right| < \frac{\sum_{\imath=1}^{\infty} |\varrho_\imath - b_\imath|}{1 - \sum_{\imath=1}^{\infty} b_\imath} < \frac{\zeta(3 + 2\wp - \hbar)}{(2 + 2\wp)} = 1 - \aleph,$$

if $\aleph$ is supplied by (15.29).
Hence, by definition, $\vartheta \in \sigma_p(\hbar, \wp, \aleph)$ for $\aleph$ indicated by (15.29), this concludes the explanation. □

Bibliography

[1] Ahuja, O.P., Murugusundaramoorthy, G. and Magesh, N. 2008. Integral means for uniformly convex and starlike mappings associated with generalized hypergeometric mappings, *J. Inequal Pure. Appl. Math.* 8:4: Art.118, 9 pp.

[2] Bharathi, R., Parvatham, R. and Swaminathan, A. 1997. On subclass of uniformly convex mappings and corresponding class of starlike functions, *Tamkang J. Math.* 28:1: 17–32.

[3] Clunie, J. 1959. On meromorphic schlicht functions, *J. London Math. Soc.* 34: 215–216.

[4] Goodman, A.W. 1957. Univalent functions and non-analytic curves, *Proc. Amer. Math. Soc.* 8: 598–601.

[5] Goodman, A.W. 1991. On uniformly convex functions, *Ann. Pol. Math.* 56: 87–92.

[6] Goodman, A.W. 1991. On Uniformly starlike functions, *J. Math. Anal. Appl.* 155: 364–370.

[7] Juneja, O.P. and Reddy, T.R. 1985. Meromorphic starlike univalent functions with positive coefficients, *Ann. Univ. Maiiae Curie Sklodowska, Sect A* 39: 65–76.

[8] Kanas, S. and Wisniowska, A. 1999. Conic regions and k-uniformly convexity, *Comput. Appl. Math.* 105: 327–336.

[9] Kanas, S. and Wisniowska, A. 2000. Conic domains and starlike functions, *Rev. Roum. Math. Pures Appl.* 45: 647–657.

[10] Murugusundarmurthy, G. and Magesh, N. 2010. Certain subclass of starlike mappings of complex order involving generalized hypergeometric mappings, *Int. J. Math. Math.Sci.* Art.ID 178605, 12 pp.

[11] Pommerenke, Ch. 1963. On meromorphic starlike functions, *Pacfic J. Math.* 13: 221–235.

[12] Ronning, F. 1991. On starlike functions associated with parabolic regions, *Ann. Univ. Mariae Curie-Sklodowska Sect. A,* 45: 117–122.

[13] Ronning, F. 1993. Unitypely convex functions and a corresponding class of starlike mappings, *Proc. Amer. Math. Soc.* 118: 189–196.

[14] Ronning, F. 1995. Integral representations of bounded starlike functions, *Ann. Pol. Math.* 60:3: 298–297.

[15] Royster, W.C. 1963. Meromorphic starlike multivalent functions, *Trans. Amer. Math.Soc.* 107: 300–308.

[16] Ruscheweyh, St. 1981. Neighbourhoods of univalent functions, *Proc. Amer Math. Soc.* 81: 521–527.

[17] Silverman, H. 1975. Univalent functions with negative coefficients, *Proc. Amer. Math. Soc.* 51: 109–116.

[18] Venkateswarlu, B., Thirupathi Reddy, P. and Rani, N. 2019. Certain subclass of meromorphically uniformly convex functions with positive coefficients, *Mathematica (Cluj)* 61:84:1: 85–97 .

[19] Venkateswarlu, B., Thirupathi Reddy, P. and Rani, N. 2019. On new subclass of meromorphically convex functions with positive coefficients, *Surveys Math. Appl.* 14: 49–60.

[20] Venkateswarlu, B., Thirupathi Reddy, P., Meng, C. and Shilpa, R.M. 2020. A new subclass of meromorphic functions with positive coeficients defned by Bessel functions, *Note di Math.* 40:1: 13–25.

Chapter 16

Sharp coefficient bounds and solution of the Fekete-Szegö problem for a certain subclass of bi-univalent functions associated with the Chebyshev polynomials

Amol Bhausaheb Patil

16.1	Introduction	279
	16.1.1 Bi-univalent function	280
	16.1.2 Subordination	281
	16.1.3 Chebyshev polynomials	282
	16.1.4 The function class $\mathcal{CH}_\Sigma(\lambda, \mu, x)$	282
16.2	Coefficient estimates for the class $\mathcal{CH}_\Sigma(\lambda, \mu, x)$	284
	16.2.1 Some immediate consequences of the theorem	287
	Bibliography	288

16.1 Introduction

Geometric function theory (GFT) is a core branch of Complex Analysis and the study of univalent and multivalent functions is a fascinating branch of it. The theory of univalent functions is properly classified under the GFT, mainly due to the interconnection between the geometric behavior and the analytic characteristics of the function. The famous Bieberbach conjecture (1916) was the basic source of motivation for researchers to accelerate the development of this subject, which was finally settled positively by de Branges in 1985. The researchers Duren [7], Goodman [10,11], Nehari [17], etc., obtained a number of interesting results and open problems in line with the Bieberbach conjecture.

DOI: 10.1201/9781003330868-16

16.1.1 Bi-univalent function

In 1967, Lewin [15] extended the theory of univalent functions to bi-univalent functions and proved that the first Taylor-Maclaurin coefficient $|a_2| < 1.51$ for a function belongs to the bi-univalent function class Σ. Initially, Netanyahu [18], Jenson and Waadeland [12], Goodman [11], Brannan and Clunie [2] (also [3,24]), Styer and Wright [23], Kedzierawski [14], Tan [25] etc. offered a concrete basis for the study of bi-univalent function theory. The development in the field of bi-univalent functions has been accelerated considerably due to the seminal research paper of Srivastava et al. [22]. In recent years, the coefficient inequalities or bounds of functions of various subclasses of bi-univalent functions have been comprehensively studied by many researchers viz. [9,13,19–21]. However, the problem of estimating sharp Taylor-Maclaurin coefficients for the class Σ of bi-univalent functions remains unsolved.
Let $\mathbb{N} := \{1, 2, 3, \cdots\}$, $\mathbb{R} := (-\infty, \infty)$ and $\mathbb{C} := x+iy$, $(x, y \in \mathbb{R}, i = \sqrt{-1})$ be the set of natural numbers, real numbers, and complex numbers, respectively. Let $\mathcal{A}$ denote the class of functions analytic in the open unit disk:

$$\mathfrak{D} := \{z : z \in \mathbb{C}, |z| < 1\},$$

which are normalized by $f(0) = 0$ and $f'(0) = 1$ and have the form

$$f(z) = z + \sum_{n=2}^{\infty} a_n z^n, \quad (n \in \mathbb{N}, a_n \in \mathbb{C}, z \in \mathfrak{D}). \tag{16.1}$$

Also we denote by $\mathcal{S}$ the subclass of $\mathcal{A}$ containing functions which are univalent in $\mathfrak{D}$. From the Koebe One-Quarter Theorem (see [7]), it is clear that every function $f \in \mathcal{S}$ given by 16.1 has an inverse f^{-1} such that $f^{-1}(f(z)) = z$, $(z \in \mathfrak{D})$ and

$$f(f^{-1}(w)) = w, \ (|w| < r_0(f), r_0(f) \geq 1/4),$$

which has a series expansion (Taylor-Maclaurin series) of the form

$$f^{-1}(w) = g(w) = w + b_2 w^2 + b_3 w^3 + \cdots \tag{16.2}$$

and hence we have

$$w = f(g(w)) = f\left(w + b_2 w^2 + b_3 w^3 + \cdots\right)$$

which, by using equation (16.1) yields

$$w = \left(w + b_2 w^2 + b_3 w^3 + ..\right) + a_2 \left(w + b_2 w^2 + ..\right)^2 + a_3 \left(w + b_2 w^2 + ..\right)^3 ...$$

After further simplifications, it becomes

$$w = w + (a_2 + b_2)w^2 + (a_3 + b_3 + 2a_2 b_2)w^3 + (a_4 + b_4 + a_2 b_2^2 + 2a_2 b_3 + 3a_3 b_2)w^4 + ...$$

which, by comparing the initial coefficients gives

$$b_2 = -a_2, \ b_3 = 2a_2^2 - a_3, \ b_4 = 5a_2 a_3 - 5a_2^3 - a_4.$$

Substituting these values in equation (16.2), we get

$$g(w) = w + (-a_2)w^2 + (2a_2^2 - a_3)w^3 + (5a_2a_3 - 5a_2^3 - a_4)w^4 + \cdots. \quad (16.3)$$

Definition 16.1.1. *If a function $f \in \mathcal{S}$ and its inverse f^{-1} both are analytic and univalent in $\mathfrak{D}$, then the function f is said to be bi-univalent in $\mathfrak{D}$.*

If $f(z) \in \mathcal{S}$ is given by (16.1), then $f'(0) = 1$ and thus, according to the Schwarz's lemma, either $f(z) \equiv z$ or there are two points z_1 and z_2 such that $|f(z_1)| < 1$ and $|f(z_2)| > 1$. The figure below depicts an image of the open unit disc $\mathfrak{D}$ under such a function f (for more information, see [11]). Suppose

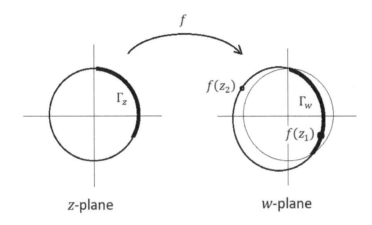

FIGURE 16.1: A function $f : \mathfrak{D} \to \mathbb{C}$ such that $f \in \mathcal{S}$ and $f'(0) = 1$

that $\Gamma_w = f(\Gamma_z)$, where Γ_w be the boundary arc of $f(\mathfrak{D})$ that lies inside $\mathfrak{D}$ in the w-plane and Γ_z is a corresponding arc on $|z| = 1$. If the function $f(z)$ is bi-univalent, it must be analytic on and across Γ_z in order for $f^{-1}(w)$ to be defined and analytic in the unit disc of the w-plane.

Let Σ represents the class of all functions that are bi-univalent. For example,

$$f_1(z) = -\log(1-z), \quad f_2(z) = \frac{z}{1-z}, \quad f_3(z) = \frac{1}{2}\log\left(\frac{1+z}{1-z}\right) \in \Sigma$$

whereas,

$$f_4(z) = \frac{z}{1-z^2}, \quad f_5(z) = z - \frac{z^2}{2}, \quad f_6(z) = \frac{z}{(1-z)^2} \notin \Sigma.$$

16.1.2 Subordination

Let the functions α and β be the members of the class $\mathcal{A}$. Then α is said to be subordinate to β in $\mathfrak{D}$, written as

$$\alpha(z) \prec \beta(z) \quad (z \in \mathfrak{D}),$$

If there is an analytic function γ in $\mathfrak{D}$, such that:

$$\gamma(0) = 0, \quad |\gamma(z)| < 1 \quad (z \in \mathfrak{D})$$

and

$$\alpha(z) = \beta(\gamma(z)) \quad (z \in \mathfrak{D}).$$

Moreover, if the function $\beta \in \mathcal{S}$, then we have:

$$\alpha(z) \prec \beta(z), \; (z \in \mathfrak{D}) \quad \text{if and only if} \quad \alpha(\mathfrak{D}) \subset \beta(\mathfrak{D}) \quad \text{and} \quad \alpha(0) = \beta(0).$$

16.1.3 Chebyshev polynomials

We know the importance of the Chebyshev polynomials due to their significant role in numerical analysis and the theory of approximations. In all, four kinds of the Chebyshev polynomials are there (see [1, 6, 8, 16], etc. for more details on Chebyshev polynomials). Of these, several researchers deal with the orthogonal polynomials due to their numerous applications. For a real variable $x = \cos\theta$ in the interval $[-1, 1]$ the degree n Chebyshev polynomials of the first and second kind are defined by

$$T_n(x) = \cos(n\theta) \text{ and } U_n(x) = \frac{\sin(n+1)\theta}{\sin\theta},$$

respectively.

16.1.4 The function class $\mathcal{CH}_\Sigma(\lambda, \mu, x)$

Now, before we define the function class $\mathcal{CH}_\Sigma(\lambda, \mu, x)$, we mention here that the expressions used in the definition of $\mathcal{CH}_\Sigma(\lambda, \mu, x)$ were used earlier by Bulut and Magesh [5], Patil and Shaba [19], Zhu [27], etc.

Definition 16.1.2. *A function $f(z) \in \mathcal{A}$ given by (16.1) is said to be in the class $\mathcal{CH}_\Sigma(\lambda, \mu, x)$ if it satisfies the following subordination conditions:*

$$(1-\lambda)\left(\frac{f(z)}{z}\right)^\mu + \lambda f'(z)\left(\frac{f(z)}{z}\right)^{\mu-1} \prec \mathcal{H}(z, x) := \frac{1}{1 - 2xz + z^2}$$

and

$$(1-\lambda)\left(\frac{g(w)}{w}\right)^\mu + \lambda g'(w)\left(\frac{g(w)}{w}\right)^{\mu-1} \prec \mathcal{H}(w, x) := \frac{1}{1 - 2xw + w^2},$$

where $z, w \in \mathfrak{D}; \lambda \geq 1; 0 \leq \mu \leq 1; x \in (1/2, 1]$ and g given by (16.3) be an extension of f^{-1} to $\mathfrak{D}$.

For $\mu = 1$, it reduces to the class $\mathcal{CH}_\Sigma(\lambda, x) \equiv \mathcal{CH}_\Sigma(\lambda, 1, x)$ containing the functions that satisfies

$$(1-\lambda)\frac{f(z)}{z} + \lambda f'(z) \prec \mathcal{H}(z, x)$$

and

$$(1-\lambda)\frac{g(w)}{w} + \lambda g'(w) \prec \mathcal{H}(w, x),$$

whereas for $\lambda = 1$, it reduces to the class $\mathcal{CH}_\Sigma(\mu, x) \equiv \mathcal{CH}_\Sigma(1, \mu, x)$ containing the functions that satisfies

$$f'(z)\left(\frac{f(z)}{z}\right)^{\mu-1} \prec \mathcal{H}(z, x)$$

and

$$g'(w)\left(\frac{g(w)}{w}\right)^{\mu-1} \prec \mathcal{H}(w, x)$$

and for $\mu = 1$ and $\lambda = 1$, it reduces to the class $\mathcal{CH}_\Sigma(x)$ containing the functions that satisfies

$$f'(z) \prec \mathcal{H}(z, x) \quad \text{and} \quad g'(w) \prec \mathcal{H}(w, x).$$

For $x = \cos\theta$ if we set $\theta \in (-\pi/3, \pi/3)$, that is, $x \in (\frac{1}{2}, 1]$, then we have

$$\mathcal{H}(z, x) = \frac{1}{1 - (2\cos\theta)z + z^2} = 1 + \sum_{n=1}^{\infty} \frac{\sin(n+1)\theta}{\sin\theta} z^n \quad (z \in \mathfrak{D}).$$

Thus

$$\mathcal{H}(z, x) = 1 + (2\cos\theta)z + \left(3\cos^2\theta - \sin^2\theta\right)z^2 + \cdots \quad (z \in \mathfrak{D}).$$

Also from [26], for $x \in (\frac{1}{2}, 1]$ we can write

$$\mathcal{H}(z, x) = 1 + U_1(x)z + U_2(x)z^2 + \cdots \quad (z \in \mathfrak{D}),$$

where $U_n(x)$ be the Chebyshev polynomials of the second kind given by

$$U_{n-1} = \frac{\sin(n\cos^{-1}x)}{\sqrt{1-x^2}} \quad (n \in \mathbb{N}),$$

with the recurrence relation

$$U_1(x) = 2x, \quad U_2(x) = 4x^2 - 1, \quad U_n(x) = 2x\,U_{n-1}(x) - U_{n-2}(x).$$

16.2 Coefficient estimates for the class $\mathcal{CH}_\Sigma(\lambda, \mu, x)$

In this section, we consider the subclass $\mathcal{CH}_\Sigma(\lambda, \mu, x)$ of the bi-univalent function class Σ involving the Chebyshev polynomials and obtain sharp bounds for the coefficients a_2, a_3 and the Fekete-Szegö functional $a_3 - \delta a_2^2$ for the functions belong to it. Moreover, some immediate consequences are mentioned as Corollaries.

Theorem 16.2.1. *For $\lambda \geq 1, 0 \leq \mu \leq 1, x \in (\frac{1}{2}, 1]$ and $\delta \in \mathbb{R}$, let a function $f(z) \in \mathcal{A}$ given by (16.1) be in the class $\mathcal{CH}_\Sigma(\lambda, \mu, x)$. Then*

$$|a_2| \leq \begin{cases} \sqrt{\frac{4x}{(2\lambda+\mu)(\mu+1)}}, & \frac{1}{2} < x \leq \frac{1+\sqrt{5}}{4} \\ \sqrt{\frac{2(4x^2-1)}{(2\lambda+\mu)(\mu+1)}}, & \frac{1+\sqrt{5}}{4} \leq x \leq 1, \end{cases} \qquad (16.4)$$

$$|a_3| \leq \begin{cases} \frac{4x}{(2\lambda+\mu)(\mu+1)}, & \frac{1}{2} < x \leq \frac{1+\sqrt{5}}{4} \\ \frac{2(4x^2-1)}{(2\lambda+\mu)(\mu+1)}, & \frac{1+\sqrt{5}}{4} \leq x \leq 1, \end{cases} \qquad (16.5)$$

$$|a_3 - \delta a_2^2| \leq \begin{cases} |1-\delta|\frac{4x}{(2\lambda+\mu)(\mu+1)}, & \frac{1}{2} < x \leq \frac{1+\sqrt{5}}{4} \\ |1-\delta|\frac{2(4x^2-1)}{(2\lambda+\mu)(\mu+1)}, & \frac{1+\sqrt{5}}{4} \leq x \leq 1. \end{cases} \qquad (16.6)$$

All of these inequalities are sharp.

Proof. Let $f \in \mathcal{CH}_\Sigma(\lambda, \mu, x)$. From Definition 16.1.2, for $z, w \in \mathfrak{D}$ we have

$$(1-\lambda)\left(\frac{f(z)}{z}\right)^\mu + \lambda f'(z)\left(\frac{f(z)}{z}\right)^{\mu-1}$$
$$= 1 + U_1(x)p(z) + U_2(x)p^2(z) + \cdots \qquad (16.7)$$

and

$$(1-\lambda)\left(\frac{g(w)}{w}\right)^\mu + \lambda g'(w)\left(\frac{g(w)}{w}\right)^{\mu-1}$$
$$= 1 + U_1(x)q(w) + U_2(x)q^2(w) + \cdots, \qquad (16.8)$$

for some analytic functions

$$p(z) = c_1 z + c_2 z^2 + c_3 z^3 + \cdots \quad (z \in \mathfrak{D}) \qquad (16.9)$$

and

$$q(w) = d_1 w + d_2 w^2 + d_3 w^3 + \cdots \quad (w \in \mathfrak{D}), \qquad (16.10)$$

such that $p(0) = 0$, $|p(z)| < 1$, $q(0) = 0$ and $|q(w)| < 1$. For such functions $p(z)$ and $q(w)$ (see Nehari [17]), we have

$$|c_1| \leq 1, \quad |d_1| \leq 1, \quad |c_2| \leq 1 - |c_1|^2, \quad |d_2| \leq 1 - |d_1|^2. \tag{16.11}$$

Using equations (16.9) and (16.10), for $z, w \in \mathfrak{D}$ we can write

$$(1-\lambda)\left(\frac{f(z)}{z}\right)^\mu + \lambda f'(z)\left(\frac{f(z)}{z}\right)^{\mu-1}$$
$$= 1 + U_1(x)c_1 z + \left[U_1(x)c_2 + U_2(x)c_1^2\right] z^2 + \cdots \tag{16.12}$$

and

$$(1-\lambda)\left(\frac{g(w)}{w}\right)^\mu + \lambda g'(w)\left(\frac{g(w)}{w}\right)^{\mu-1}$$
$$= 1 + U_1(x)d_1 w + \left[U_1(x)d_2 + U_2(x)d_1^2\right] w^2 + \cdots. \tag{16.13}$$

Equating the coefficients in (16.12) and (16.13), we get

$$(\lambda + \mu) a_2 = U_1(x) c_1, \tag{16.14}$$

$$(2\lambda + \mu)\left(a_3 + \frac{\mu - 1}{2} a_2^2\right) = U_1(x) c_2 + U_2(x) c_1^2, \tag{16.15}$$

$$-(\lambda + \mu) a_2 = U_1(x) d_1, \tag{16.16}$$

$$(2\lambda + \mu)\left(\frac{3+\mu}{2} a_2^2 - a_3\right) = U_1(x) d_2 + U_2(x) d_1^2. \tag{16.17}$$

Equation (16.14) and (16.16) gives

$$c_1 = -d_1. \tag{16.18}$$

Adding (16.15) and (16.17) we get

$$(2\lambda + \mu)(\mu + 1) a_2^2 = U_1(x)(c_2 + d_2) + U_2(x)\left(c_1^2 + d_1^2\right),$$

which, on using (16.11) and (16.18), yields

$$|a_2|^2 \leq \frac{2}{(2\lambda + \mu)(\mu + 1)} \left[U_1(x) + (U_2(x) - U_1(x))|c_1|^2\right].$$

This, according to the cases when the term $(U_2(x) - U_1(x))$ is positive or negative along with $|c_1| \leq 1$ gives us

$$|a_2|^2 \leq \begin{cases} \frac{2 U_1(x)}{(2\lambda+\mu)(\mu+1)}, & U_1(x) \geq U_2(x) \\ \frac{2 U_2(x)}{(2\lambda+\mu)(\mu+1)}, & U_1(x) \leq U_2(x), \end{cases}$$

which proves the inequality (16.4) by putting the values of $U_1(x)$ and $U_2(x)$ with $x \in (\frac{1}{2}, 1]$. Now, for the estimate on a_3, we subtract (16.17) from (16.15) to get

$$2(2\lambda + \mu)(a_3 - a_2^2) = U_1(x)(c_2 - d_2) + U_2(x)(c_1^2 - d_1^2),$$

or equivalently

$$a_3 = a_2^2 + \frac{U_1(x)(c_2 - d_2)}{2(2\lambda + \mu)} \qquad (16.19)$$
$$= \frac{2\left[U_1(x)(c_2 + d_2) + U_2(x)(c_1^2 + d_1^2)\right] + U_1(x)(\mu + 1)(c_2 - d_2)}{2(2\lambda + \mu)(\mu + 1)}$$
$$= \frac{[(\mu + 3)c_2 + (1 - \mu)d_2]U_1(x) + 4c_1^2 U_2(x)}{2(2\lambda + \mu)(\mu + 1)}.$$

This, on using (16.11) and (16.18), gives

$$|a_3| \leq \frac{2}{(2\lambda + \mu)(\mu + 1)}\left[U_1(x) + (U_2(x) - U_1(x))|c_1|^2\right],$$

which then proves the inequality (16.5) similarly as in case of a_2. Finally, for the estimate on the Fekete-Szegö functional, from equation (16.19) we have

$$a_3 - \delta a_2^2 = (1 - \delta) a_2^2 + \frac{U_1(x)(c_2 - d_2)}{2(2\lambda + \mu)}$$
$$= (1 - \delta)\frac{U_1(x)(c_2 + d_2) + 2U_2(x)c_1^2}{(2\lambda + \mu)(\mu + 1)} + \frac{U_1(x)(c_2 - d_2)}{2(2\lambda + \mu)}$$
$$= \frac{U_1(x)}{(2\lambda + \mu)}\left[\frac{(1 - \delta)(c_2 + d_2)}{(\mu + 1)} + \frac{2c_1^2(1 - \delta)U_2(x)}{(\mu + 1)U_1(x)} + \frac{(c_2 - d_2)}{2}\right]$$
$$= \frac{U_1(x)}{(2\lambda + \mu)}\left[\frac{(3 - 2\delta + \mu)c_2 + (1 - 2\delta - \mu)d_2}{2(\mu + 1)} + \frac{2c_1^2(1 - \delta)U_2(x)}{(\mu + 1)U_1(x)}\right].$$

This, on using (16.11) and (16.18), gives

$$|a_3 - \delta a_2^2| \leq \frac{U_1(x)}{(2\lambda + \mu)}\left[\frac{2|1 - \delta|(1 - |c_1|^2)}{(\mu + 1)} + \frac{2|c_1|^2|1 - \delta|U_2(x)}{(\mu + 1)U_1(x)}\right]$$
$$= \frac{2|1 - \delta|}{(2\lambda + \mu)(\mu + 1)}\left[(1 - |c_1|^2)U_1(x) + |c_1|^2 U_2(x)\right]$$
$$= \frac{2|1 - \delta|}{(2\lambda + \mu)(\mu + 1)}\left[U_1(x) + (U_2(x) - U_1(x))|c_1|^2\right],$$

which proves the inequality (16.6) similarly as in case of a_2 and a_3. Also, on putting $\delta = 0$, we get the bound on a_3 and by putting $\delta = 1$, we get the result $\left|a_3 - a_2^2\right| = 0$ or $|a_3| = \left|a_2^2\right|$.

The equations (16.7) and (16.8) with $p(z) = z$ and $q(w) = -w$ generates the bi-univalent function $f_1 \in \mathcal{CH}_\Sigma(\lambda, \mu, x)$ given by

$$f_1(z) = z + \left(\frac{2U_2(x)}{(2\lambda + \mu)(\mu + 1)}\right)^{\frac{1}{2}} z^2 + \left(\frac{2U_2(x)}{(2\lambda + \mu)(\mu + 1)}\right) z^3 + \cdots,$$

which shows the sharpness in the second part of all the three inequalities. On the other hand, the equations (16.7) and (16.8) with $p(z) = z^2$ and $q(w) = w^2$ generates the bi-univalent function $f_2 \in \mathcal{CH}_\Sigma(\lambda, \mu, x)$ given by

$$f_2(z) = z + \left(\frac{2U_1(x)}{(2\lambda + \mu)(\mu + 1)}\right)^{\frac{1}{2}} z^2 + \left(\frac{2U_1(x)}{(2\lambda + \mu)(\mu + 1)}\right) z^3 + \cdots,$$

which shows the sharpness in the first part of all the three inequalities. $\square$

16.2.1 Some immediate consequences of the theorem

For $\mu = 1$, we get the following consequence of Theorem 16.2.1.

Corollary 16.2.2. *For $\lambda \geq 1, x \in (\frac{1}{2}, 1]$ and $\delta \in \mathbb{R}$, let a function $f \in \mathcal{A}$ given by (16.1) be in the class $\mathcal{CH}_\Sigma(\lambda, x)$. Then*

$$|a_2| \leq \begin{cases} \sqrt{\frac{2x}{2\lambda+1}}, & \frac{1}{2} < x \leq \frac{1+\sqrt{5}}{4} \\ \sqrt{\frac{4x^2-1}{2\lambda+1}}, & \frac{1+\sqrt{5}}{4} \leq x \leq 1, \end{cases}$$

$$|a_3| \leq \begin{cases} \frac{2x}{2\lambda+1}, & \frac{1}{2} < x \leq \frac{1+\sqrt{5}}{4} \\ \frac{4x^2-1}{2\lambda+1}, & \frac{1+\sqrt{5}}{4} \leq x \leq 1, \end{cases}$$

$$|a_3 - \delta a_2^2| \leq \begin{cases} |1-\delta|\left(\frac{2x}{2\lambda+1}\right), & \frac{1}{2} < x \leq \frac{1+\sqrt{5}}{4} \\ |1-\delta|\left(\frac{4x^2-1}{2\lambda+1}\right), & \frac{1+\sqrt{5}}{4} \leq x \leq 1. \end{cases}$$

All of these inequalities are sharp.

For $\lambda = 1$, we get the following consequence of Theorem 16.2.1.

Corollary 16.2.3. *For $0 \leq \mu \leq 1, x \in (\frac{1}{2}, 1]$ and $\delta \in \mathbb{R}$, let a function $f \in \mathcal{A}$ given by (16.1) be in the class $\mathcal{CH}_\Sigma(\mu, x)$. Then*

$$|a_2| \leq \begin{cases} \sqrt{\frac{4x}{(\mu+1)(\mu+2)}}, & \frac{1}{2} < x \leq \frac{1+\sqrt{5}}{4} \\ \sqrt{\frac{2(4x^2-1)}{(\mu+1)(\mu+2)}}, & \frac{1+\sqrt{5}}{4} \leq x \leq 1, \end{cases}$$

$$|a_3| \leq \begin{cases} \frac{4x}{(\mu+1)(\mu+2)}, & \frac{1}{2} < x \leq \frac{1+\sqrt{5}}{4} \\ \frac{2(4x^2-1)}{(\mu+1)(\mu+2)}, & \frac{1+\sqrt{5}}{4} \leq x \leq 1, \end{cases}$$

$$|a_3 - \delta a_2^2| \leq \begin{cases} |1-\delta|\frac{4x}{(\mu+1)(\mu+2))}, & \frac{1}{2} < x \leq \frac{1+\sqrt{5}}{4} \\ |1-\delta|\frac{2(4x^2-1)}{(\mu+1)(\mu+2)}, & \frac{1+\sqrt{5}}{4} \leq x \leq 1. \end{cases}$$

All of these inequalities are sharp.

For $\lambda = 1$ and $\mu = 1$, Theorem 16.2.1 gives us the following consequence.

Corollary 16.2.4. *For $x \in (\frac{1}{2}, 1]$ and $\delta \in \mathbb{R}$, let a function $f \in \mathcal{A}$ given by (16.1) be in the class $\mathcal{CH}_\Sigma(x)$. Then*

$$|a_2| \leq \begin{cases} \sqrt{\frac{2x}{3}}, & \frac{1}{2} < x \leq \frac{1+\sqrt{5}}{4} \\ \sqrt{\frac{4x^2-1}{3}}, & \frac{1+\sqrt{5}}{4} \leq x \leq 1, \end{cases}$$

$$|a_3| \leq \begin{cases} \frac{2x}{3}, & \frac{1}{2} < x \leq \frac{1+\sqrt{5}}{4} \\ \frac{4x^2-1}{3}, & \frac{1+\sqrt{5}}{4} \leq x \leq 1, \end{cases}$$

$$|a_3 - \delta a_2^2| \leq \begin{cases} \frac{2x}{3}|1-\delta|, & \frac{1}{2} < x \leq \frac{1+\sqrt{5}}{4} \\ \frac{4x^2-1}{3}|1-\delta|, & \frac{1+\sqrt{5}}{4} \leq x \leq 1. \end{cases}$$

All of these inequalities are sharp.

Bibliography

[1] Altinkaya, Ş. and Yalçin, S. 2016. On the Chebyshev polynomial bounds for classes of univalent functions, *Khayyam J. Math.* 2:1: 1–5.

[2] Brannan, D.A. and Clunie, J.G. 1980. *Aspects of Contemporary Complex Analysis*, Academic Press, London.

[3] Brannan, D.A., Clunie, J.G. and Kirwan, W.E. 1970. Coefficient Estimates for a Class of Star-Like Functions, *Canad. J. Math.* 22: 476–485.

[4] Brannan, D.A. and Taha, T.S. 1986. On some classes of bi-univalent functions, *Stud. Univ. Babeş-Bolyai Math.* 31:2: 70–77.

[5] Bulut, S. and Magesh, N. 2016. On the sharp bounds for a comprehensive class of analytic and univalent functions by means of Chebyshev polynomials, *Khayyam J. Math.* 2:2: 194–200.

[6] Doha, E.H. 1994. The first and second kind Chebyshev coefficients of the moments of the general order derivative of an infinitely differentiable function, *Int. J. Comput. Math.* 51: 21–35.

[7] Duren, P.L. 1983. *Univalent Functions*, Grundlehren der Mathematischen Wissenschaften, Springer, New York.

[8] Dziok, J., Raina, R.K. and Sokól, J. 2015. Application of Chebyshev polynomials to classes of analytic functions, *C. R. Math. Acad. Sci. Paris* 353:5: 433—438.

[9] Frasin, B.A. and Aouf, M.K. 2011. New subclasses of bi-univalent functions, *Appl. Math. Letters* 24: 1569–1573.

[10] Goodman, A.W. 1983. *Univalent Functions*, Marinee Publ. Co., Inc., Tampa, FL., vol. I., xvii+246 pp. ISBN:0-936166-10-x.

[11] Goodman, A.W. 1979. An invitation to the study of univalent and multivalent functions, *Int. J. Math. Math. Sci.* 2: 163–186.

[12] Jensen, E. and Waadeland, H-. 1972. A coefficient inequality for biunivalent functions, *Skrifter Norske Vid. Selskab (Trondheim)* 15: 1–11.

[13] Joshi, S., Joshi, S. and Pawar, H. 2016. On some subclasses of bi-univalent functions associated with pseudo-starlike functions, *J. Egyptian Math. Soc.* 24: 522–525.

[14] Kedzierawski, A.W. 1985. Some remarks on bi-univalent functions, *Ann. Univ. Mariae Curie-Sklodowska Lublin-Polonia, XXXIX* 10: 77–81.

[15] Lewin, M. 1967. On a coefficient problem for bi-univalent functions, *Proc. Amer. Math. Soc.* 18: 63–68.

[16] Mason, J.C. 1967. Chebyshev polynomial approximations for the L-membrane eigenvalue problem, *SIAM J. Appl. Math.* 15: 172–186.

[17] Nehari, Z. 1953. *Conformal Mapping*, McGraw-Hill Book Co., New York.

[18] Netanyahu, E. 1969. The minimal distance of the image boundary from the origin and the second coefficient of a univalent function in $|z| < 1$, *Arch. Rational Mech. Anal.* 32: 100–112.

[19] Patil, A.B. and Naik, U.H. 2018. Bounds on initial coefficients for a new subclass of bi-univalent functions, *New Trends Math. Sci.* 6:1: 85–90.

[20] Porwal, S. and Darus, M. 2013. On a new subclass of bi-univalent functions, *J. Egyptian Math. Soc.* 21:3: 190–193.

[21] Srivastava, H.M. and Bansal, D. 2015. Coefficient estimates for a subclass of analytic and bi-univalent functions, *J. Egyptian Math. Soc.* 23:2: 242–246.

[22] Srivastava, H.M., Mishra, A.K. and Gochhayat, P. 2010. Certain subclasses of analytic and bi-univalent functions, *Appl. Math. Letters* 23: 1188–1192.

[23] Styer, D. and Wright, D.J. 1981. Results on bi-univalent functions, *Proc. Amer. Math. Soc.* 82:2: 243–248.

[24] Taha, T.S. *Topics in Univalent Function Theory*, (1981) Ph.D. Thesis, University of London.

[25] Tan, D.-L. 1984. Coefficient estimates for bi-univalent functions, *Chinese Ann. Math. Ser. A*, 5: 559–568.

[26] Whittaker, T. and Watson, G.N. *A Course of Modern Analysis* (1996) Reprint of the fourth (1927) edition, Cambridge Mathematical Library, Cambridge Univ. Press, Cambridge.

[27] Zhu, Y. 2007. Some starlikeness criteria for analytic functions, *J. Math. Anal. Appl.* 335: 1452--1459.

Chapter 17

Some differential sandwich theorems involving a multiplier transformation and Ruscheweyh derivative

Alb Lupaş Alina

17.1 Differential subordination and superordination 291
17.2 Strong differential subordination and superordination 300
      Bibliography 308

17.1 Differential subordination and superordination

Let $\mathcal{H}(U)$ be the class of analytic function in the open unit disc of the complex plane $U = \{z \in \mathbb{C} : |z| < 1\}$. Let $\mathcal{H}(a,n)$ be the subclass of $\mathcal{H}(U)$ consisting of functions of the form $f(z) = a + a_n z^n + a_{n+1} z^{n+1} + \ldots$.

Let $\mathcal{A}_n = \{f \in \mathcal{H}(U) : f(z) = z + a_{n+1} z^{n+1} + \ldots, z \in U\}$ and $\mathcal{A} = \mathcal{A}_1$.

Let the functions f and g be analytic in U. We say that the function f is subordinate to g, written $f \prec g$, if there exists a Schwartz function w, analytic in U, with $w(0) = 0$ and $|w(z)| < 1$, for all $z \in U$, such that $f(z) = g(w(z))$, for all $z \in U$. In particular, if the function g is univalent in U, the above subordination is equivalent to $f(0) = g(0)$ and $f(U) \subset g(U)$.

Let $\psi : \mathbb{C}^3 \times U \to \mathbb{C}$ and h be an univalent function in U. If p is analytic in U and satisfies the second order differential subordination

$$\psi(p(z), zp'(z), z^2 p''(z); z) \prec h(z), \quad z \in U, \tag{17.1}$$

then p is called a solution of the differential subordination. The univalent function q is called a dominant of the solutions of the differential subordination, or more simply a dominant, if $p \prec q$ for all p satisfying (17.1). A dominant $\widetilde{q}$ that satisfies $\widetilde{q} \prec q$ for all dominants q of (17.1) is said to be the best dominant of (17.1). The best dominant is unique up to a rotation of U.

Let $\psi : \mathbb{C}^2 \times U \to \mathbb{C}$ and h analytic in U. If p and $\psi(p(z), zp'(z), z^2 p''(z); z)$ are univalent and if p satisfies the second order

DOI: 10.1201/9781003330868-17

differential superordination

$$h(z) \prec \psi(p(z), zp'(z), z^2 p''(z); z), \quad z \in U, \tag{17.2}$$

then p is a solution of the differential superordination (17.2) (if f is subordinate to F, then F is called to be superordinate to f). An analytic function q is called a subordinant if $q \prec p$ for all p satisfying (17.2). An univalent subordinant $\tilde{q}$ that satisfies $q \prec \tilde{q}$ for all subordinates q of (17.2) is said to be the best subordinat.

Miller and Mocanu [17] obtained conditions h, q, and ψ for which the following implication holds

$$h(z) \prec \psi(p(z), zp'(z), z^2 p''(z); z) \Rightarrow q(z) \prec p(z).$$

For two functions $f(z) = z + \sum_{j=2}^{\infty} a_j z^j$ and $g(z) = z + \sum_{j=2}^{\infty} b_j z^j$ analytic in the open unit disc U, the Hadamard product (or convolution) of $f(z)$ and $g(z)$, written as $(f * g)(z)$ is defined by

$$f(z) * g(z) = (f * g)(z) = z + \sum_{j=2}^{\infty} a_j b_j z^j.$$

Definition 17.1.1. *[16] For $f \in \mathcal{A}$, $m \in \mathbb{N} \cup \{0\}$, $\lambda, l \geq 0$, the multiplier transformation $I(m, \lambda, l) f(z)$ is defined by the following infinite series*

$$I(m, \lambda, l) f(z) := z + \sum_{j=2}^{\infty} \left(\frac{1 + \lambda(j-1) + l}{1 + l} \right)^m a_j z^j.$$

Remark 17.1.2. *We have*

$$(l+1) I(m+1, \lambda, l) f(z) = (l+1-\lambda) I(m, \lambda, l) f(z) + \lambda z (I(m, \lambda, l) f(z))',$$

$z \in U$.

Remark 17.1.3. *For $l = 0$, $\lambda \geq 0$, the operator $D_\lambda^m = I(m, \lambda, 0)$ was introduced and studied by Al-Oboudi, which reduced to the Sălăgean differential operator $S^m = I(m, 1, 0)$ for $\lambda = 1$.*

Definition 17.1.4. *(Ruscheweyh [23]) For $f \in \mathcal{A}$ and $n \in \mathbb{N}$, the Ruscheweyh derivative R^n is defined by $R^n : \mathcal{A} \to \mathcal{A}$,*

$$\begin{aligned} R^0 f(z) &= f(z) \\ R^1 f(z) &= zf'(z) \\ &\cdots \\ (n+1) R^{n+1} f(z) &= z(R^n f(z))' + n R^n f(z), \quad z \in U. \end{aligned}$$

Remark 17.1.5. *If $f \in \mathcal{A}$, $f(z) = z + \sum_{j=2}^{\infty} a_j z^j$, then $R^n f(z) = z + \sum_{j=2}^{\infty} \frac{(n+j-1)!}{n!(j-1)!} a_j z^j$ for $z \in U$.*

Definition 17.1.6. ([2]) Let $\lambda, l \geq 0$ and $n, m \in \mathbb{N}$. Denote by $IR_{\lambda,l}^{m,n} : \mathcal{A} \to \mathcal{A}$ the operator given by the Hadamard product of the multiplier transformation $I(m, \lambda, l)$ and the Ruscheweyh derivative R^n,

$$IR_{\lambda,l}^{m,n} f(z) = (I(m, \lambda, l) * R^n) f(z),$$

for any $z \in U$ and each nonnegative integers m, n.

Remark 17.1.7. If $f \in \mathcal{A}$ and $f(z) = z + \sum_{j=2}^{\infty} a_j z^j$, then
$IR_{\lambda,l}^{m,n} f(z) = z + \sum_{j=2}^{\infty} \left(\frac{1+\lambda(j-1)+l}{l+1} \right)^m \frac{(n+j-1)!}{n!(j-1)!} a_j^2 z^j$, $z \in U$.

Using simple computation one obtains the next result.

Proposition 17.1.8. [1] For $m, n \in \mathbb{N}$ and $\lambda, l \geq 0$ we have

$$(n+1) IR_{\lambda,l}^{m,n+1} f(z) - n IR_{\lambda,l}^{m,n} f(z) = z \left(IR_{\lambda,l}^{m,n} f(z) \right)'. \tag{17.3}$$

The purpose of this paper is to derive the several subordination and superordination results involving a differential operator. Furthermore, we studied the results of Selvaraj and Karthikeyan [25], Shanmugam, Ramachandran, Darus and Sivasubramanian [26] and Srivastava and Lashin [27].
In order to prove our subordination and superordination results, we make use of the following known results.

Definition 17.1.9. [18] Denote by Q the set of all functions f that are analytic and injective on $\overline{U} \setminus E(f)$, where $E(f) = \{\zeta \in \partial U : \lim_{z \to \zeta} f(z) = \infty\}$, and are such that $f'(\zeta) \neq 0$ for $\zeta \in \partial U \setminus E(f)$.

Lemma 17.1.10. [18] Let the function q be univalent in the unit disc U and θ and ϕ be analytic in a domain D containing $q(U)$ with $\phi(w) \neq 0$ when $w \in q(U)$. Set $Q(z) = zq'(z) \phi(q(z))$ and $h(z) = \theta(q(z)) + Q(z)$. Suppose that

(1) Q is starlike univalent in U and

(2) $\operatorname{Re}\left(\frac{zh'(z)}{Q(z)}\right) > 0$ for $z \in U$.

If p is analytic with $p(0) = q(0)$, $p(U) \subseteq D$ and

$$\theta(p(z)) + zp'(z) \phi(p(z)) \prec \theta(q(z)) + zq'(z) \phi(q(z)),$$

then $p(z) \prec q(z)$ and q is the best dominant.

Lemma 17.1.11. [15] Let the function q be convex univalent in the open unit disc U and ν and ϕ be analytic in a domain D containing $q(U)$. Suppose that

(1) $\operatorname{Re}\left(\frac{\nu'(q(z))}{\phi(q(z))}\right) > 0$ for $z \in U$ and

(2) $\psi(z) = zq'(z)\phi(q(z))$ is starlike univalent in U.

If $p(z) \in \mathcal{H}[q(0),1] \cap Q$, with $p(U) \subseteq D$ and $\nu(p(z)) + zp'(z)\phi(p(z))$ is univalent in U and

$$\nu(q(z)) + zq'(z)\phi(q(z)) \prec \nu(p(z)) + zp'(z)\phi(p(z)),$$

then $q(z) \prec p(z)$ and q is the best subordinate.

We obtain the following results:

Theorem 17.1.12. Let $\dfrac{z \cdot IR_{\lambda,l}^{m,n+1}f(z)}{\left(IR_{\lambda,l}^{m,n}f(z)\right)^2} \in \mathcal{H}(U)$ and let the function $q(z)$ be analytic and univalent in U such that $q(z) \neq 0$, for all $z \in U$. Suppose that $\dfrac{zq'(z)}{q(z)}$ is starlike univalent in U. Let

$$Re\left(\frac{\xi}{\beta}q(z) + \frac{2\mu}{\beta}q^2(z) + 1 + z\frac{q''(z)}{q(z)} - z\frac{q'(z)}{q(z)}\right) > 0, \qquad (17.4)$$

for $\alpha, \xi, \beta, \mu \in \mathbb{C}$, $\beta \neq 0$, $z \in U$ and

$$\psi_{\lambda,l}^{m,n}(\alpha,\xi,\mu,\beta;z) := (\alpha + \beta n) + \xi \frac{z \cdot IR_{\lambda,l}^{m,n+1}f(z)}{\left(IR_{\lambda,l}^{m,n}f(z)\right)^2} + \qquad (17.5)$$

$$\mu \frac{z^2 \cdot \left(IR_{\lambda,l}^{m,n+1}f(z)\right)^2}{\left(IR_{\lambda,l}^{m,n}f(z)\right)^4} + \beta(n+2)\frac{IR_{\lambda,l}^{m,n+2}f(z)}{IR_{\lambda,l}^{m,n+1}f(z)} - 2\beta(n+1)\frac{IR_{\lambda,l}^{m,n+1}f(z)}{IR_{\lambda,l}^{m,n}f(z)}.$$

If q satisfies the following subordination

$$\psi_{\lambda,l}^{m,n}(\alpha,\beta,\mu;z) \prec \alpha + \xi q(z) + \mu(q(z))^2 + \beta\frac{zq'(z)}{q(z)}, \qquad (17.6)$$

for $\alpha, \xi, \beta, \mu \in \mathbb{C}$, $\beta \neq 0$, then

$$\frac{z \cdot IR_{\lambda,l}^{m,n+1}f(z)}{\left(IR_{\lambda,l}^{m,n}f(z)\right)^2} \prec q(z), \qquad (17.7)$$

and q is the best dominant.

Proof. Let the function p be defined by $p(z) := \dfrac{z \cdot IR_{\lambda,l}^{m,n+1}f(z)}{\left(IR_{\lambda,l}^{m,n}f(z)\right)^2}$, $z \in U$, $z \neq 0$, $f \in \mathcal{A}$. We have $p'(z) = \dfrac{IR_{\lambda,l}^{m,n+1}f(z)}{\left(IR_{\lambda,l}^{m,n}f(z)\right)^2} + \dfrac{z\left(IR_{\lambda,l}^{m,n+1}f(z)\right)'}{\left(IR_{\lambda,l}^{m,n}f(z)\right)^2} - 2z\dfrac{IR_{\lambda,l}^{m,n+1}f(z)}{\left(IR_{\lambda,l}^{m,n}f(z)\right)^2} \cdot \dfrac{\left(IR_{\lambda,l}^{m,n}f(z)\right)'}{IR_{\lambda,l}^{m,n}f(z)}.$

By using the identity (17.3), we obtain

$$\frac{zp'(z)}{p(z)} = (n+2)\frac{IR_{\lambda,l}^{m,n+2}f(z)}{IR_{\lambda,l}^{m,n+1}f(z)} - 2(n+1)\frac{IR_{\lambda,l}^{m,n+1}f(z)}{IR_{\lambda,l}^{m,n}f(z)} + n. \qquad (17.8)$$

By setting $\theta(w) := \alpha + \xi w + \mu w^2$ and $\phi(w) := \frac{\beta}{w}$, it can be easily verified that θ is analytic in $\mathbb{C}$, ϕ is analytic in $\mathbb{C}\setminus\{0\}$ and that $\phi(w) \neq 0$, $w \in \mathbb{C}\setminus\{0\}$. Also, by letting

$$Q(z) = zq'(z)\phi(q(z)) = \beta\frac{zq'(z)}{q(z)}$$

and

$$h(z) = \theta(q(z)) + Q(z) = \alpha + \xi q(z) + \mu(q(z))^2 + \beta\frac{zq'(z)}{q(z)},$$

we find that $Q(z)$ is starlike univalent in U.

We have

$$h'(z) = \xi q'(z) + 2\mu q(z) q'(z) + \beta\frac{q'(z)}{q(z)} + \beta z\frac{q''(z)}{q(z)} - \beta z\left(\frac{q'(z)}{q(z)}\right)^2$$

and

$$\frac{zh'(z)}{Q(z)} = \frac{\xi}{\beta}q(z) + \frac{2\mu}{\beta}q^2(z) + 1 + z\frac{q''(z)}{q(z)} - z\frac{q'(z)}{q(z)}.$$

We deduce that

$$\operatorname{Re}\left(\frac{zh'(z)}{Q(z)}\right) = \operatorname{Re}\left(\frac{\xi}{\beta}q(z) + \frac{2\mu}{\beta}q^2(z) + 1 + z\frac{q''(z)}{q(z)} - z\frac{q'(z)}{q(z)}\right) > 0.$$

By using (17.8), we obtain

$$\alpha + \xi p(z) + \mu(p(z))^2 + \beta\frac{zp'(z)}{p(z)}$$

$$= (\alpha + \beta n) + \xi\frac{z \cdot IR_{\lambda,l}^{m,n+1}f(z)}{\left(IR_{\lambda,l}^{m,n}f(z)\right)^2} + \mu\frac{z^2 \cdot \left(IR_{\lambda,l}^{m,n+1}f(z)\right)^2}{\left(IR_{\lambda,l}^{m,n}f(z)\right)^4}$$

$$+ \beta(n+2)\frac{IR_{\lambda,l}^{m,n+2}f(z)}{IR_{\lambda,l}^{m,n+1}f(z)} - 2\beta(n+1)\frac{IR_{\lambda,l}^{m,n+1}f(z)}{IR_{\lambda,l}^{m,n}f(z)}.$$

By using (17.6), we have

$$\alpha + \xi p(z) + \mu(p(z))^2 + \beta\frac{zp'(z)}{p(z)} \prec \alpha + \xi q(z) + \mu(q(z))^2 + \beta\frac{zq'(z)}{q(z)}.$$

By an application of Lemma 17.1.10, we have $p(z) \prec q(z)$, $z \in U$, i.e.,

$$\frac{z \cdot IR_{\lambda,l}^{m,n+1}f(z)}{\left(IR_{\lambda,l}^{m,n}f(z)\right)^2} \prec q(z),$$

$z \in U$ and q is the best dominant. $\square$

Corollary 17.1.13. Let $m, n \in \mathbb{N}$, $\lambda, l \geq 0$. Assume that (17.4) holds. If $f \in \mathcal{A}$ and

$$\psi_{\lambda,l}^{m,n}(\alpha, \beta, \mu; z) \prec \alpha + \xi \frac{1+Az}{1+Bz} + \mu \left(\frac{1+Az}{1+Bz}\right)^2 + \frac{\beta(A-B)z}{(1+Az)(1+Bz)},$$

for $\alpha, \beta, \mu, \xi \in \mathbb{C}$, $\beta \neq 0$, $-1 \leq B < A \leq 1$, where $\psi_{\lambda,l}^{m,n}$ is defined in (17.5), then

$$\frac{z \cdot IR_{\lambda,l}^{m,n+1} f(z)}{\left(IR_{\lambda,l}^{m,n} f(z)\right)^2} \prec \frac{1+Az}{1+Bz},$$

and $\frac{1+Az}{1+Bz}$ is the best dominant.

Proof. For $q(z) = \frac{1+Az}{1+Bz}$, $-1 \leq B < A \leq 1$ in Theorem 17.1.12, we get the corollary. □

Corollary 17.1.14. Let $m, n \in \mathbb{N}$, $\lambda, l \geq 0$. Assume that (17.4) holds. If $f \in \mathcal{A}$ and

$$\psi_{\lambda,l}^{m,n}(\alpha, \beta, \mu; z) \prec \alpha + \xi \left(\frac{1+z}{1-z}\right)^\gamma + \mu \left(\frac{1+z}{1-z}\right)^{2\gamma} + \frac{2\beta\gamma z}{1-z^2},$$

for $\alpha, \beta, \mu, \xi \in \mathbb{C}$, $0 < \gamma \leq 1$, $\beta \neq 0$, where $\psi_{\lambda,l}^{m,n}$ is defined in (17.5), then

$$\frac{z \cdot IR_{\lambda,l}^{m,n+1} f(z)}{\left(IR_{\lambda,l}^{m,n} f(z)\right)^2} \prec \left(\frac{1+z}{1-z}\right)^\gamma,$$

and $\left(\frac{1+z}{1-z}\right)^\gamma$ is the best dominant.

Proof. Corollary follows by using Theorem 17.1.12 for $q(z) = \left(\frac{1+z}{1-z}\right)^\gamma$, $0 < \gamma \leq 1$. □

Theorem 17.1.15. Let q be analytic and univalent in U such that $q(z) \neq 0$ and $\frac{zq'(z)}{q(z)}$ be starlike univalent in U. Assume that

$$\operatorname{Re}\left(\frac{\xi}{\beta} q(z) q'(z) + \frac{2\mu}{\beta} q^2(z) q'(z)\right) > 0, \text{ for } \xi, \beta, \mu \in \mathbb{C},\ \beta \neq 0. \quad (17.9)$$

If $f \in \mathcal{A}$, $\frac{z \cdot IR_{\lambda,l}^{m+1,n} f(z)}{\left(IR_{\lambda,l}^{m,n} f(z)\right)^2} \in \mathcal{H}[q(0), 1] \cap Q$ and $\psi_{\lambda,l}^{m,n}(\alpha, \beta, \mu; z)$ is univalent in U, where $\psi_{\lambda,l}^{m,n}(\alpha, \beta, \mu; z)$ is as defined in (17.5), then

$$\alpha + \xi q(z) + \mu (q(z))^2 + \frac{\beta z q'(z)}{q(z)} \prec \psi_{\lambda,l}^{m,n}(\alpha, \beta, \mu; z) \quad (17.10)$$

implies
$$q(z) \prec \frac{z \cdot IR_{\lambda,l}^{m,n+1} f(z)}{\left(IR_{\lambda,l}^{m,n} f(z)\right)^2}, \quad z \in U, \tag{17.11}$$

and q is the best subordinant.

Proof. Let the function p be defined by
$$p(z) := \frac{z \cdot IR_{\lambda,l}^{m,n+1} f(z)}{\left(IR_{\lambda,l}^{m,n} f(z)\right)^2},$$

$z \in U$, $z \neq 0$, $f \in \mathcal{A}$.

By setting $\nu(w) := \alpha + \xi w + \mu w^2$ and $\phi(w) := \frac{\beta}{w}$, it can be easily verified that ν is analytic in $\mathbb{C}$, ϕ is analytic in $\mathbb{C}\setminus\{0\}$ and that $\phi(w) \neq 0$, $w \in \mathbb{C}\setminus\{0\}$.
Since
$$\frac{\nu'(q(z))}{\phi(q(z))} = \frac{q'(z) q(z) [\xi + 2\mu q(z)]}{\beta},$$
it follows that $Re\left(\frac{\nu'(q(z))}{\phi(q(z))}\right) = Re\left(\frac{\xi}{\beta} q(z) q'(z) + \frac{2\mu}{\beta} q^2(z) q'(z)\right) > 0$, for $\alpha, \beta, \mu \in \mathbb{C}$, $\mu \neq 0$.

By using (17.8) and (17.10) we obtain
$$\alpha + \xi q(z) + \mu(q(z))^2 + \frac{\beta z q'(z)}{q(z)} \prec \alpha + \xi p(z) + \mu(p(z))^2 + \frac{\beta z p'(z)}{p(z)}.$$

Using Lemma 17.1.11, we have
$$q(z) \prec p(z) = \frac{z \cdot IR_{\lambda,l}^{m,n+1} f(z)}{\left(IR_{\lambda,l}^{m,n} f(z)\right)^2}, \quad z \in U,$$

and q is the best subordinant. □

Corollary 17.1.16. *Let $m, n \in \mathbb{N}$, $\lambda, l \geq 0$. Assume that (17.9) holds. If $f \in \mathcal{A}$, $\frac{z \cdot IR_{\lambda,l}^{m,n+1} f(z)}{\left(IR_{\lambda,l}^{m,n} f(z)\right)^2} \in \mathcal{H}[q(0), 1] \cap Q$ and*

$$\alpha + \xi \frac{1 + Az}{1 + Bz} + \mu \left(\frac{1 + Az}{1 + Bz}\right)^2 + \frac{\beta(A - B)z}{(1 + Az)(1 + Bz)} \prec \psi_{\lambda,l}^{m,n}(\alpha, \beta, \mu; z),$$

for $\alpha, \beta, \xi, \mu \in \mathbb{C}$, $\beta \neq 0$, $-1 \leq B < A \leq 1$, where $\psi_{\lambda,l}^{m,n}$ is defined in (17.5), then
$$\frac{1 + Az}{1 + Bz} \prec \frac{z \cdot IR_{\lambda,l}^{m,n+1} f(z)}{\left(IR_{\lambda,l}^{m,n} f(z)\right)^2},$$

and $\frac{1+Az}{1+Bz}$ is the best subordinant.

Proof. For $q(z) = \frac{1+Az}{1+Bz}$, $-1 \leq B < A \leq 1$ in Theorem 17.1.15, we get the corollary. □

Corollary 17.1.17. Let $m, n \in \mathbb{N}$, $\lambda, l \geq 0$. Assume that (17.9) holds. If $f \in \mathcal{A}$, $\frac{z \cdot IR_{\lambda,l}^{m,n+1} f(z)}{\left(IR_{\lambda,l}^{m,n} f(z)\right)^2} \in \mathcal{H}[q(0), 1] \cap Q$ and

$$\alpha + \xi \left(\frac{1+z}{1-z}\right)^\gamma + \mu \left(\frac{1+z}{1-z}\right)^{2\gamma} + \frac{2\beta\gamma z}{1-z^2} \prec \psi_{\lambda,l}^{m,n}(\alpha, \beta, \mu; z),$$

for $\alpha, \beta, \mu, \xi \in \mathbb{C}$, $\beta \neq 0$, $0 < \gamma \leq 1$, where $\psi_{\lambda,l}^{m,n}$ is defined in (17.5), then

$$\left(\frac{1+z}{1-z}\right)^\gamma \prec \frac{z \cdot IR_{\lambda,l}^{m,n+1} f(z)}{\left(IR_{\lambda,l}^{m,n} f(z)\right)^2},$$

and $\left(\frac{1+z}{1-z}\right)^\gamma$ is the best subordinant.

Proof. For $q(z) = \left(\frac{1+z}{1-z}\right)^\gamma$, $0 < \gamma \leq 1$ in Theorem 17.1.15, we get the corollary. □

Combining Theorem 17.1.12 and Theorem 17.1.15, we state the following sandwich theorem.

Theorem 17.1.18. Let q_1 and q_2 be analytic and univalent in U such that $q_1(z) \neq 0$ and $q_2(z) \neq 0$, for all $z \in U$, with $\frac{zq_1'(z)}{q_1(z)}$ and $\frac{zq_2'(z)}{q_2(z)}$ being starlike univalent. Suppose that q_1 satisfies (17.4) and q_2 satisfies (17.9). If $f \in \mathcal{A}$, $\frac{z \cdot IR_{\lambda,l}^{m+1,n} f(z)}{\left(IR_{\lambda,l}^{m,n} f(z)\right)^2} \in \mathcal{H}[q(0), 1] \cap Q$ and $\psi_{\lambda,l}^{m,n}(\alpha, \beta, \mu; z)$ is as defined in (17.5) univalent in U, then

$$\alpha + \xi q_1(z) + \mu (q_1(z))^2 + \frac{\beta z q_1'(z)}{q_1(z)} \prec \psi_{\lambda,l}^{m,n}(\alpha, \beta, \mu; z) \prec$$

$$\alpha + \xi q_2(z) + \mu (q_2(z))^2 + \frac{\beta z q_2'(z)}{q_2(z)},$$

for $\alpha, \beta, \mu, \xi \in \mathbb{C}$, $\beta \neq 0$, implies

$$q_1(z) \prec \frac{z \cdot IR_{\lambda,l}^{m,n+1} f(z)}{\left(IR_{\lambda,l}^{m,n} f(z)\right)^2} \prec q_2(z),$$

and q_1 and q_2 are respectively the best subordinant and the best dominant.

For $q_1(z) = \frac{1+A_1z}{1+B_1z}$, $q_2(z) = \frac{1+A_2z}{1+B_2z}$, where $-1 \leq B_2 < B_1 < A_1 < A_2 \leq 1$, we have the following corollary.

Corollary 17.1.19. Let $m, n \in \mathbb{N}$, $\lambda, l \geq 0$. Assume that (17.4) and (17.9) hold. If $f \in \mathcal{A}$, $\frac{z \cdot IR_{\lambda,l}^{m,n+1} f(z)}{\left(IR_{\lambda,l}^{m,n} f(z)\right)^2} \in \mathcal{H}[q(0), 1] \cap Q$ and

$$\alpha + \xi \frac{1+A_1z}{1+B_1z} + \mu \left(\frac{1+A_1z}{1+B_1z}\right)^2 + \frac{\beta(A_1 - B_1)z}{(1+A_1z)(1+B_1z)} \prec \psi_{\lambda,l}^{m,n}(\alpha, \beta, \mu; z)$$

$$\prec \alpha + \xi \frac{1+A_2z}{1+B_2z} + \mu \left(\frac{1+A_2z}{1+B_2z}\right)^2 + \frac{\beta(A_2 - B_2)z}{(1+A_2z)(1+B_2z)},$$

for $\alpha, \beta, \mu, \xi \in \mathbb{C}$, $\beta \neq 0$, $-1 \leq B_2 \leq B_1 < A_1 \leq A_2 \leq 1$, where $\psi_{\lambda,l}^{m,n}$ is defined in (17.5), then

$$\frac{1+A_1z}{1+B_1z} \prec \frac{z \cdot IR_{\lambda,l}^{m,n+1} f(z)}{\left(IR_{\lambda,l}^{m,n} f(z)\right)^2} \prec \frac{1+A_2z}{1+B_2z},$$

hence $\frac{1+A_1z}{1+B_1z}$ and $\frac{1+A_2z}{1+B_2z}$ are the best subordinant and the best dominant, respectively.

For $q_1(z) = \left(\frac{1+z}{1-z}\right)^{\gamma_1}$, $q_2(z) = \left(\frac{1+z}{1-z}\right)^{\gamma_2}$, where $0 < \gamma_1 < \gamma_2 \leq 1$, we have the following corollary.

Corollary 17.1.20. Let $m, n \in \mathbb{N}$, $\lambda, l \geq 0$. Assume that (17.4) and (17.9) hold. If $f \in \mathcal{A}$, $\frac{z \cdot IR_{\lambda,l}^{m,n+1} f(z)}{\left(IR_{\lambda,l}^{m,n} f(z)\right)^2} \in \mathcal{H}[q(0), 1] \cap Q$ and

$$\alpha + \xi \left(\frac{1+z}{1-z}\right)^{\gamma_1} + \mu \left(\frac{1+z}{1-z}\right)^{2\gamma_1} + \frac{2\beta\gamma_1 z}{1-z^2} \prec \psi_{\lambda,l}^{m,n}(\alpha, \beta, \mu; z)$$

$$\prec \alpha + \xi \left(\frac{1+z}{1-z}\right)^{\gamma_2} + \mu \left(\frac{1+z}{1-z}\right)^{2\gamma_2} + \frac{2\beta\gamma_2 z}{1-z^2},$$

for $\alpha, \beta, \mu, \xi \in \mathbb{C}$, $\beta \neq 0$, $0 < \gamma_1 < \gamma_2 \leq 1$, where $\psi_{\lambda,l}^{m,n}$ is defined in (17.5), then

$$\left(\frac{1+z}{1-z}\right)^{\gamma_1} \prec \frac{z \cdot IR_{\lambda,l}^{m,n+1} f(z)}{\left(IR_{\lambda,l}^{m,n} f(z)\right)^2} \prec \left(\frac{1+z}{1-z}\right)^{\gamma_2},$$

hence $\left(\frac{1+z}{1-z}\right)^{\gamma_1}$ and $\left(\frac{1+z}{1-z}\right)^{\gamma_2}$ are the best subordinant and the best dominant, respectively.

17.2 Strong differential subordination and superordination

Consider $U = \{z \in \mathbb{C} : |z| < 1\}$ the unit disc of the complex plane, the closed unit disc of the complex plane $\overline{U} = \{z \in \mathbb{C} : |z| \leq 1\}$ and the class of analytic functions in $U \times \overline{U}$ denoted $\mathcal{H}(U \times \overline{U})$.
Let

$$\mathcal{A}^*_{n\zeta} = \{f \in \mathcal{H}(U \times \overline{U}) : f(z,\zeta) = z + a_{n+1}(\zeta) z^{n+1} + \ldots, \ z \in U, \ \zeta \in \overline{U}\},$$

where $a_k(\zeta)$ are holomorphic functions in $\overline{U}$ for $k \geq 2$, for $n = 1$ we denote this class with $\mathcal{A}^*_\zeta$, and $\mathcal{H}^*[a, n, \zeta] =$

$$\{f \in \mathcal{H}(U \times \overline{U}) : f(z,\zeta) = a + a_n(\zeta) z^n + a_{n+1}(\zeta) z^{n+1} + \ldots, \ z \in U, \ \zeta \in \overline{U}\},$$

for $a \in \mathbb{C}$ and $n \in \mathbb{N}$, $a_k(\zeta)$ are holomorphic functions in $\overline{U}$ for $k \geq n$.
J.A. Antonino and S. Romaguera defined in [14] the notion of strong differential subordinations, which was developed by G.I. Oros and Gh. Oros in [21,22].

Definition 17.2.1. *[21] Let $f(z,\zeta)$, $H(z,\zeta)$ analytic in $U \times \overline{U}$. The function $f(z,\zeta)$ is said to be strongly subordinate to $H(z,\zeta)$ if there exists a function w analytic in U, with $w(0) = 0$ and $|w(z)| < 1$ such that $f(z,\zeta) = H(w(z),\zeta)$ for all $\zeta \in \overline{U}$. In such a case we write $f(z,\zeta) \prec\prec H(z,\zeta)$, $z \in U, \zeta \in \overline{U}$.*

Remark 17.2.2. *[21] (i) Since $f(z,\zeta)$ is analytic in $U \times \overline{U}$, for all $\zeta \in \overline{U}$, and univalent in U, for all $\zeta \in \overline{U}$, Definition 17.2.1 is equivalent to $f(0,\zeta) = H(0,\zeta)$, for all $\zeta \in \overline{U}$, and $f(U \times \overline{U}) \subset H(U \times \overline{U})$.*
(ii) If $H(z,\zeta) \equiv H(z)$ and $f(z,\zeta) \equiv f(z)$, the strong subordination becomes the usual notion of subordination.

In studying the strong differential subordinations we will use the following lemma.

Lemma 17.2.3. *[19] Let the function q be univalent in $U \times \overline{U}$ and θ and ϕ be analytic in a domain D containing $q(U \times \overline{U})$ with $\phi(w) \neq 0$ when $w \in q(U \times \overline{U})$. Set $Q(z,\zeta) = zq'_z(z,\zeta)\phi(q(z,\zeta))$ and $h(z,\zeta) = \theta(q(z,\zeta)) + Q(z,\zeta)$. Suppose that*

(1) Q is starlike univalent in $U \times \overline{U}$ and

(2) $\mathrm{Re}\left(\frac{zh'_z(z,\zeta)}{Q(z,\zeta)}\right) > 0$ for $z \in U$, $\zeta \in \overline{U}$.

If p is analytic with $p(0,\zeta) = q(0,\zeta)$, $p(U \times \overline{U}) \subseteq D$ and

$$\theta(p(z,\zeta)) + zp'_z(z,\zeta)\phi(p(z,\zeta)) \prec\prec \theta(q(z,\zeta)) + zq'_z(z,\zeta)\phi(q(z,\zeta)),$$

then $p(z,\zeta) \prec\prec q(z,\zeta)$ and q is the best dominant.

As a dual notion of strong differential subordination G.I. Oros has introduced and developed the notion of strong differential superordinations in [20].

Definition 17.2.4. *[20] Let $f(z,\zeta)$, $H(z,\zeta)$ analytic in $U \times \overline{U}$. The function $f(z,\zeta)$ is said to be strongly superordinate to $H(z,\zeta)$ if there exists a function w analytic in U, with $w(0) = 0$ and $|w(z)| < 1$, such that $H(z,\zeta) = f(w(z),\zeta)$, for all $\zeta \in \overline{U}$. In such a case we write $H(z,\zeta) \prec\prec f(z,\zeta)$, $z \in U$, $\zeta \in \overline{U}$.*

Remark 17.2.5. *[20] (i) Since $f(z,\zeta)$ is analytic in $U \times \overline{U}$, for all $\zeta \in \overline{U}$, and univalent in U, for all $\zeta \in \overline{U}$, Definition 17.2.4 is equivalent to $H(0,\zeta) = f(0,\zeta)$, for all $\zeta \in \overline{U}$, and $H(U \times \overline{U}) \subset f(U \times \overline{U})$.*
(ii) If $H(z,\zeta) \equiv H(z)$ and $f(z,\zeta) \equiv f(z)$, the strong superordination becomes the usual notion of superordination.

Definition 17.2.6. *[11] We denote by Q^* the set of functions that are analytic and injective on $\overline{U} \times \overline{U} \backslash E(f,\zeta)$, where $E(f,\zeta) = \{y \in \partial U : \lim_{z \to y} f(z,\zeta) = \infty\}$, and are such that $f'_z(y,\zeta) \neq 0$ for $y \in \partial U \times \overline{U} \backslash E(f,\zeta)$. The subclass of Q^* for which $f(0,\zeta) = a$ is denoted by $Q^*(a)$.*

We have need the following lemma to study the strong differential superordinations.

Lemma 17.2.7. *[19] Let the function q be convex univalent in $U \times \overline{U}$ and ν and ϕ be analytic in a domain D containing $q(U \times \overline{U})$. Suppose that*

(1) $\operatorname{Re}\left(\frac{\nu'_z(q(z,\zeta))}{\phi(q(z,\zeta))}\right) > 0$ for $z \in U$, $\zeta \in \overline{U}$ and

(2) $\psi(z,\zeta) = zq'_z(z,\zeta)\phi(q(z,\zeta))$ is starlike univalent in $U \times \overline{U}$.

If $p(z,\zeta) \in \mathcal{H}^*[q(0,\zeta),1,\zeta] \cap Q^*$, with $p(U \times \overline{U}) \subseteq D$ and $\nu(p(z,\zeta)) + zp'_z(z)\phi(p(z,\zeta))$ is univalent in $U \times \overline{U}$ and

$$\nu(q(z,\zeta)) + zq'_z(z,\zeta)\phi(q(z,\zeta)) \prec\prec \nu(p(z,\zeta)) + zp'_z(z,\zeta)\phi(p(z,\zeta)),$$

then $q(z,\zeta) \prec\prec p(z,\zeta)$ and q is the best subordinant.

The author extended in [9] and [10] the multiplier transformation ([3]) and respectively, Ruscheweyh derivative ([23]) to the new class of analytic functions $\mathcal{A}_{n\zeta}^*$ introduced in [22].

Definition 17.2.8. *[9] For $f \in \mathcal{A}_\zeta^*$, $m \in \mathbb{N} \cup \{0\}$, $\lambda, l \geq 0$, the operator $I(m,\lambda,l): \mathcal{A}_\zeta^* \to \mathcal{A}_\zeta^*$ is defined by the following infinite series*

$$I(m,\lambda,l)f(z,\zeta) := z + \sum_{j=2}^{\infty} \left(\frac{1+\lambda(j-1)+l}{l+1}\right)^m a_j(\zeta) z^j.$$

Remark 17.2.9. *[9] The operator $I(m, \lambda, l)$ verifies the property*

$$(l+1) I(m+1, \lambda, l) f(z, \zeta) = [l + 1 - \lambda] I(m, \lambda, l) f(z, \zeta)$$
$$+ \lambda z \left(I(m, \lambda, l) f(z, \zeta) \right)', \zeta \in \overline{U}, \ z \in U.$$

Definition 17.2.10. *[10] For $f \in \mathcal{A}_\zeta^*$, $m \in \mathbb{N}$, the operator R^m is defined by $R^m : \mathcal{A}_\zeta^* \to \mathcal{A}_\zeta^*$,*

$$R^0 f(z, \zeta) = f(z, \zeta)$$
$$R^1 f(z, \zeta) = z f'(z, \zeta)$$
$$\ldots$$
$$(m+1) R^{m+1} f(z, \zeta) = z (R^m f(z, \zeta))' + m R^m f(z, \zeta), \quad \zeta \in \overline{U}, \ z \in U.$$

Remark 17.2.11. *[10] If $f \in \mathcal{A}_\zeta^*$, $f(z, \zeta) = z + \sum_{j=2}^\infty a_j(\zeta) z^j$, then*

$$R^m f(z, \zeta) = z + \sum_{j=2}^\infty C_{m+j-1}^m a_j(\zeta) z^j, \ \zeta \in \overline{U}, \ z \in U.$$

The author also extended and studied in [8, 12] the differential operator obtained as a convolution product (Hadamard product) of multiplier transformation and Ruscheweyh derivative [4, 5] to the class $\mathcal{A}_\zeta^*$.

Definition 17.2.12. *[8] Let $\lambda, l \geq 0$ and $m \in \mathbb{N}$. Denote by $IR_{\lambda,l}^m$ the extended operator given by the Hadamard product of the extended multiplier transformation $I(m, \lambda, l)$ and the extended Ruscheweyh derivative R^m, $IR_{\lambda,l}^m : \mathcal{A}_\zeta^* \to \mathcal{A}_\zeta^*$,*

$$IR_{\lambda,l}^m f(z, \zeta) = (I(m, \lambda, l) * R^m) f(z, \zeta).$$

Remark 17.2.13. *[8] If $f \in \mathcal{A}_\zeta^*$, $f(z, \zeta) = z + \sum_{j=2}^\infty a_j(\zeta) z^j$, then*

$$IR_{\lambda,l}^m f(z, \zeta) = z + \sum_{j=2}^\infty \left(\frac{1 + \lambda(j-1) + l}{l+1} \right)^m C_{m+j-1}^m a_j^2(\zeta) z^j, \zeta \in \overline{U}, \ z \in U.$$

Remark 17.2.14. *For $l = 0$, $\lambda \geq 0$, we obtain the operator DR_λ^n studied in [7] and for $l = 0$ and $\lambda = 1$, we obtain the operator SR^n studied in [6].*

Using simple computation one obtains the next result.

Proposition 17.2.15. *For $m, n \in \mathbb{N}$ and $\lambda, l \geq 0$ we have*

$$(n+1) IR_{\lambda,l}^{m,n+1} f(z, \zeta) - n IR_{\lambda,l}^{m,n} f(z, \zeta) = z \left(IR_{\lambda,l}^{m,n} f(z, \zeta) \right)_z'. \quad (17.12)$$

Similar to the results from section 1 we obtain the following results for the extended operator:

Theorem 17.2.16. Let $\frac{z \cdot IR_{\lambda,l}^{m,n+1}f(z,\zeta)}{\left(IR_{\lambda,l}^{m,n}f(z,\zeta)\right)^2} \in \mathcal{H}(U \times \overline{U})$ and let the function $q(z,\zeta)$ be analytic and univalent in $U \times \overline{U}$ such that $q(z,\zeta) \neq 0$, for all $z \in U$, $\zeta \in \overline{U}$. Suppose that $\frac{z(q(z,\zeta))'_z}{q(z,\zeta)}$ is starlike univalent in $U \times \overline{U}$. Let

$$Re\left(\frac{\xi}{\beta}q(z,\zeta) + \frac{2\mu}{\beta}q^2(z,\zeta) + 1 + z\frac{q''_{z^2}(z,\zeta)}{q'_z(z,\zeta)} - z\frac{q'_z(z,\zeta)}{q(z,\zeta)}\right) > 0, \quad (17.13)$$

for $\alpha, \xi, \beta, \mu \in \mathbb{C}$, $\beta \neq 0$, $z \in U$, $\zeta \in \overline{U}$ and

$$\psi_{\lambda,l}^{m,n}(\alpha, \xi, \mu, \beta; z, \zeta) := (\alpha + \beta n) + \xi \frac{z \cdot IR_{\lambda,l}^{m,n+1}f(z,\zeta)}{\left(IR_{\lambda,l}^{m,n}f(z,\zeta)\right)^2}$$

$$+ \mu \frac{z^2 \cdot \left(IR_{\lambda,l}^{m,n+1}f(z,\zeta)\right)^2}{\left(IR_{\lambda,l}^{m,n}f(z,\zeta)\right)^4}$$

$$+ \beta(n+2)\frac{IR_{\lambda,l}^{m,n+2}f(z,\zeta)}{IR_{\lambda,l}^{m,n+1}f(z,\zeta)} - 2\beta(n+1)\frac{IR_{\lambda,l}^{m,n+1}f(z,\zeta)}{IR_{\lambda,l}^{m,n}f(z,\zeta)}. \quad (17.14)$$

If q satisfies the following strong differential subordination

$$\psi_{\lambda,l}^{m,n}(\alpha, \beta, \mu; z, \zeta) \prec\prec \alpha + \xi q(z,\zeta) + \mu(q(z,\zeta))^2 + \beta\frac{zq'_z(z,\zeta)}{q(z,\zeta)}, \quad (17.15)$$

for $\alpha, \xi, \beta, \mu \in \mathbb{C}$, $\beta \neq 0$, then

$$\frac{z \cdot IR_{\lambda,l}^{m,n+1}f(z,\zeta)}{\left(IR_{\lambda,l}^{m,n}f(z,\zeta)\right)^2} \prec\prec q(z,\zeta), \quad (17.16)$$

and q is the best dominant.

Proof. Let the function p be defined by $p(z,\zeta) := \frac{z \cdot IR_{\lambda,l}^{m,n+1}f(z,\zeta)}{\left(IR_{\lambda,l}^{m,n}f(z,\zeta)\right)^2}$, $z \in U$, $\zeta \in \overline{U}$, $z \neq 0$, $f \in \mathcal{A}_\zeta^*$. We have

$$p'_z(z,\zeta) = \frac{IR_{\lambda,l}^{m,n+1}f(z,\zeta)}{\left(IR_{\lambda,l}^{m,n}f(z,\zeta)\right)^2} + \frac{z\left(IR_{\lambda,l}^{m,n+1}f(z,\zeta)\right)'_z}{\left(IR_{\lambda,l}^{m,n}f(z,\zeta)\right)^2}$$

$$- 2z\frac{IR_{\lambda,l}^{m,n+1}f(z,\zeta)}{\left(IR_{\lambda,l}^{m,n}f,\zeta\right)^2} \cdot \frac{\left(IR_{\lambda,l}^{m,n}f(z,\zeta)\right)'_z}{IR_{\lambda,l}^{m,n}f(z,\zeta)}.$$

By using the identity (17.12), we obtain

$$\frac{zp'_z(z,\zeta)}{p(z,\zeta)} = (n+2)\frac{IR_{\lambda,l}^{m,n+2}f(z,\zeta)}{IR_{\lambda,l}^{m,n+1}f(z,\zeta)} - 2(n+1)\frac{IR_{\lambda,l}^{m,n+1}f(z,\zeta)}{IR_{\lambda,l}^{m,n}f(z,\zeta)} + n. \quad (17.17)$$

By setting $\theta(w) := \alpha + \xi w + \mu w^2$ and $\phi(w) := \frac{\beta}{w}$, it can be easily verified that θ is analytic in $\mathbb{C}$, ϕ is analytic in $\mathbb{C}\setminus\{0\}$ and that $\phi(w) \neq 0$, $w \in \mathbb{C}\setminus\{0\}$. Also, by letting $Q(z,\zeta) = zq'_z(z)\phi(q(z,\zeta)) = \beta\frac{zq'_z(z,\zeta)}{q(z,\zeta)}$ and $h(z,\zeta) = \theta(q(z,\zeta)) + Q(z,\zeta) = \alpha + \xi q(z,\zeta) + \mu(q(z,\zeta))^2 + \beta\frac{zq'_z(z,\zeta)}{q(z,\zeta)}$, we find that $Q(z,\zeta)$ is starlike univalent in $U \times \overline{U}$.

We have

$$h'_z(z,\zeta) = \xi q'_z(z,\zeta) + 2\mu q(z,\zeta) q'_z(z,\zeta) + \beta\frac{q'_z(z,\zeta)}{q(z,\zeta)}$$

$$+ \beta z\frac{q''_{z^2}(z,\zeta)}{q(z,\zeta)} - \beta z\left(\frac{q'_z(z,\zeta)}{q(z,\zeta)}\right)^2$$

and

$$\frac{zh'_z(z,\zeta)}{Q(z,\zeta)} = \frac{\xi}{\beta}q(z,\zeta) + \frac{2\mu}{\beta}q^2(z,\zeta) + 1 + z\frac{q''_{z^2}(z,\zeta)}{q(z,\zeta)} - z\frac{q'_z(z,\zeta)}{q(z,\zeta)}.$$

We deduce that

$$\operatorname{Re}\left(\frac{zh'_z(z,\zeta)}{Q(z,\zeta)}\right)$$

$$= \operatorname{Re}\left(\frac{\xi}{\beta}q(z,\zeta) + \frac{2\mu}{\beta}q^2(z,\zeta) + 1 + z\frac{q''_{z^2}(z,\zeta)}{q(z,\zeta)} - z\frac{q'_z(z,\zeta)}{q(z,\zeta)}\right) > 0.$$

By using (17.17), we obtain

$$\alpha + \xi p(z,\zeta) + \mu(p(z,\zeta))^2 + \beta\frac{zp'_z(z,\zeta)}{p(z,\zeta)}$$

$$= (\alpha + \beta n) + \xi\frac{z \cdot IR_{\lambda,l}^{m,n+1}f(z,\zeta)}{\left(IR_{\lambda,l}^{m,n}f(z,\zeta)\right)^2} + \mu\frac{z^2 \cdot \left(IR_{\lambda,l}^{m,n+1}f(z,\zeta)\right)^2}{\left(IR_{\lambda,l}^{m,n}f(z,\zeta)\right)^4}$$

$$+ \beta(n+2)\frac{IR_{\lambda,l}^{m,n+2}f(z,\zeta)}{IR_{\lambda,l}^{m,n+1}f(z,\zeta)} - 2\beta(n+1)\frac{IR_{\lambda,l}^{m,n+1}f(z,\zeta)}{IR_{\lambda,l}^{m,n}f(z,\zeta)}.$$

By using (17.15), we have

$$\alpha + \xi p(z,\zeta) + \mu(p(z,\zeta))^2 + \beta\frac{zp'_z(z,\zeta)}{p(z,\zeta)}$$

$$\prec\prec \alpha + \xi q(z,\zeta) + \mu(q(z,\zeta))^2 + \beta\frac{zq'_z(z,\zeta)}{q(z,\zeta)}.$$

By an application of Lemma 17.2.3, we have $p(z,\zeta) \prec\prec q(z,\zeta)$, $z \in U$, $\zeta \in \overline{U}$, i.e., $\frac{z \cdot IR_{\lambda,l}^{m,n+1}f(z,\zeta)}{\left(IR_{\lambda,l}^{m,n}f(z,\zeta)\right)^2} \prec\prec q(z,\zeta)$, $z \in U$, $\zeta \in \overline{U}$, and q is the best dominant. $\square$

Corollary 17.2.17. *Let $m, n \in \mathbb{N}$, $\lambda, l \geq 0$. Assume that (17.13) holds. If $f \in \mathcal{A}_\zeta^*$ and*

$$\psi_{\lambda,l}^{m,n}(\alpha, \beta, \mu; z, \zeta) \prec\prec \alpha + \xi \frac{\zeta + Az}{\zeta + Bz} + \mu \left(\frac{\zeta + Az}{\zeta + Bz}\right)^2 + \frac{\beta(A-B)\zeta z}{(\zeta + Az)(\zeta + Bz)},$$

for $\alpha, \beta, \mu, \xi \in \mathbb{C}$, $\beta \neq 0$, $-1 \leq B < A \leq 1$, where $\psi_{\lambda,l}^{m,n}$ is defined in (17.14), then

$$\frac{z \cdot IR_{\lambda,l}^{m,n+1} f(z, \zeta)}{\left(IR_{\lambda,l}^{m,n} f(z, \zeta)\right)^2} \prec\prec \frac{\zeta + Az}{\zeta + Bz},$$

and $\frac{\zeta + Az}{\zeta + Bz}$ is the best dominant.

Proof. For $q(z, \zeta) = \frac{\zeta + Az}{\zeta + Bz}$, $-1 \leq B < A \leq 1$ in Theorem 17.2.16, we get the corollary. □

Corollary 17.2.18. *Let $m, n \in \mathbb{N}$, $\lambda, l \geq 0$. Assume that (17.13) holds. If $f \in \mathcal{A}_\zeta^*$ and*

$$\psi_{\lambda,l}^{m,n}(\alpha, \beta, \mu; z, \zeta) \prec\prec \alpha + \xi \left(\frac{\zeta + z}{\zeta - z}\right)^\gamma + \mu \left(\frac{\zeta + z}{\zeta - z}\right)^{2\gamma} + \frac{2\beta\gamma\zeta z}{\zeta^2 - z^2},$$

for $\alpha, \beta, \mu, \xi \in \mathbb{C}$, $0 < \gamma \leq 1$, $\beta \neq 0$, where $\psi_{\lambda,l}^{m,n}$ is defined in (17.14), then

$$\frac{z \cdot IR_{\lambda,l}^{m,n+1} f(z, \zeta)}{\left(IR_{\lambda,l}^{m,n} f(z, \zeta)\right)^2} \prec\prec \left(\frac{\zeta + z}{\zeta - z}\right)^\gamma,$$

and $\left(\frac{\zeta + z}{\zeta - z}\right)^\gamma$ is the best dominant.

Proof. Corollary follows by using Theorem 17.2.16 for $q(z, \zeta) = \left(\frac{\zeta + z}{\zeta - z}\right)^\gamma$, $0 < \gamma \leq 1$. □

Theorem 17.2.19. *Let q be analytic and univalent in $U \times \overline{U}$ such that $q(z, \zeta) \neq 0$ and $\frac{zq_z'(z, \zeta)}{q(z, \zeta)}$ be starlike univalent in $U \times \overline{U}$. Assume that*

$$\operatorname{Re}\left(\frac{\xi}{\beta} q(z, \zeta) q_z'(z, \zeta) + \frac{2\mu}{\beta} q^2(z, \zeta) q_z'(z, \zeta)\right) > 0, \text{ for } \xi, \beta, \mu \in \mathbb{C}, \beta \neq 0.$$

(17.18)

If $f \in \mathcal{A}_\zeta^*$, $\frac{z \cdot IR_{\lambda,l}^{m+1,n} f(z, \zeta)}{(IR_{\lambda,l}^{m,n} f(z, \zeta))^2} \in \mathcal{H}[q(0, \zeta), 1, \zeta] \cap Q^*$ and $\psi_{\lambda,l}^{m,n}(\alpha, \beta, \mu; z, \zeta)$ is univalent in $U \times \overline{U}$, where $\psi_{\lambda,l}^{m,n}(\alpha, \beta, \mu; z, \zeta)$ is as defined in (17.14), then

$$\alpha + \xi q(z, \zeta) + \mu (q(z, \zeta))^2 + \frac{\beta z q_z'(z, \zeta)}{q(z, \zeta)} \prec\prec \psi_{\lambda,l}^{m,n}(\alpha, \beta, \mu; z, \zeta) \quad (17.19)$$

implies and q is the best subordinant.

Proof. Let the function p be defined by $p(z,\zeta) := \dfrac{z \cdot IR_{\lambda,l}^{m,n+1} f(z,\zeta)}{\left(IR_{\lambda,l}^{m,n} f(z,\zeta)\right)^2}$, $z \in U$, $\zeta \in \overline{U}$, $z \neq 0$, $f \in \mathcal{A}_\zeta^*$.

By setting $\nu(w) := \alpha + \xi w + \mu w^2$ and $\phi(w) := \dfrac{\beta}{w}$ it can be easily verified that ν is analytic in $\mathbb{C}$, ϕ is analytic in $\mathbb{C}\backslash\{0\}$ and that $\phi(w) \neq 0$, $w \in \mathbb{C}\backslash\{0\}$. Since

$$\frac{\nu_z'(q(z,\zeta))}{\phi(q(z,\zeta))} = \frac{q_z'(z,\zeta) q(z,\zeta) [\xi + 2\mu q(z,\zeta)]}{\beta},$$

it follows that

$$Re\left(\frac{\nu_z'(q(z,\zeta))}{\phi(q(z,\zeta))}\right) = Re\left(\frac{\xi}{\beta} q(z,\zeta) q_z'(z,\zeta) + \frac{2\mu}{\beta} q^2(z,\zeta) q_z'(z,\zeta)\right) > 0$$

for $\alpha, \beta, \mu \in \mathbb{C}$, $\mu \neq 0$.

By using (17.17) and (17.19), we obtain

$$\alpha + \xi q(z,\zeta) + \mu (q(z,\zeta))^2 + \frac{\beta z q_z'(z,\zeta)}{q(z,\zeta)} \prec\prec \alpha + \xi p(z,\zeta) + \mu (p(z,\zeta))^2 + \frac{\beta z p_z'(z,\zeta)}{p(z,\zeta)}.$$

Using Lemma 17.2.7, we have

$$q(z,\zeta) \prec\prec p(z,\zeta) = \frac{z \cdot IR_{\lambda,l}^{m,n+1} f(z,\zeta)}{\left(IR_{\lambda,l}^{m,n} f(z,\zeta)\right)^2}, \quad z \in U, \ \zeta \in \overline{U},$$

and q is the best subordinant. □

Corollary 17.2.20. *Let $m, n \in \mathbb{N}$, $\lambda, l \geq 0$. Assume that (17.18) holds. If $f \in \mathcal{A}_\zeta^*$, $\dfrac{z \cdot IR_{\lambda,l}^{m,n+1} f(z,\zeta)}{\left(IR_{\lambda,l}^{m,n} f(z,\zeta)\right)^2} \in \mathcal{H}[q(0,\zeta), 1, \zeta] \cap Q^*$ and*

$$\alpha + \xi \frac{\zeta + Az}{\zeta + Bz} + \mu \left(\frac{\zeta + Az}{\zeta + Bz}\right)^2 + \frac{\beta(A-B)\zeta z}{(\zeta + Az)(\zeta + Bz)} \prec\prec \psi_{\lambda,l}^{m,n}(\alpha, \beta, \mu; z, \zeta),$$

for $\alpha, \beta, \xi, \mu \in \mathbb{C}$, $\beta \neq 0$, $-1 \leq B < A \leq 1$, where $\psi_{\lambda,l}^{m,n}$ is defined in (17.14), then

$$\frac{\zeta + Az}{\zeta + Bz} \prec\prec \frac{z \cdot IR_{\lambda,l}^{m,n+1} f(z,\zeta)}{\left(IR_{\lambda,l}^{m,n} f(z,\zeta)\right)^2},$$

and $\dfrac{\zeta + Az}{\zeta + Bz}$ is the best subordinant.

Proof. For $q(z,\zeta) = \dfrac{\zeta + Az}{\zeta + Bz}$, $-1 \leq B < A \leq 1$ in Theorem 17.2.19, we get the corollary. □

Corollary 17.2.21. *Let $m, n \in \mathbb{N}$, $\lambda, l \geq 0$. Assume that (17.18) holds. If $f \in \mathcal{A}_\zeta^*$, $\dfrac{z \cdot IR_{\lambda,l}^{m,n+1} f(z,\zeta)}{\left(IR_{\lambda,l}^{m,n} f(z,\zeta)\right)^2} \in \mathcal{H}[q(0,\zeta), 1, \zeta] \cap Q^*$ and*

$$\alpha + \xi \left(\frac{\zeta + z}{\zeta - z}\right)^\gamma + \mu \left(\frac{\zeta + z}{\zeta - z}\right)^{2\gamma} + \frac{2\beta\gamma\zeta z}{\zeta^2 - z^2} \prec\prec \psi_{\lambda,l}^{m,n}(\alpha, \beta, \mu; z, \zeta),$$

for $\alpha, \beta, \mu, \xi \in \mathbb{C}$, $\beta \neq 0$, $0 < \gamma \leq 1$, where $\psi_{\lambda,l}^{m,n}$ is defined in (17.14), then

$$\left(\frac{\zeta+z}{\zeta-z}\right)^\gamma \prec\prec \frac{z \cdot IR_{\lambda,l}^{m,n+1} f(z,\zeta)}{\left(IR_{\lambda,l}^{m,n} f(z,\zeta)\right)^2},$$

and $\left(\frac{\zeta+z}{\zeta-z}\right)^\gamma$ is the best subordinant.

Proof. For $q(z,\zeta) = \left(\frac{\zeta+z}{\zeta-z}\right)^\gamma$, $0 < \gamma \leq 1$ in Theorem 17.2.19, we get the corollary. □

Combining Theorem 17.2.16 and Theorem 17.2.19, we state the following Sandwich theorem.

Theorem 17.2.22. *Let q_1 and q_2 be analytic and univalent in $U \times \overline{U}$ such that $q_1(z,\zeta) \neq 0$ and $q_2(z,\zeta) \neq 0$, for all $z \in U$, $\zeta \in \overline{U}$, with $\frac{z(q_1(z,\zeta))'_z}{q_1(z,\zeta)}$ and $\frac{z(q_2(z,\zeta))'_z}{q_2(z,\zeta)}$ being starlike univalent. Suppose that q_1 satisfies (17.13) and q_2 satisfies (17.18). If $f \in \mathcal{A}_\zeta^*$, $\frac{z \cdot IR_{\lambda,l}^{m+1,n} f(z,\zeta)}{\left(IR_{\lambda,l}^{m,n} f(z,\zeta)\right)^2} \in \mathcal{H}[q(0,\zeta), 1, \zeta] \cap Q^*$ and $\psi_{\lambda,l}^{m,n}(\alpha, \beta, \mu; z, \zeta)$ is as defined in (17.14) univalent in $U \times \overline{U}$, then*

$$\alpha + \xi q_1(z,\zeta) + \mu(q_1(z,\zeta))^2 + \frac{\beta z(q_1(z,\zeta))'_z}{q_1(z,\zeta)} \prec\prec \psi_{\lambda,l}^{m,n}(\alpha, \beta, \mu; z, \zeta)$$

$$\prec\prec \alpha + \xi q_2(z,\zeta) + \mu(q_2(z,\zeta))^2 + \frac{\beta z(q_2(z,\zeta))'_z}{q_2(z,\zeta)},$$

for $\alpha, \beta, \mu, \xi \in \mathbb{C}$, $\beta \neq 0$, implies

$$q_1(z,\zeta) \prec\prec \frac{z \cdot IR_{\lambda,l}^{m,n+1} f(z,\zeta)}{\left(IR_{\lambda,l}^{m,n} f(z,\zeta)\right)^2} \prec\prec q_2(z,\zeta),$$

and q_1 and q_2 are respectively the best subordinant and the best dominant.

For $q_1(z,\zeta) = \frac{\zeta+A_1 z}{\zeta+B_1 z}$, $q_2(z,\zeta) = \frac{\zeta+A_2 z}{\zeta+B_2 z}$, where $-1 \leq B_2 < B_1 < A_1 < A_2 \leq 1$, we have the following corollary.

Corollary 17.2.23. *Let $m, n \in \mathbb{N}$, $\lambda, l \geq 0$. Assume that (17.13) and (17.18) hold. If $f \in \mathcal{A}_\zeta^*$, $\frac{z \cdot IR_{\lambda,l}^{m,n+1} f(z,\zeta)}{\left(IR_{\lambda,l}^{m,n} f(z,\zeta)\right)^2} \subset \mathcal{H}[q(0,\zeta), 1, \zeta] \cap Q^*$ and*

$$\alpha + \xi \frac{\zeta+A_1 z}{\zeta+B_1 z} + \mu \left(\frac{\zeta+A_1 z}{\zeta+B_1 z}\right)^2 + \frac{\beta(A_1-B_1)\zeta z}{(\zeta+A_1 z)(\zeta+B_1 z)} \prec\prec \psi_{\lambda,l}^{m,n}(\alpha, \beta, \mu; z, \zeta)$$

$$\prec\prec \alpha + \xi \frac{\zeta+A_2 z}{\zeta+B_2 z} + \mu \left(\frac{\zeta+A_2 z}{\zeta+B_2 z}\right)^2 + \frac{\beta(A_2-B_2)\zeta z}{(\zeta+A_2 z)(\zeta+B_2 z)},$$

for $\alpha, \beta, \mu, \xi \in \mathbb{C}$, $\beta \neq 0$, $-1 \leq B_2 \leq B_1 < A_1 \leq A_2 \leq 1$, where $\psi_{\lambda,l}^{m,n}$ is defined in (17.14), then

$$\frac{\zeta + A_1 z}{\zeta + B_1 z} \prec\prec \frac{z \cdot IR_{\lambda,l}^{m,n+1} f(z,\zeta)}{\left(IR_{\lambda,l}^{m,n} f(z,\zeta)\right)^2} \prec\prec \frac{\zeta + A_2 z}{\zeta + B_2 z},$$

hence $\frac{\zeta + A_1 z}{\zeta + B_1 z}$ and $\frac{\zeta + A_2 z}{\zeta + B_2 z}$ are the best subordinant and the best dominant, respectively.

For $q_1(z,\zeta) = \left(\frac{\zeta+z}{\zeta-z}\right)^{\gamma_1}$, $q_2(z,\zeta) = \left(\frac{\zeta+z}{\zeta-z}\right)^{\gamma_2}$, where $0 < \gamma_1 < \gamma_2 \leq 1$, we have the following corollary.

Corollary 17.2.24. Let $m, n \in \mathbb{N}$, $\lambda, l \geq 0$. Assume that (17.13) and (17.18) hold. If $f \in \mathcal{A}_\zeta^*$, $\frac{z \cdot IR_{\lambda,l}^{m,n+1} f(z,\zeta)}{\left(IR_{\lambda,l}^{m,n} f(z,\zeta)\right)^2} \in \mathcal{H}\left[q(0,\zeta), 1, \zeta\right] \cap Q^*$ and

$$\alpha + \xi \left(\frac{\zeta+z}{\zeta-z}\right)^{\gamma_1} + \mu \left(\frac{\zeta+z}{\zeta-z}\right)^{2\gamma_1} + \frac{2\beta\gamma_1 \zeta z}{\zeta^2 - z^2} \prec\prec \psi_{\lambda,l}^{m,n}(\alpha, \beta, \mu; z, \zeta)$$

$$\prec\prec \alpha + \xi \left(\frac{\zeta+z}{\zeta-z}\right)^{\gamma_2} + \mu \left(\frac{\zeta+z}{\zeta-z}\right)^{2\gamma_2} + \frac{2\beta\gamma_2 \zeta z}{\zeta^2 - z^2},$$

for $\alpha, \beta, \mu, \xi \in \mathbb{C}$, $\beta \neq 0$, $0 < \gamma_1 < \gamma_2 \leq 1$, where $\psi_{\lambda,l}^{m,n}$ is defined in (17.14), then

$$\left(\frac{\zeta+z}{\zeta-z}\right)^{\gamma_1} \prec\prec \frac{z \cdot IR_{\lambda,l}^{m,n+1} f(z,\zeta)}{\left(IR_{\lambda,l}^{m,n} f(z,\zeta)\right)^2} \prec\prec \left(\frac{\zeta+z}{\zeta-z}\right)^{\gamma_2},$$

hence $\left(\frac{\zeta+z}{\zeta-z}\right)^{\gamma_1}$ and $\left(\frac{\zeta+z}{\zeta-z}\right)^{\gamma_2}$ are the best subordinant and the best dominant, respectively.

Bibliography

[1] Alb Lupaş, A. 2015. Some Differential Sandwich Theorems using a multiplier transformation and Ruscheweyh derivative, *Electronic J. Math. Appl. (EJMIA)* 1:2: 76–86.

[2] Alb Lupaş, A. 2016. About some differential sandwich theorems using a multiplier transformation and Ruscheweyh derivative, *J. Comput. Anal. Appl.* 21:7: 1218–1224.

[3] Alb Lupaş, A. 2011. A new comprehensive class of analytic functions defined by multiplier transformation, *Math. Comput. Modell.* 54: 2355–2362.

[4] Alb Lupaş, A. 2010. A note on a certain subclass of analytic functions defined by multiplier transformation, *J. Comput. Anal. Appl.* 12:1-B: 369–373.

[5] Alb Lupaş, A. 2010. A note on differential superordinations using a multiplier transformation and Ruscheweyh derivative, *Studia Universitatis Babes-Bolyai. Math.* LV:3: 3–20.

[6] Alb Lupaş, A. 2011. Certain strong differential subordinations using Sălăgean and Ruscheweyh operators, *Adv. Appl. Math. Anal.* 6:1: 27–34.

[7] Alb Lupaş, A. 2013. A note on strong differential subordinations using a generalized Sălăgean operator and Ruscheweyh operator, *Communications Math. Analysis, Acta Universitatis Apulensis* 34: 105–114.

[8] Alb Lupaş, A. 2011. Certain strong differential subordinations using a multiplier transformation and Ruscheweyh operator, *Int. J. Open Problems Complex Anal.* 3:1: 1–8.

[9] Alb Lupaş, A., Oros, G.I. and Oros, Gh. 2012. A note on special strong differential subordinations using multiplier transformation, *J. Comput. Anal. Appl.* 14:2: 261–265.

[10] Alb Lupaş, A. 2012. On special strong differential subordinations using a generalized Sălăgean operator and Ruscheweyh derivative, *J. Concrete Appl. Math.* 10:1-2: 17–23.

[11] Alb Lupaş, A. 2014. On special strong differential superordinations using Sălăgean and Ruscheweyh operators, *J. Adv. Appl. Comput. Math.* 1:1: 1–7.

[12] Alb Lupaş, A. 2021. Applications of a Multiplier Transformation and Ruscheweyh Derivative for Obtaining New Strong Differential Subordinations, *Symmetry* 13:1312: https://doi.org/10.3390/ sym13081312.

[13] Al-Oboudi, F.M. 2004. On univalent functions defined by a generalized Sălăgean operator, *Int. J. Math. Math. Sci.* 27: 1429–1436.

[14] Antonino, J.A. and Romaguera, S. 1994. Strong differential subordination to Briot-Bouquet differential equations, *J. Diff. Equ.* 114: 101–105.

[15] Bulboacă, T. 2017. Classes of first order differential superordinations, *Demonstratio Math.* 35:2: 287–292.

[16] Cătaş, A. *On certain class of p-valent functions defined by new multiplier transformations*, Proceedings Book of the International Symposium on Geometric Function Theory and Applications, August 20–24, 2007, TC Istanbul Kultur University, Turkey, 241–250.

[17] Miller, S.S. and Mocanu, P.T. 2003. Subordinants of Differential Superordinations, *Complex Variables* 48:10: 815–826.

[18] Miller, S.S. and Mocanu, P.T. 2000. *Differential Subordinations: Theory and Applications*, Marcel Dekker Inc., New York.

[19] Miller, S.S. and Mocanu, P.T. 2007. Briot-Bouquet differential superordinations and sandwich theorems, *J. Math. Anal. Appl.* 329:1: 237–335.

[20] Oros, G.I. 2009. Strong differential superordination, *Acta Universitatis Apulensis*, 19: 101–106.

[21] Oros, G.I. and Oros, Gh. 2009. Strong differential subordination, *Turkish J. Math.* 33: 249–257.

[22] Oros, G.I. 2012. On a new strong differential subordination, *Acta Universitatis Apulensis* 32: 243–250.

[23] Ruscheweyh, St. 1975. New criteria for univalent functions, *Proc. Amet. Math. Soc.* 49: 109–115.

[24] Sălăgean, G. St. 1983. *Subclasses of univalent functions*, Lecture Notes in Math., Springer Verlag, Berlin, 1013: 362–372.

[25] Selvaraj, C. and Karthikeyan, K.T. 2009. Differential Subordination and Superordination for Analytic Functions Defined Using a Family of Generalized Differential Operators, *An. St. Univ. Ovidius Constanta* 17:1: 201–210.

[26] Shanmugan, T.N., Ramachandran, C., Darus, M. and Sivasubramanian, S. 2007. Differential sandwich theorems for some subclasses of analytic functions involving a linear operator, *Acta Math. Univ. Comenianae* 16:2: 287–294.

[27] Srivastava, H.M. and Lashin, A.Y. 2005. Some applications of the Briot-Bouquet differential subordination, *JIPAM. J. Inequal. Pure Appl. Math.* 6:2: Article 41, 7 pp. (electronic).

Chapter 18

A study on self similar, nonlinear and complex behavior of the spread of COVID-19 in India

Dibakar Das

Sankalpa Chowdhury

Gourab Das

Anuska Chanda

Swapnesh Khamaru

Koushik Ghosh

18.1	Introduction	312
18.2	On the importance of the tests performed	314
18.3	Theory	315
	18.3.1 Calculation of moving averages	315
	18.3.2 Calculation of Hurst exponent by finite variance scaling method	316
	18.3.3 Estimation of fractal dimension by Higuchi's method	317
	18.3.4 Multifractal analysis by multifractal detrended fluctuation analysis	318
	18.3.5 Analysis for non-linearity using delay vector variance method	320
	18.3.6 0-1 test for chaos detection	322
	18.3.7 Mathematical aspects of self-organized criticality	323
18.4	Data	323
18.5	Results	324
	Bibliography	330

18.1 Introduction

The disease COVID-19 was first reported on 29 December, 2019 from Wuhan City of Hubei province in China and on 07 January, 2020 a new strain of coronavirus was isolated [46]. The viral strain was first named as 2019-nCoV (full form: 2019-novel coronavirus) by the World Health Organization (WHO) on 12 January, 2020 and later on 11 February, 2020 International Committee on Taxonomy of Virus (ICTV) renamed this virus as Severe Acute Respiratory Syndrome coronavirus 2 (SARS-CoV-2), as the virus is genetically very much related to the virus accountable for the SARS outbreak of 2003. On 21 January, 2020 WHO indicated possibility of human to human transmission and on 30 January a critical public-health emergency of international concern was raised and on 11 March, 2020 comprehending the acuteness and immensity in the outbreak WHO declared COVID-19 as pandemic [45].

The first case of COVID-19 reported in India was on 30 January, 2020 from Thissur in the State of Kerala [4] and only 3 cases were reported till 02 March, 2020 [7]. India initially showed very gradual growth for a long period but at a later stage huge growth was seen in terms of both daily new confirmed cases and daily deaths. In 2021, India faced a second wave of COVID-19 cases with a huge number of daily cases [7]. We have considered 24 June, 2021 as the last date for our present analysis since the trend of the daily new confirmed cases exhibits a valley like structure after that with a nagging decay for next six months till almost the end of December, 2021.

Epidemiology is a broad area of interest. The dynamics of a disease was initially thought to be governed by deterministic processes but a deeper understanding of how disease spread clearly indicated the presence of stochastic nature [3]. Time series analysis constitutes forecasting, modeling and understanding the dynamics of a system [44]. Time series analysis is a useful aspect of contemporary epidemiology [2, 40]. Statistical epidemiologists are interested in forecasting the spread of a disease, the priori to such computation is analysing the underlying dynamics and processes that affect the system. The understanding of governing dynamics has been enriched by various developments in physical and mathematical paradigms such as fractal, chaos and criticality [38].

Nature and natural systems are governed by complex laws inheriting nonlinearities in them. Fluctuations in nature are decorated by fractal and chaotic behaviors. To account for shortcoming in Euclidean geometry, Fractal geometry was introduced as a concept by Felix Hausdorff in the year 1918 and the term "fractal", was coined by Mandelbrot [32]. The concept of chaos present in various contexts was introduced in mathematics by Tien-Yien Li and James Yorke [29]. The recent advances in fractal theory was done by Mandelbrot and a path-breaking stream was formed when Mandelbrot observed patterns in chaotic signals, thus the concept of chaos was revisited, notable advances were made by Lorenz and Mandelbrot in the field [31,33]. Mandelbrot during his research on chaotic signals identified underlying patterns, as he broadened his view, Mandelbrot made an observation "*clouds are not spheres, mountains are not cones, coastlines are not circles, bark is not smooth, nor does lightning travel in a straight line*" [32]. This led to the introduction of fractals in mathematics to quantitatively analyse geometrical patterns in nature. The word fractal was coined by Mandelbrot in 1975 from a Latin word "fractus" which means "broken", interestingly the word is also used to describe a ragged cloud. Although fractal geometry is a subset of Euclidean space, a layman approach to describe fractal is a mathematical set whose dimension is non-integer [32] and generally exceeding its topological dimension [33]. Fractals may exhibit scale invariance [11,32], but it is not a necessity. Hence, fractals may not necessarily show scale-invariance and self-similarity. Over broad sense the concept of fractals considers detailed patterns repeating itself.

Chaos is a phenomenon seen in dynamical systems with states of disorder and irregularity in them. Chaos in mathematics is broadly segmented into deterministic and stochastic chaos. Stochastic chaos is affected by nonlinear scale invariant characteristics. The presence of nonlinearity in a time series gives a probable presence of chaos in the system. Deterministic chaotic systems exhibit high sensitivity to initial conditions, this in turn means exponential growth of the sensitive factor present in the system. For deterministic chaos even infinitesimal perturbations introduced in initial states of a system leads to huge changes in final state. On the contrary, however non-definable and random a deterministic chaos may seem to be, in practice the governing dynamics of these systems are determined by some or other physical or mathematical laws predictable over shorter ranges of action. The degrees of freedom of such systems are limited to a few. On short time scales, deterministic chaotic systems can be fairly simply comprehended to understand its dynamics and their study reveals the interaction between order and absolute chaos inherent in the system. Stochastic chaotic systems i.e., systems with nonlinear scale invariant dynamics, generally show infinite degrees of freedom.

The concept of self organized criticality (self organized criticality) can be established in dynamical systems with internal mechanisms that translates to statistical stationarity when attacked by perturbations, this concept was crucial in giving explanatory answers in open and extended driven systems exhibiting avalanche like energy dissipation [5]. Systems with self organized criticality properties in general show scale invariance in distribution of relaxation events with sufficient competency to reach statistical stationarity due to self-regulatory internal mechanisms. A significant characteristic of systems exhibiting self organized criticality is the existence of power law in the observed magnitude of events.

In the present work we analysed the time series of daily new cases for various factors. The data has been pre-processed by using a dynamic 7-point moving average to remove random fluctuations in the time series. Initially we performed scaling analysis to understand the inherent memory of the time series. Henceforth, we have tested for self similarity. After finding the monofractal dimension by Higuchi method [19], we felt the need to understand the multifractal properties. The Delay Vector Variance method has been used to identify if the time series of daily new cases exhibits non-linearity. The possibility of chaos was examined by 0-1 chaos test [13]. In the end the presence of self-organized criticality was assessed by the presence of power law distribution in the cumulative distribution. All the tests performed have significant contributions in the understanding of governing dynamics of the disease.

18.2 On the importance of the tests performed

The prediction in various analyses such as forecasting is heavily affected by long memory processes [18]. Generally the auto-correlation function decay in short memory is of exponential nature while for long memory it is very gradual. Hurst [21] analyzed the hydrological properties of river Nile to control irregular flows and regular floods. He thereafter introduced a computational static called 'Adjusted Range' [21], at present more commonly called as Rescaled Range Analysis [34], the exponent thus calculated was later coined as 'Hurst exponent'. The procedures for estimation Hurst exponent were later modified with the introduction of fractal theory by Mandelbrot [33, 34]. Estimating the Hurst exponent for epidemiological data has been performed in [41].

Fractals can be represented as equations that are nowhere differentiable. This way fractals are broadly segmented as monofractal and multifractal with respect to self-similarity in change of scales. Monofractals have homogeneous distribution of singularity exponent, hence have self-similar properties over the complete scale for the considered time series [6]. When a single power law $F^{-\alpha}$ (where F represents frequency) describes the power spectrum of the considered time series, it may show fractal nature, where α is the irregularity index. Such power laws approximate the memory for self-similar series. The method of estimating fractal dimension was innovated by Higuchi by relating fractal length $L(k)$ with the time interval k in a power law relation $L(k) \propto k^{-D}$, where index D is the computed fractal dimension [19]. Higuchi also investigated the relation between α and D formulated $D = (5 - \alpha)/2$. The value of D lies in $1 < D < 2$ for statistically self-similar curves if they are embedded in a plane, while D is identical to topological dimension for smooth rectifiable curves. The Higuchi method is statistically significant and stable even for non-stationary time series [28]. Higuchi method has been used in epidemic data in [8, 41]. Eventually with the expansion of self-similar time series, the global self-similarity may diminish and local self-similar properties may emerge for different ranges of scales. Complex systems like natural processes inherit inhomogeneous distributions in their time series. Multifractal Detrended Fluctuation Analysis is a mathematical tool widely used in various analytical processes. MFDFA is used to understand the multifractal behaviors of a time series [23, 36]. Multifractality has been established in epidemic spread models in [20] and considered for the case of China in [30].

A stochastic process may be governed by nonlinear dynamical equations. Previous methods in understanding nonlinearity in time series such as Kaplan [24] and by Kennel et al. [26], Kaplan's $\delta - \epsilon$ method [25], Correlation Exponents (COR) [17], etc. were statistically insignificant due to certain limitations [1]. Delay Vector Variance is a novel method introduced by Gautama et al. [12] to investigate nonlinearity and determinism in time series.

In the recent era with increase in computational strength and speed, the procedures for visualizing chaos have been modified heavily. The "0-1 chaos test" is an important tool for detection of chaos, with the advantage of single integer output [13]. The test for nonlinearity and chaos in epidemic data was performed in [9].

The self organized criticality phenomenon in the contemporary context of time series requires fractality as a necessary condition, thereafter the self organized criticality is established in systems that show power law relation in its cumulative distribution. Self organized criticality has been established in various physical organizations, the introduction of self organized criticality in epidemic dynamics was done in [16] and later revisited in [39]. The presence of self organized criticality was investigated in COVID-19 spread in [8] and has been revisited for pan India context, in the present work.

18.3 Theory

The time series collected from the website Covid19India is denoted by d_t, where $t = 1, 2, ..., N$. This series is initially treated with 7-point moving average technique which generates a new time series $x(t)$, where $t = 1, 2, ..., N$. Thus, time series $x(t)$ is used for the calculation in other sub-sections.

18.3.1 Calculation of moving averages

Irregular fluctuations in reported data is caused by difference in biomedical reaction in different patients or delay in reporting. These can be eradicated by an odd window moving average. But, if the window size is too large, then the moving average distorts the dynamics. In the present daily data of new confirmed cases of COVID-19, we can find seven-day cycles in the long run with weekly minima usually on Monday (sometimes on Sunday also) due to less number of tests during weekends. For this reason to detrend these week long cyclic behavior we have employed seven point moving average on the present data.

The standard practice of $(2\omega + 1)$-point moving average setup (where ω is a positive integer) the sample size gets reduced by 2ω, this was solved in [8] in the following method:

Let d_t be the raw data and $x(t)$ be the moving average data, where $(t = 1, 2, ..., N)$, such that:

$$x(t) = \begin{cases} d_t & \text{, for } t = 1 \text{ and } N \\ \frac{1}{3}(d_{t-1} + d_t + d_{t+1}) & \text{, for } t = 2 \text{ and } (N-1) \\ \frac{1}{5}(d_{t-2} + d_{t-1} + d_t + d_{t+1} + d_{t+2}) & \text{, for } t = 3 \text{ and } (N-2) \\ \frac{1}{7}(d_{t-3} + d_{t-2} + d_{t-1} + d_t + \\ \quad d_{t+1} + d_{t+2} + d_{t+3}) & \text{, for } t = 4, 5, ..., (N-3) \end{cases}$$
(18.1)

18.3.2 Calculation of Hurst exponent by finite variance scaling method

Memory analysis plays an important role in understanding the viability and accuracy of modeling or forecasting a dynamical process, by understanding low frequency variation and trends. The memory is a quantitative estimation of the rate of change of correlation function, i.e., the statistical dependence between two points in a stochastic series, as the separation between them increases. A strong dependence with the past points gives us long memory processes, while in the short memory process the dependence is only visible in the neighborhood. Finite Variance Scaling Method is an efficient technique for performing Scaling Analysis on time series [21] and a well known procedure is Standard Deviation Analysis (SDA). A sequence of cumulative standard deviation $D(t_j)$ is calculated using FVSM for partial time series $x(t)$ $(t = (1, 2, ..., j)$, where $(j = 1, 2, ...N)$ as given by:

$$D(t_j) = \left[\frac{1}{j}\sum_{t=1}^{j} x^2(t) - \left\{\frac{1}{j}\sum_{t=1}^{j} x(t)\right\}^2\right]^{\frac{1}{2}}, \quad j = 1, 2, 3, ..., N \quad (18.2)$$

Eventually a self-similar time series may exhibit a power law relation [21] for $D(t)$ in the following way:

$$D(t) \sim t^H \quad (18.3)$$

The regression analysis of log-log plot of $D(t)$ vs t using the best fitted straight line approach gives us the term indexed H named as the Hurst exponent. For time series obeying laws of self-similarity the value of Hurst exponent calculated using FVSM lies in the range of [0,1]. The Hurst exponent can be considered a reliable statistical tool to identify the correlational properties without inspecting for nonlinearities in a time series. A comparative analysis is provided for various Hurst exponent values that exist and the specific behavior of the time series exhibiting the memory.

Hurst exponent value within the range $0 < H < 0.5$ is indicative of mean-reverting or anti-persistent time series, i.e., the series possesses a short memory. A time series exhibiting short memory will have revertions in magnitude in the future.

$H = 0.5$, coined as "Joseph effect" indicates a time series exhibiting Brownian motion or true random walk. Time series displaying Hurst exponents of 0.5 or its close approximate are hard to predict. This indicates that in this situation, there is either null or very feeble autocorrelation.

Hurst exponent value within range $0.5 < H < 1$ is indicative of persistent time series possessing long memory. A time series with long memory indicates slow decay in the autocorrelation of the time series which means that an increase will be followed by an increase more frequently and vice versa.

$H = 1$ signifies a trend following smooth time series with a high degree of autocorrelation. In such time series an uptrend or increase is followed by the same and vice versa for the complete sequence of finite time series within some confidence level. These time series theoretically indicate an infinite growth of epidemic leading to a very fatal situation, although this is not a general case, if a disease is highly contagious and prevention measures aren't adequate, the possibility of this phase remains, but, only for limited time windows.

18.3.3 Estimation of fractal dimension by Higuchi's method

Fractal analysis can be performed in various ways to estimate the fractal dimension of a time series. Higuchi method is an important and efficient tool for finding the fractal dimension.

From the time series $x(t)$, a new time series X_k^m is constructed as below:

$$X_k^m = \{X(m), X(m+k), X(m+2k), ..., X(m + [(N-m)/k] \cdot k)\}, \quad (18.4)$$

where m ($m = 1, 2, ..., k$) indicates initial time with an integer value while another integer k indicates time interval, while the term in the [.] indicates the greatest integer following the Gaussian notation. Following the above procedures, we obtain k sets of new time series from initial time series.

As defined by Higuchi, the length of curve for each time series X_k^m is calculated as:

$$L_m(k) = \frac{1}{k}\left\{\sum_{i=1}^{[(N-m)/k]} |X(m+ik) - X(m + (i-1)k)|\right\}\left(\frac{N-1}{[(N-m)/k] \cdot k}\right), \quad (18.5)$$

where the normalization is performed by the term $(\{N-1\}/[(N-m)/k] \cdot k)$ present in the length equation.

The length of curve $\langle L(k) \rangle$ is given by the mean of k sets of $L_m(k)$ for each time interval k. For time series to be fractal, the length of curve $\langle L(k) \rangle$ follows a inverse power law relation for different time interval as follows:

$$\langle L(k) \rangle \sim k^{-D}, \quad (18.6)$$

where the term indexed as D is a measure of the fractal dimension of a time series. The slope of the best fitted straight line against the log-log plot of $\langle L(k) \rangle$ vs k gives us an estimate of D using simple regression analysis and lies in the interval [1,2] for a 1-dimensional time series [19].

18.3.4 Multifractal analysis by multifractal detrended fluctuation analysis

Detrended Fluctuation Analysis (DFA) [35, 37] is an important statistical tool. Multifractal Detrended Fluctuation Analysis (MFDFA) an extension of DFA and it is an useful tool to understand the multifractal properties of a time series. MFDFA is used to eliminate polynomial trends from the system to introspect on the scaling properties of the non-stationary time series considered. The time series $x(t)$, where $t = 1, 2, .., N$ is assumed to be of compact support or $x(t) = 0$, for a finitely small number of samples only. The steps for computation of MFDFA are given as [23]:

1. The time series $x(t)$ ($t = 1, 2, ..., N$) is used to determine the profile $Y(j)$ defined as:

$$Y(j) = \sum_{t=1}^{j} (x(t) - \langle x \rangle), \qquad j = 1, 2, 3, ..., N, \qquad (18.7)$$

where $\langle x \rangle$ is the mean of the time series $x(t)$ and it can be easily verified that $Y(N) = 0$.

2. $N_s = int(N/s)$ is calculated and thereafter the profile $Y(j)$ into N_s non-overlapping segments of s equal lengths. A general observation in performing the above shows that a short tail is always disregarded due to the fact that generally s is not a factor of N. We repeat the same procedure from the opposite end and thereby fulfill the tail preservation. Hence, $2N_s$ non-overlapping segments are obtained.

3. For each $2N_s$ segment, we compute the local trends by means of least square fitting. The variance is calculated by

$$F^2(s, v) = \frac{1}{s} \sum_{j=1}^{s} \{Y[(v-1)s + j] - y_v(j)\}^2 \qquad (18.8)$$

for each segment v, $v = 1, 2, ..., N_s$ and

$$F^2(s, v) = \frac{1}{s} \sum_{j=1}^{s} \{Y[N - (v - N_s)s + j] - y_v(j)\}^2 \qquad (18.9)$$

for $v = N_s + 1, N_s + 2, ..., 2N_s$.

The fitting polynomial $y_v(j)$ in the segment v is subtracted from the profile eliminates the trend in the time series. The order of the trend which needs to be removed from the time series is determined by the order of the polynomial $y_v(j)$. The trends of m_{th} order in the profile eliminate the $(m-1)_{th}$ order trends in the original time series and is called MFDFAm. Hence in linear, quadratic, cubic polynomial fitting is called MFDFA1, MFDFA2, MFDFA3, respectively for higher order polynomial fitting [22]. In the present scenario we have considered MFDFA2 to remove at most the quadratic trends.

4. The mean $F^2(s,v)$ of all the segments provide the q_{th} order fluctuation function $F_q(s)$ defined as

$$F_q(s) = \left\{ \frac{1}{2N_s} \sum_{v=1}^{2N_s} [F^2(s,v)]^{\frac{q}{2}} \right\}^{\frac{1}{q}}, \quad (18.10)$$

where moment q can assume any non-zero real value. To account the diverging exponent when $q \to 0$ we take a logarithmic average to find $F_q(s)$ at $q \to 0$:

$$F_0(s) = \exp\left\{ \frac{1}{4N_s} \sum_{v=1}^{2N_s} \ln|F^2(s,v)| \right\}. \quad (18.11)$$

Hence, we find $F_q(s)$ for $s \geq m+2$. For understanding the multifractality, we look into the dependence of $F_q(s)$ on s i.e. the time scale for specific values of q.

5. The slope of log-log plots of $F_q(s)$ v/s s for different values of q gives us the scaling behavior of the fluctuation function $F_q(s)$. We see if a long range power law correlation is present in the analyzed time series, thus for sufficiently large values of s we may find that the scaling of fluctuation $F_q(s)$ follows a power-law relation:

$$F_q(s) \sim s^{h(q)}. \quad (18.12)$$

The right hand index of the equation gives us an exponent $h(q)$ which depends on the variable q.

In the case of stationary time series, the exponent calculated by MFDFA2, $h(2)$ it is established that the same is identical to the standard Hurst exponent, H [43]. We may thus consider that the exponent $h(q)$ generalizes the standard Hurst exponent and thus, the name generalized Hurst exponent is commonly used, we use the term generalized Hurst exponent hereafter. Thus, the generalized Hurst exponent becomes an important tool to distinguish between time series. For monofractal time series the generalized Hurst exponent is equal to standard Hurst exponent for all values of q i.e., $h(q) = H$ and time series for which $h(q)$ decreases monotonously as a function of q, the corresponding time series is multifractal. Segments with small fluctuations are characterized by negative values of $h(q)$, q is dominated by segments v holding small variances $F^2(s,v)$, the negative $h(q)$, q are formed by scaling of the segments with small variances. While segments with large fluctuations are characterized by positive values of $h(q)$, q dominated by the segments v, holding large variances $F^2(s,v)$.

Here, we compute the values of $h(q)$ for different values of q and analyze them to understand if the considered time series is multifractal.

18.3.5 Analysis for non-linearity using delay vector variance method

A time series $x(t)$ of length N is embedded by time delay and can be represented in 'phase space'. For a set of delay vectors of a given dimension are obtained by performing the embedding of the time delay. A mth dimensional delay vector, the DVs can be represented as $x(k) = [x_{k-m\tau}, \ldots, x_{k-\tau}]$, where the term τ denotes time lag and $k = 1, 2, \ldots, N$. We group the DVs in a way so that they are within a certain Euclidean distance to DV $x(k)$ with a corresponding target to the next sample. These groups are denoted by λ_k. The Euclidean distances are varied with respect to the pair-wise distribution of distances between DVs in a standardized manner. The mean target variance σ^{*2} is computed over all sets of λ_k for an optimal embedding dimension m. The optimal embedding dimension (m) is chosen for the minimal target variance σ^{*2}, thereafter we examine the complete range of pairwise distance for the variations of the standardized distance. We present the algorithm for Delay Vector Variance (DVV) method:

1. We define the pairwise Euclidean distance between the DVs, and calculate the mean and standard deviation. The distance is given by

$$d(t,j) = ||x(t) - x(j)|| \qquad (t \neq j). \tag{18.13}$$

2. We generate sets of $\lambda_k(\tau_d)$, for $k = 1, 2, \ldots, N$ so that they satisfy the equation given as:

$$\lambda_k(\tau_d) = \{x(t) | ||x(k) - x(t)|| \leq \tau_d\}. \tag{18.14}$$

So, $\lambda_k(\tau_d)$ is chosen on the basis that the DVs lie closer to $x(t)$ than the distance τ_d where $[\min\{0, (\mu_d - n_d\sigma_d)\} \leq \tau_d \leq (\mu_d + n_d\sigma_d)]$, and n_d is a parameter controlling the span selected for DVV analysis.

3. We compute the variance of target counterparts $\sigma_k^2(\tau_d)$ for each set $\lambda_k(\tau_d)$. The target variance σ_k^2 is calculated from the mean of all sets of $\lambda_k(\tau_d)$ and normalized by the the variance of the time series $x(t)$, $\sigma^{*2}(\tau_d)$ given as:

$$\sigma_s^{*2}(\tau_d) = \frac{(\frac{1}{N})\sum_{k=1}^{N}\sigma_k^2(\tau_d)}{\sigma_x^2}. \tag{18.15}$$

The measured target variance gives us the unpredictability. The reliability of estimating the true population variance we require sufficient enough points for calculating the sample variance. To attain the reliability in the measurement of variance, we consider that the set of $\lambda_k(\tau_d)$ must atleast have 30 DVs [12]. This gives us N_0, the minimum set size.

Due to the standardization of the distance axis, τ_d is substituted by $\frac{\tau_d - \mu_d}{\sigma_d}$, which has a unit variance and zero mean. The plot of target variance $\sigma^{*2}(\tau_d)$ as a function of the standard distance gives us the DVV plots. Since the stochastic component incorporates largest quantity of noise, it will have large $\sigma^{*2}(\tau_d)$. If the signal posses strong deterministic components, the value of $\sigma^{*2}(\tau_d)$ will be small. Since all DVs considered are in the same set and the variance of the target matches to the variance of the time series for maximum span, the DVV shall converge to unity, the span parameter n_d is updated by increasing if the criteria is not met.

4. Using iterated amplitude adjusted Fourier transform (IAAFT) method, we yield a number of surrogate time series which has amplitude spectra identical and approximately identical signal distribution to the original time series [27, 42]. The DVV plots for the surrogate time series are also calculated using optimal embedding dimension of time series $x(t)$.

5. The "DVV Scatter Diagram" is obtained by plotting the target variance $\sigma^{*2}(\tau_d)$ of the original time series on horizontal axis and on vertical axis, we plot the mean of the surrogate time series, on the standardized distance axis.

6. If the DVV plots of surrogate time series are similar to the DVV plot of the original time series we see that the 'DVV Scatter Diagram' coincides with the bisector line and the original time series is linear in nature. Whereas if the DVV plots of surrogate time series are not similar to the DVV plot of original time series the 'DVV Scatter Diagram' deviates from the bisector line, and we can identify the original time series as non-linear.

7. To give a quantitative estimate of the deviation from bisector line signifying nonlinearity, we compute the root mean square error (RMSE) between the average $\sigma_s^{*2}(\tau_d)$ of the surrogate time series and the $\sigma^{*2}(\tau_d)$ of the original time series, as:

$$\text{RMSE} = \left[\text{mean} \left\{ \sigma^{*2}(\tau_d) - \frac{\sum_{k=1}^{N_s} \sigma_{s,k}^{*2}(\tau_d)}{N_s} \right\}^2 \right]^{\frac{1}{2}}, \quad (18.16)$$

where we take the mean over all spans of τ_d valid in all DVV plots of the surrogate series and $\sigma_{s,k}^{*2}(\tau_d)$ is the target variance at span τ_d for the k_{th} surrogate [1].

18.3.6 0-1 test for chaos detection

The 0-1 chaos test is a binary test invented and later modified by Gottawald and Melbourne [13]. It is an important statistical tool due to its computational simplicity over the Lyapunov exponent and the phase-space reconstruction for detection of chaos. The 0-1 test takes an input time series and gives the output in the form of "0" or "1" which is reliable and robust [14] for the inspection of deterministic chaos in experimental data [10]. The time series $x(t)$ can be tested for the presence of chaos following the steps given by [15]:

1. From the original time series a Fourier transformed series is constructed as,

$$p_n = \sum_{t=1}^{n} x(t) e^{itc}, \qquad (18.17)$$

where i represents the imaginary unit and multiple values of c are randomly chosen. Here, we consider 100 values of c chosen randomly in the range $[\pi/5, 4\pi/5]$.

2. The smooth mean square displacement denoted by $D_c(t)$ is computed as

$$D_c(n) = \frac{1}{N-m} \sum_{t=1}^{N-m} |p_{t+n} - p_t|^2 - \langle x \rangle^2 \frac{1 - \cos nc}{1 - \cos c}, \qquad (18.18)$$

where $\langle x \rangle \equiv \frac{1}{N} \sum_{t=1}^{N} x(t)$ defines the average of considered time series and $s \leq m = N/10 \ll N$

A system with underlying dynamics of deterministic chaos, the $D_c(n)$ scales linearly with n, i.e., $D_c(n) \sim n$ and we thus see p_n exhibits a Brownian motion in the complex plane. For a system with regular dynamics $D_c(n)$ is seen a bounded function of n, i.e., $D_c(n)$ does not increase with n.

The test is improved by the modification of $D_c(n)$ to $D^*(n)$ defined below which makes the test more robust to weak noise corrupted signals and therefore also to weakly chaotic dynamics.

$$D_c^*(n) = D_c(n) + \alpha V_{damp}(n), \qquad (18.19)$$

where $\alpha V_{damp}(n)$ is given as $V_{damp}(n) = \langle x^2 \rangle \sin(\sqrt{2}n)$. The magnitude of α, coefficient of $\alpha V_{damp}(n)$ determines the sensitivity of the test to weak chaos inherent from the sensitivity to weak noise.

3. The asymptotic growth rate K_c is obtained in order to assess the linear growth strength defined as

$$K_c = \operatorname{corr}(n, D^*(n)). \qquad (18.20)$$

The measured quantity provides the strength of correlation of $D^*{}_c(n)$ with linear growth.

4. The value of K is the binary output of the 0-1 test,

$$K = \text{median}(K_c). \tag{18.21}$$

Practically the test in general does not return a "0" or "1", if the estimated value of K is close to "1" we may judge the dynamics as chaotic otherwise we may judge the dynamics as regular i.e., the value of K close to "0".

18.3.7 Mathematical aspects of self-organized criticality

A dynamical system exhibiting self organized criticality is given by the presence of power law in cumulative magnitude of observed events [5]. A property t of a dynamical system with distribution of events $x(t)$ exhibiting a the property $x(t) \sim t^{-\zeta}$ is analyzed using the cumulative distribution as follows [8]:

$$\bar{X}(t) = \int_t^M x(t)dt, \tag{18.22}$$

where M denotes the maximal event occurring in the data. For finite values of M, we may find

$$\bar{X}(t) \propto t^{-\zeta+1}[1 - (\frac{t}{M})^{\zeta-1}] \tag{18.23}$$

For large sample denoted by sufficiently large values of M, we can modify the equation without loss of generality in the form:

$$\bar{X}(t) \sim t^{-\zeta+1}. \tag{18.24}$$

The slope of log-log plots for $\bar{X}(t)$ vs t, determined by the best fitted straight line gives us the value of $-\zeta + 1$ and therefore, the value of ζ. If the value of ζ lies in the right hand neighborhood of 1, we may confirm the presence of self-organized criticality.

18.4 Data

In the present study, the daily new confirmed cases of COVID-19 in India. We can visualize the data and its moving average in FIGURE 18.1 and the data has been retrieved from Covid19India on 24 June, 2021. The data considered for various tests have been described in Table 18.1.

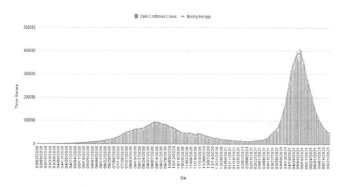

FIGURE 18.1: Daily New Confirmed Cases and 7-point moving average in India (Days are mentioned in mm/dd/yyyy format)

TABLE 18.1: Description of the time series taken:

Time Series	Range of Data	Sample size (in days)
Daily new cases in India	2 March, 2020 to 24 June, 2021 (02/03/2020-24/06/2020)	480
First Wave	02/03/2020-08/02/2021	344
Growth Window1	02/03/2020-14/09/2020	197
Declining Window1	15/09/2020-08/02/2021	147
Second Wave	09/02/2021-24/06/2021	136
Growth Window2	09/02/2021-05/05/2021	86
Declining Window2	06/05/2021-24/06/2021	50

18.5 Results

We have analyzed the time series of daily new confirmed cases in India for Hurst exponent, monofractal dimension, multifractal property, nonlinearity, chaos, and self-organized criticality. The results for Standard Deviation Analysis, Higuchi method, Multi Fractal Detrended Fluctuation Analysis, Delayed Vector Variance method, 0-1 chaos test and value of ζ for understanding self organized criticality are given in Tables 18.2 to 18.6 and their figurative outputs have been given in Figures 18.2 to 18.9.

For the Hurst exponent, we have fragmented the time series into four parts, two growing and two declining windows, one from each wave. The Hurst

TABLE 18.2: Results for scaling analysis

Time series considered	H (Hurst exponent value)
Growth Window1 (First Wave)	0.357
Declining Window1 (First Wave)	0.229
Growth Window2 (Second Wave)	0.421
Growth Window2 (Second Wave)	0.227

TABLE 18.3: Monofractal dimension and generalized Hurst exponent

Time series considered	D (Monofractal Dimension)	$h(2)$
Complete Time Series	1.610722521	0.779
First Wave	1.159152718	0.918
Second Wave	1.140204302	0.71

TABLE 18.4: Results for DVV method for detection of nonlinearity

Time series	R.M.S.E.
Complete Time Series	0.589

TABLE 18.5: Outputs of 0-1 Chaos test

Time series considered	$K = median(K_c)$
Complete time series	−0.0795
First Wave	0.0696
Second Wave	0.0253

TABLE 18.6: Results for self organized criticality

Time series considered	ζ
Complete time series	1.152

exponent that signifies the memory of a time series shows that overall all the window gives an antipersistent type memory. For the growth window in the second wave, the memory is inclined toward Brownian motion. In both cases, we find that the Hurst exponent of declining window is smaller than the Hurst exponent for growing window. For inspection of self similarity, we considered two techniques. For monofractal analysis we have used the Higuchi method. The monofractal analysis for the complete time series, first wave and second wave give D strictly greater than topological dimension i.e., 1. Hence, monofractality is established in the time series. The generalized Hurst exponent has been found to be greater than 0.5 in all the three cases considered i.e., complete time series, first wave and second wave. The differences in generalized

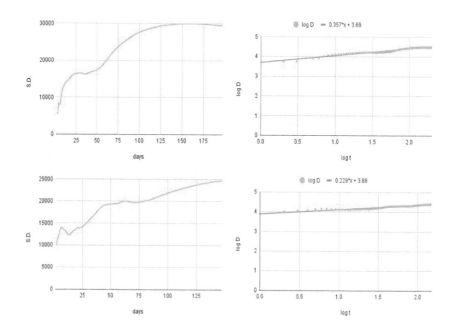

FIGURE 18.2: a. $D(t)$ vs. t and b. $\log D(t)$ vs. $\log t$ for Growth Window1. c. $D(t)$ vs. t and d. $\log D(t)$ vs. $\log t$ for Decline Window1.

Hurst exponent in different sections indicate the presence of multifractality in the time series. The RMSE and Scatter diagram in DVV analysis give us clear evidence of nonlinearity in the time series. The results of 0-1 test indicate the absence of deterministic chaos in the time series. Overall with anti-persistent memory presence of fractality, non-linearity and the absence of chaos we may ascertain that long term forecasting is possible with limitations. Additionally we have analyzed the profile of self organized criticality for pan India context. With respect to previous studies the present value of ζ has moved nearer to 1, indicating that there might be a possibility of introduction of self organized criticality afterward. The self organized criticality is not a static parameter like others, it changes with the change of the dynamics of the disease and in the current scenario the system does not exhibit self organized criticality, but the introduction of self organized criticality is a race between vaccination drive and mutation of the virus. Thus, fast and efficient vaccination drive is an important step in minimizing the possibility of introduction of self organized criticality in the governing dynamics of COVID-19, hence controlling and reducing the pandemic.

A Study on Self Similar, Nonlinear 327

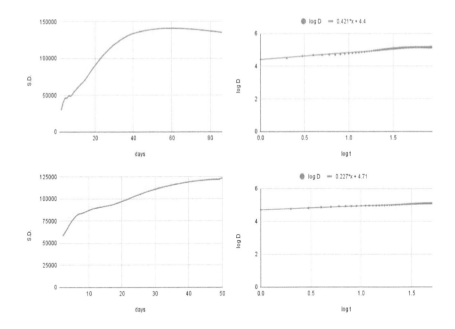

FIGURE 18.3: a. $D(t)$ vs. t and b. $\log D(t)$ vs. $\log t$ for Growth Window2. c. $D(t)$ vs. t and d. $\log D(t)$ vs. $\log t$ for Decline Window2.

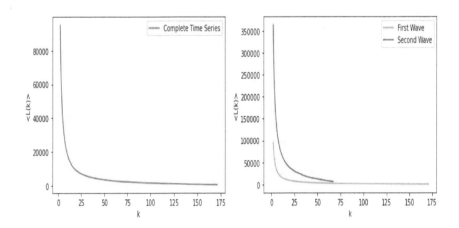

FIGURE 18.4: $\langle L(k) \rangle$ vs. k for a. complete time series b. First and Second Wave.

328 Advances in Mathematical Analysis and its Applications

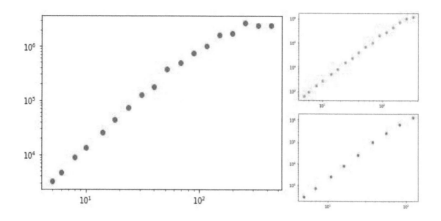

FIGURE 18.5: $h(q)$ vs. q for a. complete time series b. First and Second Wave.

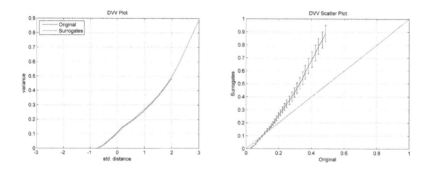

FIGURE 18.6: DVV Plot and DVV scatter diagram for complete series.

A Study on Self Similar, Nonlinear   329

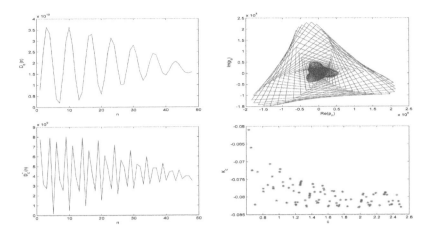

FIGURE 18.7: a. $D_c(n)$ vs. n, b. p_n in complex plane, c. $D_c^*(n)$ vs. n and d. K_c vs. c for complete time series.

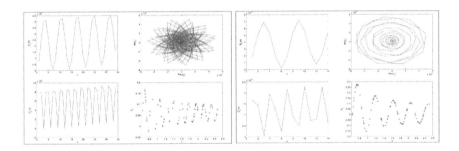

FIGURE 18.8: a. $D_c(n)$ vs. n, b. p_n in complex plane, c. $D_c^*(n)$ vs. n and d. K_c vs. c for i. First wave ii. Second wave

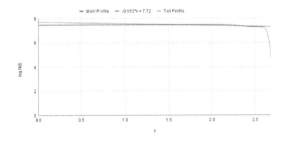

FIGURE 18.9: Profile of self organized criticality.

Bibliography

[1] Ahmed, I. 2010. Detection of nonlinearity and stochastic nature in time series by delay vector variance method, *Int. J. Engineering & Technology*, 10:2: 22–27.

[2] Allard, R. 1998. Use of time-series analysis in infectious disease surveillance, *Bull. World Health Organization*, 76:4: 327–333.

[3] Alonso, D., McKane, A.J. and Pascual, M. 2007. Stochastic amplification in epidemics, *J. The Royal Society Interface*, 4:14: 575–582.

[4] Andrews, M.A., Areekal, B., Rajesh, K.R., Krishnan, J., Suryakala, R., Krishnan, B., Muraly, C.P. and Santhosh, P.V. 2020. First confirmed case of COVID-19 infection in India: A case report, *Indian J. Med. Res.* 151:5: https://www.ijmr.org.in/text.asp.2020/151/5/490/285776.

[5] Bak, P., Tang, C. and Wiesenfeld, K. 1988. Self-organized criticality, *Phys. Rev. A*, 38:1:364.

[6] Bunde, A. and Havlin, S. 1996. (eds.), *Fractals and Disordered Systems*, 2nd ed, Springer, Berlin.

[7] COVID19INDIA, Covid19India, https://www.covid19india.org/ 2020-21, Accessed on: 25 June 2021.

[8] Das, D., Chowdhury, S., Khamaru, S., Chanda, A. and Ghosh, K. 2021. Investigation of self-organized criticality in daily new confirmed cases of covid-19 in some major affected Indian states, *Bull. Calcutta Math. Soc.* 113:1: 31–52.

[9] Ellner, S., Gallant, A.R. and Theiler, J. 1995. Detecting nonlinearity and chaos in epidemic data, Epidemic models: their structure and relation to data, https://www.osti.gov/servlets/purl/10179849.

[10] Falconer, I., Gottwald, G.A., Melbourne, I. and Wormnes, K. 2007. Application of the 0-1 test for chaos to experimental data, *SIAM J. Appl. Dynamical Systems*, 6:2: 395–402.

[11] Falconer, K. 2003. *Fractal Geometry: Mathematical Foundations and Applications, Second Edition*, John Wiley & Sons, Ltd.

[12] Gautama, T., Mandic, D.P. and van Hulle, M.M. 2004. The delay vector variance method for detecting determinism and nonlinearity in time series, *Physica D: Nonlinear Phenomena*, 190:3-4: 167–176.

[13] Gottwald, G.A. and Melbourne, I. 2004. A new test for chaos in deterministic systems, *Proc. Royal Society of London. Series A: Mathematical, Physical and Engineering Sciences*, 460:2042: 603–611.

[14] Gottwald, G.A. and Melbourne, I. 2008. Comment on "reliability of the 0-1 test for chaos", *Physical Review E*, 77:2: 028201, doi.org/10.1103/PhysRevE.77.028201.

[15] Gottwald, G.A. and Melbourne, I. 2009. On the implementation of the 0–1 test for chaos, *SIAM J. Appl. Dynamical Systems*, 8:1: 129–145.

[16] Grassberger, P. 1983. On the critical behavior of the general epidemic process and dynamical percolation, *Mathematical Biosciences*, 63:2: 157–172.

[17] Grassberger, P. and Procaccia, I. 2004. Measuring the strangeness of strange attractors, *The Theory of Chaotic Attractors*, 9: 170–189.

[18] Graves, T., Franzke, C.L.E., Watkins, N.W., Gramacy, R.B. and Tindale, E. 2017. Systematic inference of the long-range dependence and heavy-tail distribution parameters of ARFIMA models, *Physica A: Statistical Mechanics Appl.* 473: 60–71.

[19] Higuchi, T. 1988. Approach to an irregular time series on the basis of the fractal theory, *Physica D: Nonlinear Phenomena*, 31:2: 277–283.

[20] Holdsworth, A.M., Kevlahan, N.K.-R. and Earn, D.J.D. 2012. Multifractal signatures of infectious diseases, *J. Royal Society Interface*, 9:74: 2167–2180.

[21] Hurst, H.E. 1951. Long-Term Storage Capacity of Reservoirs, *Trans. Amer. Soc. Civil Engineers*, 116:1: doi.org/10.1061/TACEAT.0006518.

[22] Kantelhardt, J.W., Bunde, E.K., Rego, H.H.A., Havlind, S., and Bunde, A. 2001. Detecting long-range correlations with detrended fluctuation analysis, *Physica A: Statistical Mechanics Appl.* 295:3-4: 441–454.

[23] Kantelhardt, J.W., Zschiegner, S.A., Bunde, E.K., Havlind, S., Bunde, A. and Stanley, H.E. 2002. Multifractal detrended fluctuation analysis of nonstationary time series, *Physica A: Statistical Mechanics Appl.* 316:1-4: 87–114.

[24] Kaplan, D.T. 1994. Exceptional events as evidence for determinism, *Physica D: Nonlinear Phenomena* 73:1-2: 38–48.

[25] Kaplan, D.T. 1997. Nonlinearity and nonstationarity: the use of surrogate data in interpreting fluctuations, *Studies Health Tech. Informa.* 215–281.

[26] Kennel, M.B., Brown, R., Henry, and Abarbanel, D.I. 1992. Determining embedding dimension for phase-space reconstruction using a geometrical construction, *Physical review A*, 45:6: 3403.

[27] Kugiumtzis, D. 1999. Test your surrogate data before you test for nonlinearity, *Physical Review E*, 60:3: 2808.

[28] Torre, F De la, González-Trejo, J.I., Real-Ramí rez, C.A. and Hoyos-Reyes, L.F. 2013. Fractal dimension algorithms and their application to time series associated with natural phenomena, *J. Phys.: Conf. Ser.* 475: 012002.

[29] Li, T-Y and Yorke, J.A. 1975. Period Three Implies Chaos, *The Amer. Math. Monthly*, 82:10: 985–992.

[30] Long, Y., Chen, Y. and Li, Y. 2020. Multifractal scaling analyses of the spatial diffusion pattern of COVID-19 pandemic in Chinese mainland.

[31] Lorenz, E.N. and Haman, K. 1996. The essence of chaos, *Pure Appl. Geophysics* 147:3: 598–599.

[32] Mandelbrot, B.B. 1983. *The Fractal Geometry of Nature*, Macmillan.

[33] Mandelbrot, B.B. 2004. *Fractals and Chaos*, Springer, Berlin.

[34] Mandelbrot, B.B. and Wallis, J.R. 1969. Robustness of the rescaled range R/S in the measurement of noncyclic long run statistical dependence, *Water Resources Research*, 5:5: 967–988.

[35] Ossadnik, S.M., Buldyrev, S.V., Goldberger, A.L., Havlin, S., Mantegna, R.N., Peng, C.K., Simons, M. and Stanley, H.E. 1994. Correlation approach to identify coding regions in DNA sequences, *Biophysical J.* 67:1: 64–70.

[36] Oswiecimka, P., Kwapien, J. and Drozdz, S. 2006. Wavelet versus detrended fluctuation analysis of multifractal structures, *Physical Review E*, 74:1:016103: 1–17.

[37] Peng, C.-K., Buldyrev, S.V., Havlin, S., Simons, M., Stanle, H.E. and Goldberger, A.L. 1994. Mosaic organization of DNA nucleotides, *Physical Review E*, 49:2: 1685–1689.

[38] Philippe, P. 1993. Chaos, Population Biology, and Epidemiology: Some Research Implications, *Human Biology* 65:4: 525–546.

[39] Rhodes, C.J., Jensen, H.J. and Anderson, R.M. 1997. On the critical behaviour of simple epidemics, *Proc. R. Soc. Lond. B.* 264: 1639–-1646.

[40] Saha, G., Rakshit, K., Ghosh, K. and Chaudhuri, K. 2019. A New Proposal on the Relation between Irregularity Index and Scaling Index in a Non-stationary Self-affine Signal obeying Fractional Brownian Motion, *Bull. Calcutta Math. Soc.* 111:1: 79–86.

[41] Samadder, S. and Ghosh, K. 2021. Analysis of Self-Similarity, Memory and Variation in Growth Rate of COVID-19 Cases in Some Major Impacted Countries, *J. Physics: Conference Series* 1797:1: 012010.

[42] Schreiber, T. and Schmitz, A. 2000. Surrogate time series, *Physica D: Nonlinear Phenomena* 142:3-4: 346–382.

[43] Shimizu, Y., Thurner, S. and Ehrenberger, K. 2002. Multifractal spectra as a measure of complexity in human posture, *Fractals* 10:1:: 103–116.

[44] Sprott, J.C. 2003. *Chaos and Time Series Analysis*, Oxford University Press.

[45] WHO, "Novel Coronavirus–China", https://www.who.int/csr/don/12-january-2020-novel-coronavirus-china/en/, 12 January 2020, Accessed on: 25 June 2021.

[46] Zhu, N., Zhang, D., Wang, W., Li, X., Yan, B., Song, J., Zhao, X., Huang, B., Shi, W., Lu, R., Niu, P., Zhan, F., Ma, X., Wang, D., Xu, W., Wu, G., Gao, G.F. and Tan, W. 2020. A Novel Coronavirus from Patients with Pneumonia in China, 2019, N Engl J. Med. 382:8: 727–733.

Index

$\mathcal{A}$-statistical relative, 49
$\mathcal{A}$-statistically convergent, 50
$A-$density, 49
B-continuous, 146
$B(u,v)$-transform, 59
BK-space, 113
P-asymptotically equal, 12
P-strongly asymptotically
    equivalent, 12
RH-regular, 50
Δ_2−condition, 61
$\Gamma^{\mathcal{I}_3}$stat.c., 28
β-connected set, 232
η-convex, 207
η-quasi-convex, 207
$\mathcal{A}$-statistically bounded, 141
$\mathcal{A}$-statistically convergent, 140
$\mathcal{I}$-statistically convergent, 23
$\mathcal{I}_3$-minimal closed set, 28
$\mathcal{I}_3$-statistically pre-Cauchy, 27
$\mathcal{I}_3$-density zero, 23
$\mathcal{I}_3$-nonthin subsequence, 25
$\mathcal{I}_3$-stat.c.p., 24
$\mathcal{I}_3$-statistical cluster points, 24
$\mathcal{I}_3$-statistically bounded, 25
$\mathcal{I}_3$-statistically convergent, 24
$\mathcal{I}_3$-thin subsequence, 25
$\mathcal{I}_3$nont.ss., 25
$\mathcal{I}_3$statb, 25
$n-$normed space, 58
q-binomial coefficient, 115
q-calculus, 115
q-factorial, 115
q-multivariable Lagrange
    polynomials, 130

q-number, 115
q-shifted factorial, 130
$\mathcal{O}$-regularly varying, 97

$\mathcal{I}_3$-convergent, 23
matrix mapping, 58

admissible ideal, 23
almost convergent, 60
asymptotic analysis, 5
asymptotic density, 44
auxiliary index function, 96

Bögel bounded, 146
Banach limit, 59
Banach space, 40
Bell polynomials, 172
Bennett's factorization, 77
biconvex set, 232
bidiagonal matrix, 86
bounded, 58, 71
Boyd-Wong type contraction, 159
Bregman distance function, 245

canonical pre-image, 52
Cauchy, 58
Cauchy sequence, 158
Cesàro matrix, 72
Chaos, 313
Chebyshev polynomial, 283
class $\mathsf{ARV_s}$, 104
class $\mathsf{ERV_f}$ of Matuszewska, 99
class $\mathsf{ERV_s}$, 99
complementary function, 60
complete, 53
continuous, 158

converge regularly, 48
converge statistically uniformly relative, 47
converge uniformly, 39
converge uniformly relative, 41
converges relatively uniformly, 41
convex, 206, 231
Copson matrix, 75
Covid-19, 326

density, 22
diagonal matrix, 93
diagonal sequence, 3
difference double sequence, 54
difference sequence spaces, 40
differential transformation, 171
double sequence, 2, 49

embedable, 99
equal almost all, 45
Euler matrix, 114
exponent of convergence, 14
exponential bivariational inequality, 244
exponentially pseudo β-biconvex function, 240
exponentially log-convex, 232
exponentially affine biconvex, 232
exponentially log-biconvex, 233
exponentially affine convex function, 231
exponentially biconvex function, 232
exponentially convex function, 231
exponentially pseudo-biconvex, 240
exponentially quasi biconvex, 233
exponentially quasi-convex, 231

finite diagonal sum, 3
finite Pringsheim sum, 3
four dimensional matrix, 140

Gamma matrix, 73
generalize Hilbert's inequality, 87
Generalized Boolean Sum, 130

generalized Cesàro matrix, 92
generating function, 130
GPF-integral operator, 213

h-convex, 206
Hölder matrix, 73
Hadamard differential operator, 212
Hadamard product, 302
Hardy's inequality, 80
Hausdorff matrix, 73
Hellinger-Toeplitz theorem, 75
Higuchi method, 324
Hilbert matrix, 72
Hilbert's inequality, 80
Hilbert's norm, 72

index function, 96
interpolative Kannan type contraction, 156
inverse differential transformation, 171

k-fractionals operator, 214
Katugampola fractional integral, 212
Katugampola's fractional operator, 214
Korovkin theorem, 133

lacunary $\mathcal{I}_3$-statistical cluster points, 32
lacunary triple sequence, 32
Landau-Hurwicz sequence, 3
lower Matuszewska index, 103
Lyapunov exponent, 322

m-concave, 206
m-convex, 206
matrix mapping, 114
max-converges, 2
max-strong asymptotic equivalence, 12
measure of noncompactness, 191
meromorphically convex, 272
modulus of continuity, 47

monotone, 52
Multifractal Detrended
    Fluctuation Analysis, 318
multiplier space, 119

Nörlund means, 58
norm, 71
norm statistically convergent, 22
normal, 52

Orlicz function, 60

paranorm, 60
pointwise convergent, 39
power series method , 132
power series summability method, 132
Pringsheim convergence, 2
Pringsheim limit, 2, 50, 140
Pringsheim sense, 133

quasi-convex, 206
quasi-partial b-metric, 157

rapidly varying, 100
rapidly varying of index, 101
rapidly varying sequences, 102
regular, 132, 140
regular relative uniform
    convergent, 53
regularly converges, 2
regularly varying, 96
relatively uniformly convergent, 52
Riemann-Liouville fractional
    integral, 208
root mean square error (RMSE), 321
Ruscheweyh derivative, 302

Sakaguchi mapping, 254
scale function, 50
Schauder basis, 118
Schauder's fixed point theorem, 194
Schur's theorem, 76
selection hypothesis, 4

Selection principles, 4
Seneta functions, 98
Seneta sequences, 98
sequence space, 60
sequentially continuous, 158
shift operator, 59
shifting distance function, 194
skew symmetric, 236
slowly varying functions, 96
solid, 52
starlike, 253
statistical cluster point, 22
statistical convergence, 23
statistical limit point, 22
statistically Cauchy, 45
statistically convergent, 22, 44, 49
statistically null, 49
statistically pointwise convergent, 46
statistically relatively uniform, 50, 51
statistically relatively uniform
    convergent, 46
statistically uniform convergent, 46
Stochastic chaos, 313
strongly subordinate, 300
strongly superordinate, 301
subordinant, 292
sum-converges, 2
sum-strong asymptotic
    equivalence, 13
symmetric, 52

total paranorm, 60
translationally rapidly varying
    double sequences, 5
translationally rapidly varying
    functions, 102
translationally rapidly varying
    sequences, 5, 102

uniform convergence, 49
uniformly convex mapping, 267

weighted mean matrix, 81
Wong function, 157